Algebras of Pseudodifferential Operators

Mathematics and Its Applications (*Soviet Series*)

Volume 43

B. A. Plamenevskii

Leningrad Electrical Engineering Institute,
Leningrad, U.S.S.R.

Algebras of Pseudodifferential Operators

KLUWER ACADEMIC PUBLISHERS

DORDRECHT / BOSTON / LONDON

Library of Congress Cataloging in Publication Data

Plamenevskiĭ, B. A.
[Algebry psevdodifferent͡sial'nykh operatorov. English]
Algebras of pseudodifferential operators / by B.A. Plamenevskii ; translated from the Russian by R.A.M. Hoksbergen.
p. cm. -- (Mathematics and its applications. Soviet series ; 43)
Translation of: Algebry psevdodifferent͡sial'nykh operatorov.
Includes bibliographical references.
ISBN 0-7923-0231-1
1. Pseudodifferential operators. 2. C*-algebras. I. Title. II. Series: Mathematics and its applications (Kluwer Academic Publishers). Soviet series ; 43.
QA329.7.P5313 1989
515'.7242--dc20 89-19849

ISBN 0-7923-0231-1

Published by Kluwer Academic Publishers,
P.O. Box 17, 3300 AA Dordrecht, The Netherlands.

Kluwer Academic Publishers incorporates
the publishing programmes of
D. Reidel, Martinus Nijhoff, Dr W. Junk and MTP Press.

Sold and distributed in the U.S.A. and Canada
by Kluwer Academic Publishers,
101 Philip Drive, Norwell, MA 02061, U.S.A.

In all other countries, sold and distributed
by Kluwer Academic Publishers Group,
P.O. Box 322, 3300 AH Dordrecht, The Netherlands.

Printed on acid-free paper

This is the translation of the original work
АЛГЕБРЫ ПСЕВДОДИФФЕРЕНЦИАЛЬНЫХ ОПЕРАТОРОВ
Published by Nauka, Moscow, © 1986

Translated from the Russian by R. A. M. Hoksbergen

Printed in The Netherlands

SERIES EDITOR'S PREFACE

'Et moi, ..., si j'avait su comment en revenir, je n'y serais point allé.'

Jules Verne

The series is divergent; therefore we may be able to do something with it.

O. Heaviside

One service mathematics has rendered the human race. It has put common sense back where it belongs, on the topmost shelf next to the dusty canister labelled 'discarded non-sense'.

Eric T. Bell

Mathematics is a tool for thought. A highly necessary tool in a world where both feedback and non-linearities abound. Similarly, all kinds of parts of mathematics serve as tools for other parts and for other sciences.

Applying a simple rewriting rule to the quote on the right above one finds such statements as: 'One service topology has rendered mathematical physics ...'; 'One service logic has rendered computer science ...'; 'One service category theory has rendered mathematics ...'. All arguably true. And all statements obtainable this way form part of the raison d'être of this series.

This series, *Mathematics and Its Applications*, started in 1977. Now that over one hundred volumes have appeared it seems opportune to reexamine its scope. At the time I wrote

> "Growing specialization and diversification have brought a host of monographs and textbooks on increasingly specialized topics. However, the 'tree' of knowledge of mathematics and related fields does not grow only by putting forth new branches. It also happens, quite often in fact, that branches which were thought to be completely disparate are suddenly seen to be related. Further, the kind and level of sophistication of mathematics applied in various sciences has changed drastically in recent years: measure theory is used (non-trivially) in regional and theoretical economics; algebraic geometry interacts with physics; the Minkowsky lemma, coding theory and the structure of water meet one another in packing and covering theory; quantum fields, crystal defects and mathematical programming profit from homotopy theory; Lie algebras are relevant to filtering; and prediction and electrical engineering can use Stein spaces. And in addition to this there are such new emerging subdisciplines as 'experimental mathematics', 'CFD', 'completely integrable systems', 'chaos, synergetics and large-scale order', which are almost impossible to fit into the existing classification schemes. They draw upon widely different sections of mathematics."

By and large, all this still applies today. It is still true that at first sight mathematics seems rather fragmented and that to find, see, and exploit the deeper underlying interrelations more effort is needed and so are books that can help mathematicians and scientists do so. Accordingly MIA will continue to try to make such books available.

If anything, the description I gave in 1977 is now an understatement. To the examples of interaction areas one should add string theory where Riemann surfaces, algebraic geometry, modular functions, knots, quantum field theory, Kac-Moody algebras, monstrous moonshine (and more) all come together. And to the examples of things which can be usefully applied let me add the topic 'finite geometry'; a combination of words which sounds like it might not even exist, let alone be applicable. And yet it is being applied: to statistics via designs, to radar/sonar detection arrays (via finite projective planes), and to bus connections of VLSI chips (via difference sets). There seems to be no part of (so-called pure) mathematics that is not in immediate danger of being applied. And, accordingly, the applied mathematician needs to be aware of much more. Besides analysis and numerics, the traditional workhorses, he may need all kinds of combinatorics, algebra, probability, and so on.

In addition, the applied scientist needs to cope increasingly with the nonlinear world and the extra mathematical sophistication that this requires. For that is where the rewards are. Linear models are honest and a bit sad and depressing: proportional efforts and results. It is in the non-linear world that infinitesimal inputs may result in macroscopic outputs (or vice versa). To appreciate what I am hinting at: if electronics were linear we would have no fun with transistors and computers; we would have no TV; in fact you would not be reading these lines.

There is also no safety in ignoring such outlandish things as nonstandard analysis, superspace and anticommuting integration, p-adic and ultrametric space. All three have applications in both electrical engineering and physics. Once, complex numbers were equally outlandish, but they frequently proved the shortest path between 'real' results. Similarly, the first two topics named have already provided a number of 'wormhole' paths. There is no telling where all this is leading - fortunately.

Thus the original scope of the series, which for various (sound) reasons now comprises five sub-series: white (Japan), yellow (China), red (USSR), blue (Eastern Europe), and green (everything else), still applies. It has been enlarged a bit to include books treating of the tools from one subdiscipline which are used in others. Thus the series still aims at books dealing with:

- a central concept which plays an important role in several different mathematical and/or scientific specialization areas;
- new applications of the results and ideas from one area of scientific endeavour into another;
- influences which the results, problems and concepts of one field of enquiry have, and have had, on the development of another.

Does an analyst need to know about such things as algebras, Noetherianess, homology, and representation theory? The answer would appear to be an emphatic yes. (He also needs to know about topology (and not only in the context of topological vector spaces, but also differential topology).) And the more complicated the setting of the problems in the analysis concerned, the more this is needed to sort out the phenomena.

It has been said before (by myself, for example), but bears to be repeated: "Mathematics is a tool for thought" (at least as long as our brains are rather spectacularly finite in their abilities to instantly oversee all the consequences and implications of, say, a certain set of assumptions). This implies that any distinct part of mathematics provides tools for thought for any other. Mostly, mathematicians are not slow to point out the importance and applicability of their science to other fields; I wonder whether we follow our own advice with equal vigour intra-mathematically.

One fairly recently created tool, of considerable interest in itself and of vast applicability, is that of algebras of pseudodifferential operators. The present book is devoted to that topic; more specifically it is devoted to the case of the theory of pseudodifferential operators on manifolds of higher dimension ($\geqslant 2$) with isolated singularities. As such, it is unique with virtually no overlap with other existing monographs; a more than worthy addition to this series.

Pseudo-differential operators are a topic from analysis, of course; in the book they are studied from the point of view of C^*-algebras, and the attentive and interested reader will become aware of the importance and power of algebraic, topological and, perhaps especially, representation theoretic considerations in this field.

The shortest path between two truths in the real domain passes through the complex domain.

J. Hadamard

La physique ne nous donne pas seulement l'occasion de résoudre des problèmes ... elle nous fait pressentir la solution.

H. Poincaré

Never lend books, for no one ever returns them; the only books I have in my library are books that other folk have lent me.

Anatole France

The function of an expert is not to be more right than other people, but to be wrong for more sophisticated reasons.

David Butler

Bussum, September 1989

Michiel Hazewinkel

Table of Contents

Introduction

The theory of pseudodifferential operators (ΨDOs), created in the last two-three decades, has become both an object of application and a generally used instrument in various branches of mathematics, such as mathematical physics, functional analysis, and topology. Several monographs have been devoted to this theory ([40], [67], [68], [69], [73], [74], etc.); in them pseudodifferential operators arise in the 'smooth' case - with smooth symbols (coefficients), on smooth manifolds. In all branches mentioned above, quite a lot of problems appear related to the study of pseudodifferential operators with singularities. In [9], [12], [17], [41], [71] one-dimensional singular integral operators with discontinuous coefficients on composite contours are treated. The corresponding problems for pseudodifferential operators on manifolds of dimensions $n \geqslant 2$ have not been reflected upon in monographs. The present book is devoted to this very theme. The results expounded in it form, it seems, the core of the theory of pseudodifferential operators with discontinuous symbols, which is being created. In essence there is no overlap with other books.

We consider algebras of pseudodifferential operators on manifolds with a finite number of isolated singularities. The symbols of the operators are allowed to have discontinuities at a finite number of points. The set of points of discontinuity depends on the operator, and by closing the algebras, operators whose symbols have an everywhere dense set of singularities arise. Different versions of symbolic calculus are considered, and necessary and sufficient conditions for a pseudodifferential operator to be Fredholm are clarified.

One of the basic aims of the book is the study of pseudodifferential operators from the point of view of the theory of C^*-algebras. The scalar pseudodifferential operators of order zero and with smooth symbols on a smooth manifold $\mathfrak{M}$ without boundary generate an algebra $\mathcal{Q}$, which after factorization by the ideal $\mathcal{K}$ of compact operators becomes a commutative algebra. Hence all irreducible representations of the quotient algebra $\mathcal{Q}/\mathcal{K}$ are one-dimensional, and its spectrum (the set of equivalence classes of irreducible representations) can be identified with the space of maximal ideals, which coincides with the bundle $S^*(\mathfrak{M})$ of unit cotangent vectors. For an $A \in \mathcal{A}$, $[A] \in \mathcal{A}/\mathcal{K}$ denotes its residue class, and if π runs through the spectrum of $\mathcal{A}/\mathcal{K}$, then the map $\pi \mapsto \pi[A]$ can be realized as the function $S^*(\mathfrak{M}) \ni \pi \mapsto \Phi(\pi)$, where Φ is the symbol of A.

The situation becomes more complicated if the manifold has a boundary or singularities and (or) if the symbols of the operators become discontinuous. The corresponding quotient algebra $\mathcal{A}/\mathcal{K}$ is, in general, not commutative. Among the irreducible representations of $\mathcal{A}/\mathcal{K}$ there appear infinite-dimensional ones. In this book we describe the spectra of algebras generated by pseudodifferential operators: all equivalence classes of irreducible representations are listed, a topology on the spectrum is elucidated (the so-called Jacobson topology), and a realization of the irreducible representations is given, i.e. a map $\hat{\mathcal{A}} \ni \pi \mapsto \pi(A)$, where $A \in \mathcal{A}$ and $\hat{\mathcal{A}}$ is the spectrum of $\mathcal{A}$. The dependence of the spectrum on the kind of discontinuity of the symbols and on the choice of function space is studied.

In the first Chapter we introduce and study the integral transform $E(\lambda)$ on the $(n-1)$-dimensional sphere S^{n-1}. The Fourier transform can be decomposed in a product of three operators: the Mellin transform, $E(\lambda)$, and the inverse Mellin transform. We compute the values of $E(\lambda)$ on spherical functions, and indicate a space on which $E(\lambda)$ acts as a continuous operator. The results of this Chapter are constantly used in the sequel.

Pseudodifferential operators with discontinuous symbols are naturally regarded in spaces with weighted norms. In the second Chapter we define the 'weighted' classes $H^s_\beta(\mathbb{R}^m)$, $H^s_\beta(\mathbb{R}^m, \mathbb{R}^{m-n})$. On these spaces we study the Fourier transform and operators of convolution with a homogeneous function. We prove boundedness theorems for these operators, and derive special representations for convolutions in terms of $E(\lambda)$ and the Mellin transform. These representations play an important part in the study of algebras of pseudodifferential operators. The third Chapter is devoted to the special class of meromorphic operator-functions which are called meromorphic pseudodifferential operators. It will later turn out that meromorphic pseudodifferential operators are values of 'operator symbols'. The algebras of these symbols is isomorphic to the quotient algebra $\mathcal{A}/\mathcal{K}$. Using meromorphic pseudodifferential operators the irreducible representations of $\mathcal{A}$ are realized (in Chapters 5, 6). The first three Chapters form the technical basis for the subsequent exposition.

In the fourth Chapter we study pseudodifferential operators with discontinuous symbols on Euclidean space $\mathbb{R}^n$ and on manifolds. In particular, we introduce the algebra $\mathfrak{S}$ of operator symbols and establish the isomorphism $\mathfrak{S} \approx \mathcal{Q}/\mathcal{K}$. On the manifolds we allow for isolated singularities - 'conical' points. In defining pseudodifferential operators on a smooth manifold one commonly uses the fact that locally a pseudodifferential operator on a manifold coincides with a pseudodifferential operator on a Euclidean space. In a neighborhood of a singular point similar considerations are not immediately applicable. Thus we have to define pseudodifferential operators on a manifold with conical points in a special manner. This is done in several stages. First we introduce meromorphic pseudodifferential operators on $\mathbb{R}^n$, and then on a cone. The result of the Mellin transform applied to a meromorphic pseudodifferential operator will be pseudodifferential operator on a cone. After this, a pseudodifferential operator on a manifold with conical singularities is glued together using partition of unity.

The spectrum of a C^*-algebra of pseudodifferential operators is described in the last two Chapters, the fifth and sixth. In Chapter 5 we consider algebras generated by pseudodifferential operators on closed manifolds, and in Chapter 6 - on manifolds with boundary. The necessary information from the theory of C^*-algebras is given in §1 of Chapter 5.

The contents of the book does not exhaust all known material concerning pseudodifferential operators with discontinuous symbols. Aside remained the theory of boundary value problems (which is sufficiently well-developed for elliptic pseudodifferential operators with isolated singularities), algebras with multidimensional discontinuities in the symbols (see [61]), and some scattered results. Some articles concerned with these problems are included in the list of references.

Chapter 1

Integral transforms on a sphere

In this Chapter we introduce and study the operator E(λ), which acts on the space of functions on the sphere S^{n-1}. It is used in all subsequent Chapters.

§1. The generalized kernels $(xy)_{\pm}^{\mu}$, $(\pm xy + i0)^{\mu}$

Let $x = (x_1, \cdots, x_n)$, $y = (y_1, \cdots, y_n)$ be unit vectors in $\mathbb{R}^n$, let $xy = x_1 y_1 + \cdots + x_n y_n$, and let S^{n-1} be the $(n-1)$-dimensional unit sphere with center at the coordinate origin. We introduce for $\mathrm{Re}\,\mu > -1$ the following transforms on functions u of class $C^{\infty}(S^{n-1})$:

$$(\mathrm{I}_{\mu}^{\pm} u)(x) = \int_{S^{n-1}} (xy)_{\pm}^{\mu} u(y) dy. \tag{1.1}$$

Here dy is the volume element on S^{n-1}, and, as usual, $t_{+}^{\mu} = 0$ for $t \leqslant 0$, $t_{+}^{\mu} = e^{\mu \ln t}$ if $t > 0$, and $t_{-}^{\mu} = (-t)_{+}^{\mu}$.

P r o p o s i t i o n 1.1. *The maps* $\mathrm{I}_{\mu}^{\pm} : C^{\infty}(S^{n-1}) \to C^{\infty}(S^{n-1})$ *are continuous. The operator-function* $\mu \mapsto \mathrm{I}_{\mu}^{\pm}$ *is analytic in the halfplane* $\mathrm{Re}\,\mu > -1$. *Each of these functions can be analytically extended onto the whole* μ*-plane, except for the points* $\mu = -1, -2, \cdots$, *which are poles of the first order.*

P r o o f. Let $\{\eta_j\}$ be a partition of unity on S^{n-1}, i.e. a finite collection of non-negative functions from $C^{\infty}(S^{n-1})$ such that $\Sigma \eta_j = 1$. It suffices to verify the statement for the operators $\eta_j \mathrm{I}_{\mu}^{\pm}$. We may assume that on the support $\mathrm{supp}\,\eta_j$ of η_j the angular coordinates form a regular coordinate net. Denote by g_x some rotation of S^{n-1} transforming the point $\mathfrak{N} = (0, \cdots, 0, 1)$ to the point x. Determine a family $\{g_x\}$, $x \in \operatorname{supp}\eta_j$, so that the matrix entries, giving rotations, are smooth functions of the angular coordinates of x. We have

$$(\eta_j I_\mu^{\pm} u)(x) = \eta_j(x) \int_{S^{n-1}} (g_x(\mathfrak{N})y)_\pm^\mu u(y)dy =$$

$$= \eta_j(x) \int_{S^{n-1}} (\mathfrak{N} g_x^{-1}(y))_\pm^\mu u(y)dy = \eta_j(x) \int_{S^{n-1}} (\mathfrak{N} z)_\pm^\mu u(g_x(z))dz.$$

Denote by $\theta_1, \cdots, \theta_{n-1}$ the angular coordinates of a point $z \in S^{n-1}$, related to Cartesian coordinates $(z_1, \cdots, z_n)$ by

$$\begin{aligned} z_1 &= \sin\theta_{n-1} \cdots \sin\theta_2 \sin\theta_1, \\ z_2 &= \sin\theta_{n-1} \cdots \sin\theta_2 \cos\theta_1, \\ &\cdots \quad \cdots \quad \cdots \quad \cdots \\ z_{n-1} &= \sin\theta_{n-1} \cos\theta_{n-2}, \\ z_{n-1} &= \cos\theta_{n-1}, \end{aligned} \tag{1.2}$$

where $0 \leqslant \theta_1 < 2\pi$, $0 \leqslant \theta_k \leqslant \pi$, $k \neq 1$. Introduce the function $v_x(\theta_{n-1}) = \eta_j(x) \times \int u(g_x(z))(\sin\theta_{n-2})^{n-3} \cdots \sin\theta_2 \, d\theta_1 \cdots d\theta_{n-2}$. Then

$$(\eta_j I_\mu^+ u)(x) = \int_0^{\pi/2} (\cos\theta_{n-1})^\mu (\sin\theta_{n-1})^{n-2} v_x(\theta_{n-1}) d\theta_{n-1}.$$

Putting $t = \cos\theta_{n-1}$, $w_{x,+}(t) = v_x(\arccos t)$, we extend the function $\mu \mapsto \eta_j I_\mu^+ u$ to the halfplane $\operatorname{Re}\mu > -k-1$ using the formulas

$$(\eta_j I_\mu^+ u)(x) = \int_0^1 t^\mu (1-t^2)^{(n-3)/2} w_{x,+}(t) dt = \tag{1.3}$$

$$= \int_0^1 t^\mu (1-t^2)^{(n-3)/2} \left\{ w_{x,+}(t) - \sum_{q=0}^{k-1} \frac{t^q}{q!} w_{x,+}^{(q)}(0) \right\} dt +$$

$$+ \sum_{q=0}^{k-1} \frac{1}{q!} w_{x,+}^{(q)}(0) \frac{1}{2} B \left(\frac{\mu+q+1}{2}, \frac{n-1}{2} \right).$$

Recall that the beta-function B can be expressed in terms of the gamma-function Γ, $B(\lambda, \nu) = \Gamma(\lambda)\Gamma(\nu)/\Gamma(\lambda+\nu)$, and that $\Gamma(\mu)$ is a meromorphic function on the whole μ-plane. The poles of the Γ-function are simple, are located at the points $\mu=0,-1,\cdots$, and

$$\operatorname{res}\Gamma(\mu)\,|_{\mu=-k}=\frac{(-1)^k}{k!}. \tag{1.4}$$

Put now $t=-\cos\theta_{n-1}$, $w_{x,-}(t)=v_x(\pi-\arccos t)$, and define the extension of $\mu\mapsto\eta_j\,\mathrm{I}^-_{\mu}u$ to the halfplane $\operatorname{Re}\mu>-k-1$ by

$$(\eta_j\,\mathrm{I}^-_{\mu}u)(x)=\int_0^1 t^{\mu}(1-t^2)^{(n-3)/2}w_{x,-}(t)dt= \tag{1.5}$$

$$=\int_0^1 t^{\mu}(1-t^2)^{(n-3)/2}\left\{w_{x,-}(t)-\sum_{q=0}^{k-1}\frac{t^q}{q!}w^{(q)}_{x,-}(0)\right\}dt+$$

$$+\sum_{q=0}^{k-1}\frac{1}{q!}w^{(q)}_{x,-}(0)\frac{1}{2}B\left(\frac{\mu+q+1}{2},\frac{n-1}{2}\right).$$

In view of formulas (1.3), (1.5) and the properties of the Γ-function indicated, the proposition can be immediately verified. ■

For $\operatorname{Re}\mu>-1$ and for $u\in C^{\infty}(S^{n-1})$ we introduce the operators

$$(\mathrm{J}^{\pm}_{\mu}u)(x)=\int_{S^{n-1}}(\pm xy+i0)^{\mu}u(y)dy, \tag{1.6}$$

where $(xy+i0)^{\mu}=(xy)^{\mu}_{+}+e^{i\mu\pi}(xy)^{\mu}_{-}$, $(-xy+i0)^{\mu}=e^{i\mu\pi}(xy)^{\mu}_{+}+(xy)^{\mu}_{-}$.

P r o p o s i t i o n 1.2. *The maps* $\mathrm{J}^{\pm}_{\mu}\colon C^{\infty}(S^{n-1})\to C^{\infty}(S^{n-1})$ *are continuous. The operator-function* $\mu\mapsto\mathrm{J}^{\pm}_{\mu}$ *can be analytically extended to the whole* μ*-plane.*

P r o o f. By proposition 1.1 it suffices to convince ourselves of holomorphy of the extensions of $\mu\mapsto\mathrm{J}^{\pm}_{\mu}$, which are obtained using the extensions of $\mu\mapsto\mathrm{I}^{\pm}_{\mu}$. For this we need to take into account formulas (1.3) and (1.5), and note that the singularities mutually cancel each other. ■

§2. The operator E(λ), its relation with the Fourier and Mellin transform

We define for arbitrary complex λ except $\lambda = i(k+n/2)$, $k = 0, 1, \cdots$, the following operators on functions u from $C^\infty(S^{n-1})$:

$$(\mathrm{E}(\lambda)u)(x) = \frac{1}{(2\pi)^{n/2}} e^{i\frac{\pi}{2}(i\lambda+n/2)} \Gamma(i\lambda+n/2) \times \tag{2.1}$$

$$\times \int_{S^{n-1}} (-xy+i0)^{-i\lambda-n/2} u(y)dy.$$

Proposition 1.2 implies that the operator-function $\lambda \mapsto \mathrm{E}(\lambda)$: $C^\infty(S^{n-1}) \to C^\infty(S^{n-1})$ is analytic everywhere except at the points indicated, at which it has poles of the first order. The residue at the pole $\lambda = i(k+n/2)$ is a finite-dimensional operator; using (1.4) we obtain

$$\operatorname{res} \mathrm{E}(\lambda)u\,|_{\lambda=i(k+n/2)} = \frac{(-i)^{k+1}}{(2\pi)^{n/2}} \sum_{|\gamma|=k} \frac{1}{\gamma!} x^\gamma \int_{S^{n-1}} y^\gamma u(y)dy, \tag{2.2}$$

where $x,y \in \mathbb{R}^n$, $|x| = |y| = 1$, and, as usual, $x^\gamma = x_1^{\gamma_1} \cdots x_n^{\gamma_n}$ for $x = (x_1, \cdots, x_n)$, $\gamma = (\gamma_1, \cdots, \gamma_n)$.

Let now u be a function from $C^\infty(S^{n-1})$ satisfying

$$\int_{S^{n-1}} y^\gamma u(y)dy = 0 \tag{2.3}$$

for all multi-indices γ for which $|\gamma| = k$. Formula (2.2) implies that the function $\lambda \mapsto \mathrm{E}(\lambda)u$ remains regular at the point $\lambda = i(k+n/2)$. Using the Taylor series expansion

$$(-xy+i0)^{-i\lambda-n/2} = (-xy+i0)^k +$$

$$-(-xy+i0)^k \ln(-xy+i0)(k+i\lambda+n/2) + \cdots$$

and (1.4) we obtain

$$\lim_{\lambda\to i(k+n/2)} (\mathrm{E}(\lambda)u)(x) = \tag{2.4}$$

$$= \frac{-1}{(2\pi)^{n/2}} \frac{(-i)^k}{k!} \int (xy)^k \ln(-xy+i0) u(y)dy =$$

$$= \frac{1}{(2\pi)^{n/2}} \frac{(-i)^k}{k!} \int (xy)^k \left[\ln\frac{1}{|xy|} - i\pi\theta(xy)\right] u(y)dy,$$

where $\theta(z) = 1$ in case $z > 0$ and $\theta(z) = 0$ for $z < 0$. Since

$$\frac{1}{k!}(xy)^k = \sum_{|\gamma|=k} \frac{1}{\gamma!} x^\gamma y^\gamma,$$

by (2.3) the relation (2.4) can be replaced by

$$\lim_{\lambda \to i(k+n/2)} (\mathrm{E}(\lambda)u)(x) = \tag{2.5}$$

$$= \frac{1}{(2\pi)^{n/2}} \frac{(-i)^k}{k!} \int_{S^{n-1}} (xy)^k \left\{ \ln \frac{1}{|xy|} - \frac{i\pi}{2} \operatorname{sgn}(xy) \right\} u(y) dy.$$

We will denote this operator by $\hat{\mathrm{E}}(i(k+n/2))$.

In the remainder of this section, x is an arbitrary vector from $\mathbb{R}^n$, $r = |x|$, $\phi = x/|x|$, ξ is a vector from the dual space, $\rho = |\xi|$, $\psi = \xi/|\xi|$.

We write the Fourier transform as

$$(\mathrm{F}u)(\xi) = \frac{1}{(2\pi)^{n/2}} \int e^{-i\xi x} u(x) dx,$$

and the Mellin transform (for functions from $C_0^\infty(\mathbb{R}^n \setminus 0)$) as

$$(\mathrm{M}u)(\lambda, \phi) \equiv \tilde{u}(\lambda, \phi) = \frac{1}{\sqrt{2\pi}} \int_0^\infty r^{-i\lambda - 1} u(r, \phi) dr, \quad \lambda \in \mathbb{C}.$$

For the Mellin transform the inversion formula

$$u(r, \phi) = \frac{1}{\sqrt{2\pi}} \int_{\operatorname{Im}\lambda = \tau} r^{i\lambda} \tilde{u}(\lambda, \phi) d\lambda \tag{2.6}$$

holds, as well as Parseval's equality

$$\int_{\operatorname{Im}\lambda = \tau} |\tilde{u}(\lambda, \phi)|^2 d\lambda = \int_0^\infty r^{2\beta} |u(r, \phi)|^2 dr, \quad \tau = \beta + \frac{1}{2}. \tag{2.7}$$

(It is obtained by the change of variable $r = e^t$ from the corresponding property of the one-dimensional Fourier transform.)

P r o p o s i t i o n 2.1. *For functions u of class* $C_0^\infty(\mathbb{R}^n \setminus 0)$ *the equality*

$$\mathrm{F}_{x \to \xi} u = \mathbf{M}^{-1}_{(in/2 - \lambda) \to \rho} \mathbf{E}_{\phi \to \psi}(\lambda) \mathbf{M}_{r \to (\lambda + in/2)} u \tag{2.8}$$

holds, where $\mathrm{Im}\lambda = 0$, $M^{-1}_{\nu\to\rho} = (M^{-1})_{\nu\to\rho}$.

P r o o f. Using the inversion formula (2.6) we write the Fourier transform Fu as

$$(Fu)(\rho,\psi) = \frac{1}{(2\pi)^{(n+1)/2}} \times$$

$$\times \int\int e^{-ir\rho\phi\psi} r^{n-1} dr\, d\phi \int_{-\infty}^{+\infty} r^{i\lambda - n/2} \tilde{u}(\lambda + in/2, \phi) d\lambda.$$

In order to substantiate the interchange of the integrals in this formula we introduce a parameter $\tau > 0$:

$$(Fu)(\rho,\psi) = \lim_{\tau\to+0} \frac{1}{(2\pi)^{(n+1)/2}} \int_0^{+\infty} \int_{S^{n-1}} e^{-ir\rho\phi\psi - r\tau} r^{n-1} dr d\phi \times \tag{2.9}$$

$$\times \int_{-\infty}^{+\infty} \tilde{u}(\lambda + in/2, \phi) r^{i\lambda - n/2} d\lambda =$$

$$= \lim_{\tau\to+0} \frac{1}{(2\pi)^{(n+1)/2}} \int_{-\infty}^{+\infty} \int_{S^{n-1}} \tilde{u}(\lambda + in/2, \phi) d\lambda d\phi \times$$

$$\times \int_0^{+\infty} e^{-ir\rho\phi\psi - r\tau} r^{i\lambda + n/2 - 1} dr.$$

Further,

$$\int_0^{\infty} e^{-ir\rho\phi\psi - r\tau} r^{i\lambda - 1 + n/2} dr = \tag{2.10}$$

$$= e^{i\frac{\pi}{2}(i\lambda + n/2)} (-\rho\phi\psi + i\tau)^{-i\lambda - n/2} \Gamma(i\lambda + n/2).$$

(The calculation of this integral was given in, e.g., [5].) Using the last equality and (2.9) we obtain (2.8). ■

§3. The action of E(λ) on spherical functions

Denote by $\mathcal{K}$ a sequence of integers $(k_1, k_2, \cdots, \pm k_{n-2})$ for which $m = k_0 \geqslant k_1 \geqslant \cdots \geqslant k_{n-2} \geqslant 0$, and by $Y_{m\mathcal{K}}$ the following spherical function of order m:

$$Y_{m\mathcal{K}} = A_{m\mathcal{K}} e^{\pm ik_{n-2}\psi_1} \prod_{j=0}^{n-3} (\sin\psi_{n-j-1})^{k_{j+1}} \times \tag{3.1}$$

$$\times C_{k_j - k_{j+1}}^{k_{j+1} + (n-2-j)/2} (\cos\psi_{n-j-1}),$$

where $\psi \in S^{n-1}$, $\psi_1, \cdots, \psi_{n-1}$ are the angular coordinates of ψ (cf. (1.2)), C_k^p are the Gegenbauer polynomials, and

$$A_{m\mathcal{K}}^2 = \frac{1}{\Gamma(n/2)} \times \tag{3.2}$$

$$\times \prod_{j=0}^{n-3} \frac{2^{2k_{j+1}+n-j-4}(k_j - k_{j+1})!(n-j+2k_j-2)\Gamma^2((n-j-2)/2+k_{j+1})}{\sqrt{\pi}\,\Gamma(k_j + k_{j+1} + n - j - 2)}$$

(cf., e.g., [10]). The functions $Y_{m\mathcal{K}}$ form a complete orthogonal system in the space $L_2(S^{n-1})$.

P r o p o s i t i o n 3.1. *The formula*

$$(\mathrm{E}(\lambda) Y_{m\mathcal{K}})(\phi) = \tag{3.3}$$

$$= \frac{1}{(2\pi)^{n/2}} e^{i\frac{\pi}{2}(i\lambda + n/2)} \Gamma(i\lambda + n/2) \int_{S^{n-1}} (-\phi\theta + i0)^{-i\lambda - n/2} Y_{m\mathcal{K}}(\theta) d\theta =$$

$$= (-i)^m 2^{i\lambda} \frac{\Gamma\left(\dfrac{m + i\lambda + n/2}{2}\right)}{\Gamma\left(\dfrac{m - i\lambda + n/2}{2}\right)} Y_{m\mathcal{K}}(\phi)$$

holds.

P r o o f. Consider the integral

$$\int (-\phi\theta + i0)^{\nu} Y_{m\mathcal{K}}(\theta) d\theta. \tag{3.4}$$

Denote by g_ϕ some rotation of S^{n-1} mapping the point $\mathfrak{N} = (0, \cdots, 0, 1)$ to the point ϕ. Rewrite the integral (3.4) as

$$\int(-g_\phi(\mathfrak{N})\cdot\theta+i0)^\nu Y_{m\mathcal{K}}(\theta)d\theta =$$

$$= \int Y_{m\mathcal{K}}(g_\phi\psi)(-\mathfrak{N}\psi+i0)^\nu d\psi = \int Y_{m\mathcal{K}}(g_\phi\psi)(-\cos\psi_{n-1}+i0)^\nu d\psi.$$

Under rotations of the sphere, the spherical functions transform by the rule (cf. [10]).

$$Y_{m\mathcal{K}}(g_\phi\psi) = \sum_{\mathfrak{M}} t^m_{\mathfrak{M}\mathcal{K}}(g_\phi^{-1})Y_{m\mathfrak{M}}(\psi),$$

where $t^m_{\mathfrak{M}\mathcal{K}}$ are the matrix entries of the irreducible representations of the rotation group SO(n), $g \in$ SO(n). In the sequel we only need the following properties of the functions $t^m_{\mathfrak{M}\mathcal{K}}$:

$$t^m_{0\mathcal{K}}(g_\phi^{-1}) = \tag{3.5}$$

$$= (m!\Gamma(n-1))^{1/2}(\Gamma(m+n-2)(2m+n-2))^{-1/2}Y_{m\mathcal{K}}(\phi);$$

here $0 = (0, \cdots, 0)$. Using this rule we obtain

$$\int Y_{m\mathcal{K}}(g_\phi\psi)(-\cos\psi_{n-1}+i0)^\nu d\psi = \tag{3.6}$$

$$= \sum_{\mathfrak{M}} t^m_{\mathfrak{M}\mathcal{K}}(g_\phi^{-1})\int Y_{m\mathfrak{M}}(\psi)(-\cos\psi_{n-1}+i0)^\nu d\psi.$$

We compute the last integral. Formula (3.1) implies that

$$\int Y_{m\mathfrak{M}}(\psi)(-\cos\psi_{n-1}+i0)^\nu d\psi =$$

$$= A_{m\mathfrak{M}}\int e^{\pm im_{n-2}\psi_1}\prod_{j=0}^{n-3}(\sin\psi_{n-j-1})^{m_{j+1}}\times$$

$$\times C^{m_{j+1}+(n-2-j)/2}_{m_j-m_{j+1}}(\cos\psi_{n-j-1})(-\cos\psi_{n-1}+i0)^\nu\times$$

$$\times \sin^{n-2}\psi_{n-1}\cdot\sin^{n-3}\psi_{n-2}\cdots\sin\psi_2 d\psi_1\cdots d\psi_{n-1}.$$

This integral is equal to the product of the following integrals:

$$\int_0^{2\pi} e^{\pm im_{n-2}\psi_1}d\psi_1, \tag{3.7}$$

$$\int_0^{\pi}(\sin\psi_{n-j-1})^{m_{j+1}+n-j-2}C^{m_{j+1}+(n-2-j)/2}_{m_j-m_{j+1}}(\cos\psi_{n-j-1})d\psi_{n-j-1}, \tag{3.8}$$

$1 \leqslant j \leqslant n-3,$

$$\int_0^\pi (\sin\psi_{n-1})^{m_1+n-2} C_{m-m_1}^{m_1+(n-2)/2}(\cos\psi_{n-1}) \times \tag{3.9}$$

$$\times(-\cos\psi_{n-1}+i0)^\nu d\psi_{n-1}.$$

We show that all integrals (3.7), (3.8) are distinct from zero only if $m_1 = \cdots = m_{n-2} = 0$. Clearly, the integrals (3.7) are distinct from zero only if $m_{n-2} = 0$. Suppose now that $m_{j+1} = 0$. Then the integral

$$\int_0^\pi (\sin\psi_{n-j-1})^{m_{j+1}+n-j-2} C_{m_j - m_{j+1}}^{m_{j+1}+(n-2-j)/2}(\cos\psi_{n-j-1}) d\psi_{n-j-1} =$$

$$= \int_0^\pi (\sin\psi_{n-j-1})^{n-j-2} C_{m_j}^{(n-2-j)/2}(\cos\psi_{n-j-1}) d\psi_{n-j-1}$$

is distinct from zero if and only if $m_j = 0$, which follows from the orthogonality of the Gegenbauer polynomials:

$$\int_0^\pi C_l^p(\cos\theta) C_m^p(\cos\theta)\sin^{2p}\theta\, d\theta = 0 \quad \text{for} \quad m \neq l$$

([10]), and the equation $C_0^p(\cos\theta) \equiv 1$.

Thus, $m_1 = \cdots = m_{n-2} = 0$. In this case the product of the integrals (3.7), (3.8) is equal to

$$2\pi \prod_{j=1}^{n-3} \int_0^\pi (\sin\psi)^{n-j-2} d\psi = 2\pi^{(n-1)/2} \Gamma\left(\frac{n-1}{2}\right).$$

Now we consider the integral

$$\int_0^\pi (\sin\psi)^{n-2} C_m^{(n-2)/2}(\cos\psi)(-\cos\psi + i0)^\nu d\psi =$$

$$= \int_{-1}^1 (1-t^2)^{(n-3)/2} C_m^{(n-2)/2}(t)(-t+i0)^\nu dt,$$

which coincides with the integral (3.9) for $m_1 = 0$. Taking into account that ([6], 11.1, (25))

$$C_m^{(n-2)/2}(t) = \frac{\Gamma((n-1)/2)\Gamma(m+n-2)(1-t^2)^{(3-n)/2}}{(-2)^m\Gamma(m+(n-1)/2)\Gamma(n-2)} \times$$

$$\times \frac{d^m}{dt^m}(1-t^2)^{m+(n-3)/2},$$

we integrate by parts m times, and obtain

$$\int_{-1}^{1}(1-t^2)^{(n-3)/2}C_m^{(n-2)/2}(t)(-t+i0)^\nu dt =$$

$$= \frac{\nu(\nu-1)\cdots(\nu-m+1)\Gamma((n-1)/2)\Gamma(m+n-2)}{(-2)^m m!\Gamma(n-2)\Gamma(m+(n-1)/2)} \times$$

$$\times \int_{-1}^{1}(1-t^2)^{m+(n-3)/2}(t+i0)^{\nu-m}dt.$$

Since $(t+i0)^{\nu-m} = t_+^{\nu-m} + e^{i\pi(\nu-m)}t_-^{\nu-m}$, we have

$$\int_{-1}^{1}(1-t^2)^{m+(n-3)/2}(t+i0)^{\nu-m}dt = \tag{3.10}$$

$$= (1+e^{i\pi(\nu-m)})\int_{0}^{1}t^{\nu-m}(1-t^2)^{m+(n-3)/2}dt =$$

$$= (1+e^{i\pi(\nu-m)})\frac{\Gamma(m+(n-1)/2)\Gamma((\nu-m+1)/2)}{2\Gamma((m+n+\nu)/2)}$$

Combining the equation (cf. (3.2))

$$A_{m0} = (m!\Gamma(n-2)(2m+n-2))^{1/2}(\Gamma(n+m-2)(n-2))^{-1/2}$$

and formulas (3.5), (3.6), (3.10) leads to the relation

$$\int(-\phi\theta+i0)^\nu Y_{m\mathcal{K}}(\theta)d\theta =$$

$$= \frac{\pi^{(n-1)/2}\nu\cdots(\nu-m+1)\Gamma((\nu-m+1)/2)(1+e^{i\pi(\nu-m)})}{(-2)^m\Gamma((m+n+\nu)/2)}Y_{m\mathcal{K}}(\phi).$$

Putting $\nu = -\lambda - n/2$ and calling to mind formula (2.1) we find

$$(\mathrm{E}(\lambda)Y_{m\mathcal{K}})(\phi) =$$

$$= \frac{\Gamma(-\nu)\nu\cdots(\nu-m+1)\Gamma((\nu-m+1)/2)e^{-i\nu\pi/2}(1+e^{i\pi(\nu-m)})}{(-1)^m 2^{m+n/2}\sqrt{\pi}\,\Gamma((m+n+\nu)/2)} \times$$

$$\times Y_{m\mathcal{K}}(\phi).$$

This and the equation $\Gamma(z+1) = z\Gamma(z)$ imply that

$$(\mathrm{E}(\lambda)Y_{m\mathcal{K}})(\phi) = \mu_m(\lambda)Y_{m\mathcal{K}}(\phi), \tag{3.11}$$

where

$$\mu_m(\lambda) = \tag{3.12}$$

$$= \frac{(-i)^m\Gamma(i\lambda+m+n/2)\Gamma((1-i\lambda-m-n/2)/2)}{2^{m-1+n/2}\sqrt{\pi}\,\Gamma((m-i\lambda+n/2)/2)} \times$$

$$\times \cos\frac{(i\lambda+n/2+m)\pi}{2}.$$

Applying the formula $\Gamma(1-z)\Gamma(z) = \pi/\sin\pi z$ for $z = i\lambda+m+n/2+1$, we find

$$\Gamma((1-i\lambda-m-n/2)/2) =$$

$$= \frac{\pi}{\Gamma((i\lambda+m+n/2+1)/2)\sin(\pi(i\lambda+m+n/2+1)/2)}.$$

Hence (3.12) may be rewritten as

$$\mu_m(\lambda) =$$

$$= \frac{i^m\sqrt{\pi}\,\Gamma(i\lambda+m+n/2)}{2^{m-1+n/2}\Gamma((m+n/2-i\lambda)/2)\Gamma((m+n/2+1+i\lambda)/2)}.$$

Now we use the duplication formula

$$\Gamma(2z) = 2^{2z-1}\pi^{-1/2}\Gamma(z)\Gamma(z+1/2)$$

for $z = (i\lambda+m+n/2)/2$. We have

$$\frac{\Gamma(i\lambda+m+n/2)}{\Gamma((i\lambda+m+n/2+1)/2)} =$$

$$= 2^{i\lambda+m+n/2-1}\pi^{-1/2}\Gamma((i\lambda+n/2+m)/2).$$

So that

$$\mu_m(\lambda) = (-i)^m 2^{i\lambda}\Gamma((m+n/2+i\lambda)/2)\Gamma((m+n/2-i\lambda)/2)^{-1},$$

which was required. ■

We now consider the operator $\hat{E}(i(k+n/2))$, defined by (2.5) for functions satisfying the conditions (2.3). We first note that the conditions (2.3) are equivalent to

$$\int u(x) Y_{m\mathcal{K}}(x)dx = 0, \quad 0 \leqslant m \leqslant k, \quad m \equiv k (\mathrm{mod}\, 2). \tag{3.13}$$

Indeed, any homogeneous polynomial P_k of degree k in the variables $x_1, \cdots, x_n$ has on S^{n-1} the canonical expansion

$$P_k(x) = \sum_{s-0}^{[k/2]} h_{k-2s}(x),$$

where the h_j are homogeneous harmonic polynomials of degree j (see [10]). Therefore (3.13) implies (2.3). Conversely, by (2.3) $\int u(x)x^\gamma h_{k-2s}(x)dx = 0$ for every multi-index γ for which $|\gamma| = 2s$. This clearly implies (3.13).

Proposition 3.2. *The subspace of all functions from* $C^\infty(S^{n-1})$ *satisfying* (3.13) *(or, equivalently,* (2.3)*) is invariant under the operator* $\hat{E}(i(k+n/2))$ *(see* (2.5)*). The formula*

$$(\hat{E}(i(k+n/2))Y_{m\mathcal{K}})(\phi) = (-i)^m 2^{-n/2-k} \frac{\Gamma((m-k)/2)}{\Gamma((m+n+k)/2)} Y_{m\mathcal{K}}(\phi)$$

holds.

Proof. Obtained by applying proposition 3.1. ■

§4. Operators related with the transform E(λ)

1. The operator $E(\lambda)^{-1}$. We define for arbitrary complex λ except $\lambda = -i(k+n/2)$, $k = 0, 1, \cdots,$ the following operator on functions v from $C^\infty(S^{n-1})$:

$$(E(\lambda)^{-1}v)(x) = \tag{4.1}$$

$$= \frac{1}{(2\pi)^{n/2}} e^{i\frac{\pi}{2}(n/2-i\lambda)} \Gamma(n/2-i\lambda) \int (xy+i0)^{i\lambda-n/2} v(y)dy,$$

where $x,y \in \mathbb{R}^n$, $|x| = |y| = 1$, and dy is the volume element on S^{n-1}. Proposition 1.2 implies that the function $\lambda \mapsto E(\lambda)^{-1}$: $C^\infty(S^{n-1}) \to C^\infty(S^{n-1})$ is analytic throughout the λ-plane, with the exception of the points indicated, at which it has poles of the first order. The residue at $\lambda = -i(k+n/2)$ is a finite-dimensional operator,

$$\operatorname{res}(E(\lambda)^{-1}v)|_{\lambda=-i(k+n/2)} = \tag{4.2}$$

$$= \frac{1}{(2\pi)^{n/2}} i^{k+1} \sum_{|\gamma|=k} \frac{1}{\gamma!} x^{\gamma} \int y^{\gamma} v(y)dy.$$

If a function v satisfies (2.3), then

$$\lim_{\lambda\to -i(k+n/2)} E(\lambda)^{-1}v = \frac{-1}{(2\pi)^{n/2}} \frac{i^k}{k!} \int (xy)^k \ln(xy+i0)v(y)dy.$$

We denote this operator by $\hat{E}(-i(k+n/2))^{-1}$. We have

$$\hat{E}(-i(k+n/2))^{-1}v = \tag{4.3}$$

$$= \frac{1}{(2\pi)^{n/2}} \frac{i^k}{k!} \int (xy)^k \left\{ \ln \frac{1}{|xy|} + i\frac{\pi}{2}\operatorname{sgn}(xy) \right\} v(y)dy.$$

P r o p o s i t i o n 4.1. *For* $\lambda \neq \pm i(k+n/2)$ *the operators* E(λ) *and* $E(\lambda)^{-1}$ *are inverse to each other.*

P r o o f. By formula (2.8),

$$(Fu)(\rho,\psi) = \frac{1}{\sqrt{2\pi}} \int_{-\infty}^{+\infty} \rho^{-i\lambda-n/2} E_{\phi\to\psi}(\lambda)\tilde{u}(\lambda+in/2,\phi)d\lambda.$$

We apply the inverse Fourier transform. Then

$$u(r,\theta) = \lim_{\tau\to+0} \frac{1}{(2\pi)^{(n+1)/2}} \int_0^{\infty} \int_{S^{n-1}} e^{ir\rho\theta\psi-\tau\rho} \rho^{n-1} d\rho d\psi \times$$

$$\times \int_{-\infty}^{\infty} \rho^{-i\lambda-n/2} E_{\phi\to\psi}(\lambda)\tilde{u}(\lambda+in/2,\phi)d\lambda.$$

Changing the order of integration we obtain, because of (2.10),

$$u(r,\theta) = \frac{1}{(2\pi)^{(n+1)/2}} \int_{-\infty}^{+\infty} \int_{S^{n-1}} r^{i\lambda-n/2} e^{i\frac{\pi}{2}(n/2-i\lambda)} \times$$

$$\times \Gamma(n/2-i\lambda)(\theta\psi+i0)^{i\lambda-n/2}\mathrm{E}_{\phi\to\psi}(\lambda)\tilde{u}(\lambda+in/2,\phi)d\lambda d\psi =$$

$$= \mathrm{M}^{-1}{}_{(\lambda+in/2)\to r}\mathrm{E}_{\psi\to\theta}(\lambda)^{-1}\mathrm{E}_{\phi\to\psi}(\lambda)\tilde{u}(\lambda+in/2,\phi).$$

After applying the Mellin transform $\mathrm{M}_{r\to\lambda+in/2}$ we obtain the equation

$$\tilde{u}(\lambda+in/2,\cdot) = \mathrm{E}(\lambda)^{-1}\mathrm{E}(\lambda)\tilde{u}(\lambda+in/2,\cdot),$$

which was required.

So, $\mathrm{E}(\lambda)^{-1}\mathrm{E}(\lambda) = \mathrm{I}$. By interchanging the roles of the transforms F and F^{-1} we can verify the equality $\mathrm{E}(\lambda)\ \mathrm{E}(\lambda)^{-1} = \mathrm{I}$. ■

P r o p o s i t i o n 4.2. *The formula*

$$(\mathrm{E}(\lambda)^{-1}Y_{m\mathcal{K}})(\phi) = \frac{i^m 2^{-i\lambda}\Gamma((m+n/2-i\lambda)/2)}{\Gamma((m+n/2+i\lambda)/2)}Y_{m\mathcal{K}}(\phi) \tag{4.4}$$

holds.

For the p r o o f it suffices to compare propositions 3.1 and 4.1. ■

We denote by $\mathfrak{H}_k$ the space spanned on the spherical harmonics $\{Y_{m\mathcal{K}}\}$ for which $m \leqslant k$, $m \equiv k \bmod 2$, and by $\mathfrak{H}_k^\perp$ the subspace of functions from $C^\infty(S^{n-1})$ satisfying the conditions (3.13) (or, equivalently, the conditions (2.3)). The subspace $\mathfrak{H}_k^\perp$ is invariant under the operators $\mathrm{E}(\lambda)^{\pm 1}$ for all λ, and also under the operators $\hat{\mathrm{E}}(ik+in/2)$ and $\hat{\mathrm{E}}(-ik-in/2)^{-1}$ given by (2.5) and (4.3). We yet introduce the restrictions $\hat{\mathrm{E}}(ik+in/2)^{-1}$ and $\hat{\mathrm{E}}(-ik-in/2)$ onto the subspace $\mathfrak{H}_k^\perp$.

P r o p o s i t i o n 4.3.

1) *The operators* $\hat{\mathrm{E}}(ik+in/2)^{-1}$ *and* $\hat{\mathrm{E}}(-ik-in/2)$ *are inverses of* $\hat{\mathrm{E}}(ik+in/2)$ *and* $\hat{\mathrm{E}}(-ik-in/2)^{-1}$, *respectively, and are defined by the following formulas:*

$$\hat{\mathrm{E}}(\pm i(k+n/2))^{\mp 1}v = \tag{4.5}$$

$$= \frac{1}{(2\pi)^{n/2}} i^{n+k}\Gamma(n+k)\int(\pm\phi\theta+i0)^{-n-k}v(\theta)d\theta.$$

2) *The operators* $\mathrm{E}(-ik-in/2)$ *and* $\mathrm{E}(ik+in/2)^{-1}$ *annihilate the subspace* $\mathfrak{H}_k$. *Thus,*

$$\hat{E}(\pm i(k+n/2))^{\pm 1}E(\pm i(k+n/2))^{\mp 1}u = u - \mathrm{pr}_k u,$$

where $\mathrm{pr}_k u$ *is the orthogonal projection to* $\mathfrak{H}_k$ *(in the sense of* $L_2(S^{n-1})$*) of the element u.*

P r o o f. The first assertion of the theorem follows from proposition 4.1 and the definitions of the operators (cf. (2.5) and (4.3)). It remains to convince ourselves that $E(-ik-in/2)\mathfrak{H}_k = 0$ and $E(ik+in/2)^{-1}\mathfrak{H}_k = 0$. In order to verify, e.g., the first of these equations one must use formula (3.3) and take into account that at $\lambda = -ik-in/2$ the function $\Gamma((m-i\lambda+n/2)/2)$ has a pole if $m \leqslant k$, $m \equiv k \bmod 2$. The second equation is verified using formula (4.4). ■

2. The operator E(λ)* adjoint to E(λ). We denote by $E(\lambda)^*$ the operator adjoint to $E(\lambda)$ with respect to the scalar product in $L_2(S^{n-1})$.

P r o p o s i t i o n 4.4. *The equation*

$$E(\lambda)^* = E(\bar{\lambda})^{-1} \tag{4.6}$$

holds.

P r o o f. The definition of the generalized function $(-\phi\omega+i0)^\mu$ implies that for arbitrary complex μ,

$$\int\overline{v(\phi)}d\phi\int(-\phi\omega+i0)^\mu u(\omega)d\omega =$$

$$= \int\overline{v(\phi)}d\phi\int[e^{i\pi\mu}(\phi\omega)_+^\mu+(\phi\omega)_-^\mu]u(\omega)d\omega =$$

$$= \int u(\omega)d\omega\int\overline{[e^{-i\pi\bar{\mu}}(\phi\omega)_+^{\bar{\mu}}+(\phi\omega)_-^{\bar{\mu}}]v(\phi)d\phi} =$$

$$= \int u(\omega)d\omega\int\overline{e^{-i\pi\bar{\mu}}(\phi\omega+i0)^{\bar{\mu}}v(\phi)d\phi}.$$

Since $\overline{\Gamma(u)} = \Gamma(\bar{u})$, formulas (2.1) and (4.1) imply (4.6). ■

3. The Fourier transform of a homogeneous function. Let the homogeneous function have the form $G(x) = r^a f(\phi)$, where $r = |x|$, $\phi = x/|x|$, and f is a smooth function on the sphere S^{n-1}. It suffices to calculate the Fourier transform of this homogeneous function for $\mathrm{Re}\, a > -n$; for remaining a the formula is obtained as the result of analytic extension. We have

$$\mathrm{F}G = \frac{1}{(2\pi)^{n/2}} \lim_{\tau \to +0} \int\int e^{-ir\rho\phi\psi - \tau r} f(\phi) r^{a+n-1} dr d\phi =$$

$$= \frac{1}{(2\pi)^{n/2}} \lim_{\tau \to +0} \int f(\phi) d\phi \int e^{-ir\rho\phi\psi - \tau r} r^{a+n-1} dr,$$

where $\rho = |\xi|, \psi = \xi / |\xi|$. This and (2.10) implies

$$(\mathrm{F}G)(\xi) = \frac{1}{(2\pi)^{n/2}} \rho^{-a-n} e^{i(a+n)\pi/2} \Gamma(a+n) \int (-\phi\psi + i0)^{-a-n} f(\phi) d\phi.$$

Put $\Phi(\psi) = \rho^{a+n}(\mathrm{F}G)(\xi)$. Recalling formula (2.1) we are led to

$$\Phi = \mathrm{E}(-i(a+n/2))f. \tag{4.7}$$

If $a = -n-k$, where k is a nonnegative integer, then in order for G to be a homogeneous function it is necessary and sufficient for f to satisfy (2.3). In this case (4.7) can be changed to

$$\Phi(\psi) = \hat{\mathrm{E}}_{\phi \to \psi}(ik + in/2) f(\phi) = \tag{4.8}$$

$$= \frac{1}{(2\pi)^{n/2}} \frac{(-i)^k}{k!} \int (\phi\psi)^k \left\{ \ln \frac{1}{|\phi\psi|} - i\frac{\pi}{2} \operatorname{sgn}(\phi\psi) \right\} f(\phi) d\phi.$$

In drawing this paragraph to an end we give some remarks.

R e m a r k 4.5. In the theory of singular integral operators the function f is called a *characteristic*, and Φ a *symbol* (cf. [40]). Formulas (4.7), (4.8) (as well as (2.1), (4.1), and (4.3)) express one of these functions in terms of the other, for arbitrary a. In case $k = 0$ equation (4.8) is called the *Calderón-Zygmund formula.*

R e m a r k 4.6. Extend $u_+, u_- \in C^\infty(S^{n-1})$ to $\mathbb{R}^n \setminus 0$ as (positively) homogeneous functions of degrees $i\lambda - n/2$ and $-i\lambda - n/2$, respectively, i.e. assume that $u_+(ty) = t^{i\lambda - n/2} u_+(y)$, $u_-(ty) = t^{-i\lambda - n/2} u_-(y)$ for all $y \in \mathbb{R}^n \setminus 0$ and $t > 0$. Introduce the exterior differential form $\omega(y) = y_1 dy_2 \wedge \cdots \wedge dy_n + \cdots + (-1)^{n-1} y_n dy_1 \wedge \cdots \wedge dy_{n-1}$. Denote by S an arbitrary $(n-1)$-dimensional smooth surface in $\mathbb{R}^n$ encircling the coordinate origin. Then the operators $\mathrm{E}(\lambda)^{\pm 1}$ allow the representations

$$(\mathrm{E}(\lambda)^{\pm 1}u_{\pm})(x) = \frac{1}{(2\pi)^{n/2}} e^{i\frac{\pi}{2}(n/2\pm i\lambda)} \Gamma(n/2\pm i\lambda) \times \tag{4.9}$$

$$\times \int_S (\mp xy + i0)^{-n/2\mp i\lambda} u_{\pm}(y)\omega(y).$$

Since the integrand is homogeneous of degree 0 (in y), the 'integral' is independent of the choice of S. It is clear that the operator E(λ) (resp. $\mathrm{E}(\lambda)^{-1}$) thus defined maps homogeneous functions of degree $i\lambda - n/2$ (resp. $-i\lambda - n/2$) to homogeneous functions of degree $-i\lambda - n/2$ (resp. $i\lambda - n/2$).

R e m a r k 4.7. Consider the operators $\mathrm{E}(\lambda)^{\pm 1}$ for $n = 1$. In the one-dimensional case x and y take the two values ± 1, and the 'integral' over the zero-dimensional sphere is the sum $u(1) + u(-1)$. Hence for $n = 1$ we have the formulas

$$\begin{bmatrix} (\mathrm{E}(\lambda)u)(1) \\ (\mathrm{E}(\lambda)u)(-1) \end{bmatrix} = \frac{\Gamma(1/2+i\lambda)}{\sqrt{2\pi}} \times \tag{4.10}$$

$$\times \begin{bmatrix} e^{-\frac{i\pi}{2}\left[\frac{1}{2}+i\lambda\right]} & e^{\frac{i\pi}{2}\left[\frac{1}{2}+i\lambda\right]} \\ e^{\frac{i\pi}{2}\left[\frac{1}{2}+i\lambda\right]} & e^{-\frac{i\pi}{2}\left[\frac{1}{2}+i\lambda\right]} \end{bmatrix} \begin{bmatrix} u(1) \\ u(-1) \end{bmatrix},$$

$$\begin{bmatrix} (\mathrm{E}(\lambda)^{-1}u)(1) \\ (\mathrm{E}(\lambda)^{-1}u)(-1) \end{bmatrix} = \frac{\Gamma(1/2-i\lambda)}{\sqrt{2\pi}} \times \tag{4.11}$$

$$\times \begin{bmatrix} e^{\frac{i\pi}{2}\left[\frac{1}{2}-i\lambda\right]} & e^{-\frac{i\pi}{2}\left[\frac{1}{2}-i\lambda\right]} \\ e^{-\frac{i\pi}{2}\left[\frac{1}{2}-i\lambda\right]} & e^{\frac{i\pi}{2}\left[\frac{1}{2}-i\lambda\right]} \end{bmatrix} \begin{bmatrix} u(1) \\ u(-1) \end{bmatrix}.$$

R e m a r k 4.8. Let

$$J^{+}_{\mu,\epsilon}(x) = \int_{S^{n-1}\cap\{y:|xy|>\epsilon\}} (-xy+i0)^{\mu} u(y)dy.$$

If for small positive ϵ this integral can be represented as

$$J^{+}_{\mu,\epsilon}(x) = a_\epsilon(x)+b(x)\ln\epsilon^{-1}+\sum_{k=1}^{N} a_k(x)\epsilon^{-\lambda_k},$$

where $\mathrm{Re}\lambda_k > 0$, and if a finite limit $a_0(x) = \lim_{\epsilon\to 0} a_\epsilon(x)$ exists, then we say that the principal value (of the) integral

$$\int(-xy+i0)^{\mu}u(y)dy$$

exists. By definition

$$\text{p.v.}\int(-xy+i0)^{\mu}u(y)dy = a_0(x).$$

It turns out that the principal value exists for $\mu \neq -1,-2,\cdots$, and coincides with the integral J^{+}_{μ} defined in §1 (formula (1.6)). For details see [62].

§5. The spaces $H^s(\lambda,S^{n-1})$. The operator E(λ) on the spaces $H^s(\lambda,S^{n-1})$

In this paragraph we introduce for an arbitrary real number s the spaces $H^s(\lambda,S^{n-1})$ of (generalized) functions on the sphere, with norms depending on the parameter $\lambda\in\mathbf{C}$. For each definite value of this parameter the norm in $H^s(\lambda,S^{n-1})$ is equivalent to the norm in the Sobolev-Slobodetskii space $W^s_2(S^{n-1})\equiv H^s(S^{n-1})$. It turns out that the maps $\mathrm{E}(\lambda)^{\pm 1}$: $H^s(\lambda,S^{n-1})\to H^{s\pm\mathrm{Im}\lambda}(\lambda,S^{n-1})$ are continuous.

1. The Sobolev-Slobodetskii spaces $H^s(\mathfrak{M})$ on a manifold $\mathfrak{M}$. For an arbitrary real s the space $H^s(\mathbf{R}^n)$ is defined as the completion of the set $C^\infty_0(\mathbf{R}^n)$ with respect to the norm

$$\|u;H^s(\mathbf{R}^n)\| = \left[\int|(\mathrm{F}u)(\xi)|^2(1+|\xi|^2)^s d\xi\right]^{1/2}. \tag{5.1}$$

For nonnegative integer s this norm is equivalent to the norm

$$\left[\sum_{|\alpha|\leq s}\int|(D^\alpha u)(x)|^2dx\right]^{1/2},$$

and for arbitrary $s>0$ it is equivalent to the norm

$$\left[\|u;H^{[s]}(\mathbf{R}^n)\|^2+\sum_{|\alpha|=[s]}\int_{\mathbf{R}^n}\int_{\mathbf{R}^n}\frac{|(D^\alpha u)(x)-(D^\alpha u)(y)|^2}{|x-y|^{n+2s-2[s]}}dx\,dy\right]^{1/2}.$$

Let $\mathfrak{M}$ be an n-dimensional compact manifold without boundary, of class C^∞, and let $\{U, \chi\}$ be some atlas on it, i.e. $\{U\}$ is a finite open cover of $\mathfrak{M}$ and χ: $U \to \mathbb{R}^n$ are coordinate maps. Let $\{\zeta\}$ be a partition of unity subordinate to this cover. The space $H^s(\mathfrak{M})$ is the completion of the set $C^\infty(\mathfrak{M})$ with respect to the norm

$$\|u;H^s(\mathfrak{M})\| = \left[\sum_U \|\zeta_\chi u;H^s(\mathbb{R}^n)\|^2\right]^{1/2}, \tag{5.2}$$

where $\zeta_\chi u = \zeta u \circ \chi^{-1}$ on $\chi(U)$, and $\zeta_\chi u = 0$ outside $\chi(U)$. Another partition of unity and another, equivalent, atlas lead to an equivalent norm in $H^s(\mathfrak{M})$. For $s_1 < s_2$ the space $H^{s_2}(\mathfrak{M})$ compactly belongs to $H^{s_1}(\mathfrak{M})$. The spaces $H^s(\mathfrak{M})$ and $H^{-s}(\mathfrak{M})$ are dual with respect to the scalar product in $H^0(\mathfrak{M})$.

2. Expansion in spherical harmonics. In the sequel we conveniently denote the spherical harmonics of order m ($m = 0, 1 \cdots$) by Y_{mk}. The index k enumerates the harmonics of the same order, $k = 1, \cdots, k_m$, where $k_m = (2m+n-2)\times$ $\times(m+n-3)![(n-2)!m!]^{-1}$, i.e. $k_m = O(m^{n-2})$ as $m \to \infty$. The functions Y_{mk} form a complete orthonormal system in $H^0(S^{n-1}) = L_2(S^{n-1})$. Hence for $u \in H^0(S^{n-1})$ the series $u = \Sigma_{k,m} u_{mk} Y_{mk}$ converges in $H^0(S^{n-1})$, and the Parseval equality $\|u;H^0(S^{n-1})\|^2 = \Sigma_{m,k} |u_{mk}|^2$ holds. Here $u_{mk} = (u, Y_{mk})$ (scalar product in $H^0(S^{n-1})$).

Let δ be the Laplace-Beltrami operator on S^{n-1}:

$$\delta = -\sum_{j=1}^{n} \frac{1}{q_j \sin^{n-j-1}\theta_{n-j}} \cdot \frac{\partial}{\partial \theta_{n-j}} \left[\sin^{n-j-1}\theta_{n-j} \frac{\partial}{\partial \theta_{n-j}}\right],$$

where $\theta_1, \cdots, \theta_{n-1}$ are the angular coordinates (see (1.2)), $q_1 = 1$, $q_j = (\sin\theta_{n-1} \cdots \sin\theta_{n-j+1})^2$. The Laplace operator Δ on $\mathbb{R}^n$, written in spherical coordinates, has the form

$$\Delta = \frac{\partial^2}{\partial r^2} + \frac{n-1}{r}\frac{\partial}{\partial r} - \frac{1}{r^2}\delta. \tag{5.3}$$

On $H^0(S^{n-1})$, the operator δ is nonnegative and has a selfadjoint Friedrichs extension, which we will also denote by δ. The spectrum of δ consists of the eigenvalues $\lambda_m = m(m+n-2)$, $m = 0, 1, \cdots$. To each λ_m correspond the k_m orthonormal eigenfunctions Y_{mk}. The domain of definition of a power $(I+\delta)^s$, $s > 0$, consists of only those functions v for which the series

$$\|(I+\delta)^s v; H^0(S^{n-1})\|^2 = \sum_{m,k}(1+m(m+n-2))^{2s}|v_{mk}|^2 \tag{5.4}$$

converges. The (generalized) functions v on the sphere for which (5.4) converges belong to the domain of definition of $(I+\delta)^s$ also for $s<0$.

The aim of the present paragraph is to prove the following assertion.

P r o p o s i t i o n 5.1. *For arbitrary $s \in \mathbb{R}$ the inequalities*

$$c_1\|v; H^{2s}(S^{n-1})\| \leqslant \|(I+\delta)^s v; H^0(S^{n-1})\| \leqslant c_2\|v; H^{2s}(S^{n-1})\|$$

hold.

The verification of these inequalities is subdivided into several lemmas.

L e m m a 5.2. *For $u \in C_0^\infty(\mathbb{R}^n)$ the formula*

$$(I+\delta)^s \mathrm{F}u = \mathrm{F}(I+\delta)^s u$$

holds.

P r o o f. It suffices to check $\delta \mathrm{F}u = \mathrm{F}\delta u$. Using (5.3) we obtain

$$\delta_\psi F_{x\to\xi}u(x) = \mathrm{F}_{x\to\xi}\{[(ir\rho\phi\psi)^2 + i(n-1)r\rho\phi\psi - \Delta_x r^2]u(x)\}, \tag{5.5}$$

where $r = |x|$, $\phi = x/|x|$, $\psi = \xi/|\xi|$. Since

$$\Delta(r^2u) = \frac{\partial^2}{\partial r^2}(r^2u) + \frac{n-1}{r}\frac{\partial}{\partial r}(r^2u) - \delta u,$$

it remains to convince ourselves that

$$\mathrm{F}\left\{\left[(ir\rho\phi\psi)^2 + i(n-1)r\rho\phi\psi - \frac{\partial^2}{\partial r^2}r^2 - \frac{n-1}{r}\frac{\partial}{\partial r}r^2\right]u(x)\right\} = 0.$$

For this we must integrate by parts the two last terms. ■

Let $\eta \in C^\infty(\mathbb{R})$, $\eta = 1$ on the (closed) interval [3/4, 5/4] and $\eta = 0$ outside the interval [1/2, 3/2]. For $v \in C^\infty(S^{n-1})$, put $\mathcal{V}(r\phi) = \eta(r)v(\phi)$. It is easy to check that

$$c_1\|\mathcal{V}; H^s(\mathbf{R}^n)\| \leqslant \|v\,; H^s(S^{n-1})\| \leqslant c_2\|\mathcal{V}; H^s(\mathbf{R}^n)\|, \tag{5.6}$$

with constants c_1, c_2 not depending on v.

L e m m a 5.3. *For* $0<s<1$ *the estimates*

$$c_1\|\mathcal{V}; H^{2s}(\mathbf{R}^n)\| \leqslant \|(\mathrm{I}+\delta)^s\mathcal{V}; H^0(\mathbf{R}^n)\| \leqslant c_2\|\mathcal{V}; H^{2s}(\mathbf{R}^n)\| \tag{5.7}$$

hold.

P r o o f. By Parseval's equality and Lemma 5.2,

$$\|(\mathrm{I}+\delta)^s\mathcal{V}; H^0(\mathbf{R}^n)\|^2 = \int_0^\infty \rho^{n-1}d\rho \int_{S^{n-1}} |(\mathrm{I}+\delta)^s \mathrm{F}\mathcal{V}|^2 d\psi. \tag{5.8}$$

Taking into account the assertions concerning the spectrum of the operator δ we have

$$\int |(1-\delta)^s\mathrm{F}\mathcal{V}|^2 d\psi =$$

$$= \sum_m \left[(1+m(m+n-2))^2 \sum_k |(\mathrm{F}\mathcal{V})_{mk}|^2\right]^2 \left[\sum |(\mathrm{F}\mathcal{V})_{mk}|^2\right]^{1-s}.$$

Now the Hölder inequality with exponents s^{-1} and $(1-s)^{-1}$ gives

$$\int |(\mathrm{I}+\delta)^2\mathrm{F}\mathcal{V}|^2 d\psi \leqslant \left[\int |(\mathrm{I}+\delta)\mathrm{F}\mathcal{V}|^2 d\psi\right]^s \left[\int |\mathrm{F}\mathcal{V}|^2 d\psi\right]^{1-s}. \tag{5.9}$$

Formula (5.5) implies the estimate

$$|(\mathrm{I}+\delta)\mathrm{F}\mathcal{V}|^2 \leqslant (1+\rho^2)^2 \sum_j |\mathrm{F}\chi_j\mathcal{V}|^2, \tag{5.10}$$

where $\chi_j \in C^\infty(\mathbf{R}^n)$ and the sum is finite. By (5.8) - (5.10),

$$\|(\mathrm{I}+\delta)^2\mathcal{V}; H^0(\mathbf{R}^n)\|^2 \leqslant$$

$$\leqslant \int_0^\infty (1+\rho^2)^{2s}\rho^{n-1}d\rho \left[\sum_j \int |\mathrm{F}\chi_j\mathcal{V}|^2 d\psi\right]^s \left[\int |\mathrm{F}\mathcal{V}|^2 d\psi\right]^{1-s} \leqslant$$

$$\leqslant c\int_0^\infty (1+\rho^2)^{2s}\rho^{n-1}d\rho \int \left[\sum_j |\mathrm{F}\chi_j\mathcal{V}|^2 + |\mathrm{F}\mathcal{V}|^2\right] d\psi$$

(we have used Young's inequality $a^s b^{1-s} \leqslant sa+(1-s)b$). Hence

$$\|(\mathrm{I}+\delta)^s \mathcal{V};\, H^0(\mathbb{R}^n)\|^2 \leqslant c\left[\|\mathcal{V};\, H^{2s}(\mathbb{R}^n)\|^2 + \left\|\sum_j \chi_j \mathcal{V};\, H^{2s}(\mathbb{R}^n)\right\|^2\right].$$

Since $\chi_j \in C^\infty(\mathbb{R}^n)$ and $\mathcal{V}(r,\phi) = 0$ for $r > 3/2$, we have

$$\|(\mathrm{I}+\delta)^s \mathcal{V};\, H^0(\mathbb{R}^n)\| \leqslant c\|\mathcal{V};\, H^{2s}(\mathbb{R}^n)\|. \tag{5.11}$$

One of the inequalities in (5.7) has been proved. Let us verify the second. We have

$$J = \|\mathcal{V};\, H^{2s}(\mathbb{R}^n)\|^2 \leqslant \int_0^\infty (1+\rho^2)^{2s}\rho^{n-1}d\rho \int_{S^{n-1}} |(\mathrm{I}+\delta)^{-s}\mathrm{F}(\mathrm{I}+\delta)^s\mathcal{V}|^2 d\psi.$$

The integral over the sphere is at most

$$\left[\int |(\mathrm{I}+\delta)^{-1}\mathrm{F}(\mathrm{I}+\delta)^s\mathcal{V}|^2 d\psi\right]^s \left[\int |\mathrm{F}(\mathrm{I}+\delta)^s\mathcal{V}|^2 d\psi\right]^{1-s}$$

(this estimate can be derived similarly to (5.9)). By Hölder's inequality,

$$J \leqslant \left[\int_0^\infty (1+\rho^2)^2\rho^{n-1}d\rho \int |(\mathrm{I}+\delta)^{-1}\mathrm{F}(\mathrm{I}+\delta)^s\mathcal{V}|^2 d\psi\right]^s \times \tag{5.12}$$

$$\times \left[\int_0^\infty \rho^{n-1}d\rho\left[\int |\mathrm{F}(\mathrm{I}+\delta)^s\mathcal{V}|^2 d\psi\right]\right]^{1-s} \leqslant$$

$$\leqslant \left[\int (1+|\xi|^2)^2 |\mathrm{F}(\mathrm{I}+\delta)^{s-1}\mathcal{V}|^2 d\xi\right]^s \left[\int |(\mathrm{I}+\delta)^s\mathcal{V}|^2 d\xi\right]^{1-s}.$$

It is clear that

$$\int (1+|\xi|^2)^2 |\mathrm{F}(\mathrm{I}+\delta)^{s-1}\mathcal{V}|^2 d\xi = \int |(-\Delta+\mathrm{I})(\mathrm{I}+\delta)^{s-1}\mathcal{V}|^2 dx. \tag{5.13}$$

Using (5.3) we find that the righthand side does not exceed

$$c\left[\int |(\mathrm{I}+\delta)^{s-1}v|^2 d\psi + \int |(\mathrm{I}+\delta)^{s-1}\delta v|^2 d\psi\right] \leqslant c\int |(\mathrm{I}+\delta)^s\mathcal{V}|^2 dx. \tag{5.14}$$

Combining (5.12) - (5.14) we are led to the estimate $J \leqslant \|(\mathrm{I}+\delta)^s\mathcal{V};\; H^0(\mathbb{R}^n)\|^2$, which together with (5.11) gives (5.7). ■

L e m m a 5.4. *For* $\sigma > 0$ *the inequalities*

$$c_1 \|\mathcal{V}; H^{2\sigma+2}(\mathbf{R}^n)\| \leqslant \|(I+\delta)\mathcal{V}; H^{2\sigma}(\mathbf{R}^n)\| \leqslant c_2 \|\mathcal{V}; H^{2\sigma+2}(\mathbf{R}^n)\| \tag{5.15}$$

hold.

P r o o f. Formula 5.3 and the equation $\mathcal{V} = \eta v$ imply that $(I+\delta)\mathcal{V}(r,\phi) = \chi(r) v(\phi) + (I - r^2\Delta)\mathcal{V}(r,\phi)$, where $\chi \in C_0^\infty(\mathbf{R}_+^n)$. Hence

$$\big|\, \|(I+\delta)\mathcal{V}; H^{2\sigma}(\mathbf{R}^n)\| - \|(I-r^2\Delta)\mathcal{V}; H^{2\sigma}(\mathbf{R}^n)\| \,\big| \leqslant c \|\chi v; H^{2\sigma}(\mathbf{R}^n)\|.$$

Since $\mathcal{V}$ vanishes outside the annulus $1/2 < r < 3/2$,

$$c_1 \|(I-\Delta)\mathcal{V}; H^{2\sigma}(\mathbf{R}^n)\| \leqslant$$
$$\leqslant \|(I-r^2\Delta)\mathcal{V}; H^{2\sigma}(\mathbf{R}^n)\| \leqslant c_2 \|(I-\Delta)\mathcal{V}; H^{2\sigma}(\mathbf{R}^n)\|.$$

Hence

$$\|(I+\delta)\mathcal{V}; H^{2\sigma}(\mathbf{R}^n)\| \leqslant c(\|\mathcal{V}; H^{2\sigma+2}(\mathbf{R}^n)\| + \|\chi v; H^{2\sigma}(\mathbf{R}^n)\|), \tag{5.16}$$

$$\|\mathcal{V}; H^{2\sigma+2}(\mathbf{R}^n)\| \leqslant c(\|(I+\delta)\mathcal{V}; H^{2\sigma}(\mathbf{R}^n)\| + \|\chi v; H^{2\sigma}(\mathbf{R}^n)\|). \tag{5.17}$$

Using inequalities of the type of (5.6) we obtain

$$\|\chi v; H^{2\sigma}(\mathbf{R}^n)\| \leqslant c \|v; H^{2\sigma}(S^{n-1})\| \leqslant c \|\mathcal{V}; H^{2\sigma}(\mathbf{R}^n)\|.$$

Together with (5.16) this leads to the estimate $\|(I+\delta)\mathcal{V}; H^{2\sigma}(\mathbf{R}^n)\| \leqslant c \|\mathcal{V}; H^{2\sigma+2}(\mathbf{R}^n)\|$, i.e. to the inequality on the right in (5.15).

Let us now prove the left inequality. We have

$$\|\mathcal{V}; H^{2\sigma}(\mathbf{R}^n)\|^2 = \int (1+|\xi|^2)^{2\sigma} |F\mathcal{V}|^2 d\xi \leqslant \tag{5.18}$$
$$\left[\int (1+|\xi|^2)^{2\sigma+2} |F\mathcal{V}|^2 d\xi\right]^{\sigma/(\sigma+1)} \left[\int |F\mathcal{V}|^2 d\xi\right]^{1/(\sigma+1)}.$$

The inequality $\|\mathcal{V}; H^0(\mathbf{R}^n)\| \geqslant c \|(I+\delta)\mathcal{V}; H^0(\mathbf{R}^n)\|$ and estimate (5.18) imply

$$\|\chi v; H^{2\sigma}(\mathbf{R}^n)\| \leqslant c \|\mathcal{V}; H^{2\sigma+2}(\mathbf{R}^n)\|^{\sigma/(\sigma+1)} \|(I+\delta)\mathcal{V}; H^{2\sigma}(\mathbf{R}^n)\|^{1/(\sigma+1)}.$$

By using (5.17) and Young's inequality $a^s b^{1-s} \leqslant s\epsilon^{1/s} a + (1-s) b \epsilon^{s-1}$ for $s = \sigma/(\sigma+1)$ and a sufficiently small $\epsilon > 0$ the proof is finished. ■

P r o o f o f p r o p o s i t i o n 5.1. Let initially $s \geqslant 0$. By (5.6) it suffices to convince ourselves that

$$c_1\|\mathcal{V}; H^{2s}(\mathbf{R}^n)\| \leqslant \|(I+\delta)^s\mathcal{V}; H^0(\mathbf{R}^n)\| \leqslant c_2\|\mathcal{V}; H^{2s}(\mathbf{R}^n)\|. \tag{5.19}$$

Put $s = [s]+t$. By lemma 5.3,

$$c_1\|(I+\delta)^{[s]}\mathcal{V}; H^{2t}(\mathbf{R}^n)\| \leqslant$$

$$\leqslant \|(I+\delta)^s\mathcal{V}; H^0(\mathbf{R}^n)\| \leqslant c_2\|(I+\delta)^{[s]}\mathcal{V}; H^{2t}(\mathbf{R}^n)\|.$$

Lemma 5.4 implies

$$c_1\|\mathcal{V}; H^{2s}(\mathbf{R}^n)\| \leqslant \|(I+\delta)^{[s]}\mathcal{V}; H^{2t}(\mathbf{R}^n)\| \leqslant c_2\|\mathcal{V}; H^{2s}(\mathbf{R}^n)\|.$$

This and the previous inequalities give (5.19). The case $s<0$ is obtained by transition to the dual spaces with respect to duality in $H^0(S^{n-1})$. ■

3. The operator $\mathrm{E}(\lambda)$ **on the spaces** $H^s(\lambda, S^{n-1})$. We denote by $H^s(\lambda, S^{n-1})$ the space of functions on the sphere obtained by completing the set $C^\infty(S^{n-1})$ with respect to the norm

$$\|v; H^s(\lambda, S^{n-1})\| = \left[\sum_{m,k}(1+m^2+|\lambda|^2)^s|v_{mk}|^2\right]^{1/2}. \tag{5.20}$$

Equation (5.4) and proposition 5.1 imply that for each fixed $\lambda\in\mathbf{C}$ the norm (5.20) is equivalent to the usual norm in the Sobolev-Slobodetskii space $H^s(S^{n-1})$.

P r o p o s i t i o n 5.5. *If* $\lambda\neq i(k+n/2)$ *(or* $\lambda\neq -i(k+n/2)$*), where* $k=0,1,\cdots,$ *then the map* $\mathrm{E}(\lambda)\colon H^s(\lambda, S^{n-1})\to H^{s+\mathrm{Im}\lambda}(\lambda, S^{n-1})$ *(resp.* $\mathrm{E}(\lambda)^{-1}\colon H^s(\lambda, S^{n-1})\to H^{s-\mathrm{Im}\lambda}(\lambda, S^{n-1})$*) is continuous. On every closed set* $\mathcal{F}$ *lying in a strip* $|\mathrm{Im}\lambda|<h$ *and not containing the points* $\lambda = i(k+n/2)$ *(resp.* $\lambda=-i(k+n/2)$*),* $k=0,1,\cdots,$ *the estimate*

$$\|\mathrm{E}(\lambda); H^s(\lambda, S^{n-1})\to H^{s+\mathrm{Im}\lambda}(\lambda, S^{n-1})\| \leqslant c(\mathcal{F}) \tag{5.21}$$

(resp.

$$\|\mathrm{E}(\lambda)^{-1}; H^s(\lambda, S^{n-1})\to H^{s-\mathrm{Im}\lambda}(\lambda, S^{n-1})\| \leqslant c(\mathcal{F})) \tag{5.22}$$

holds.

P r o o f. By proposition 3.1,

$$\mathrm{E}(\lambda)v = \sum_{m,k} \mu_m(\lambda) v_{mk} Y_{mk}, \tag{5.23}$$

where

$$\mu_m(\lambda) = (-i)^m 2^{i\lambda} \frac{\Gamma((m+n/2+i\lambda)/2)}{\Gamma((m+n/2-i\lambda)/2)}. \tag{5.24}$$

Put $\lambda = \sigma + i\tau$. Since $\overline{\Gamma(z)} = \Gamma(\bar{z})$, we have $|\Gamma((m+n/2-i\lambda)/2)| = |\Gamma((m+n/2+i\sigma+\tau)/2)|$. Using a well-known asymptotic formula for a quotient of Γ-functions (cf. [5]) we obtain on $\mathfrak{F}$, as $|m+i\sigma| \to \infty$,

$$|\mu_m(\lambda)| = 2^{-\tau} \left| \frac{\Gamma((m+n/2+i\sigma-\tau)/2)}{\Gamma((m+n/2+i\sigma+\tau)/2)} \right| = \tag{5.25}$$

$$= |m+i\sigma+n/2|^{-\tau}(1+O(|m+i\sigma|^{-1})).$$

Because for $\lambda \neq i(k+n/2)$ none of the functions μ_m, $m = 0, 1, \cdots$, can have a pole, (5.23) - (5.25) imply that the map $\mathrm{E}(\lambda)$: $H^s(\lambda, S^{n-1}) \to H^{s+\mathrm{Im}\lambda}(\lambda, S^{n-1})$ is continuous, and that (5.21) holds. The corresponding properties for $\mathrm{E}(\lambda)^{-1}$ are verified in a similar manner. ■

Combining propositions 4.4 and 5.5 gives

C o r o l l a r y 5.6. *For* $\mathrm{Im}\lambda = 0$ *the operators* $\mathrm{E}(\lambda)$: $H^0(S^{n-1}) \to H^0(S^{n-1})$ *are unitary.*

§6. An analog of the Paley-Wiener theorem for the operator $\mathrm{E}(\lambda)$*)

Put $S_+^{n-1} = \{x=(x', x_n) \in \mathbb{R}^n\colon \ |x| = 1, \ x_n > 0\}$, $S_-^{n-1} = \{x=(x', x_n) \in \mathbb{R}^n\colon |x| = 1, \ x_n < 0\}$. The main result in this paragraph consists, roughly speaking, of the fact that the operator $\mathrm{E}(\lambda)$ establishes an isomorphism between the set of functions with support in the halfsphere S_+^{n-1} (or in S_-^{n-1}) and the set of homogeneous functions of degree $-i\lambda - n/2$ having an analytic extension in the last coordinate to the lower (resp. upper) complex halfplane.

First we will prove several helpful propositions. Let $\omega = (\omega', \omega_n) \in \mathbb{R}^n$, $|\omega| = 1$, $\omega' \neq 0$. Put $\mathring{\omega} = \omega'/|\omega'|$, $s = \omega_n/|\omega'|$. The map $\omega \mapsto (\mathring{\omega}, s)$

*) This paragraph is used in Chapter 6 only.

transforms the sphere S^{n-1} with poles $(0,\pm 1)$ deleted onto the cylinder $S^{n-2}\times\mathbb{R}$. We denote the inverse map by κ. Clearly, $\omega = (\omega',\omega_n) = \kappa(\mathring{\omega},s) = (\mathring{\omega}(1+s^2)^{-1/2},\ s(1+s^2)^{-1/2})$.

L e m m a 6.1. *For $u\in C^\infty(S^{n-1})$ and $\lambda\in\mathbb{R}$ the equation*

$$(\mathrm{E}(\lambda)u)\circ\kappa(\mathring{\omega},s) = \tag{6.1}$$

$$= \frac{1}{(2\pi)^{3/2}}p_\lambda(s)\int_{-\infty}^{+\infty}|s|^{i(\mu-\lambda)-1/2}\Gamma(i(\lambda-\mu)+1/2)\mathrm{E}'_{\psi\to\mathring{\omega}}(\mu)d\mu\ \times$$

$$\times\int_{-\infty}^{+\infty}e^{\operatorname{sgn}(st)(\lambda-\mu-i/2)\pi/2}p_{-\lambda}(t)^{-1}|t|^{i(\mu-\lambda)-1/2}(u\circ\kappa)(\psi,t)dt$$

holds, where $p_\lambda(s) = (1+s^2)^{(i\lambda+n/2)/2}$ and $\mathrm{E}'(\lambda)$ is 'the operator $\mathrm{E}(\lambda)$ on the sphere $S^{n-2} = \{(\omega',\omega_n)\in S^{n-1}:\omega_n=0\}$', i.e. $\mathrm{E}'(\lambda)$ is defined by formula (2.1), in which n must be replaced by $n-1$.

P r o o f. We denote by F' and $\mathfrak{F}$ the Fourier transform with respect to the coordinates $x' = (x_1,\cdots,x_{n-1})$ and x_n, respectively.

Let $f\in C_0^\infty(\mathbb{R}_+)$, $\tilde{f}(\lambda+in/2) = 1$. Applying proposition 2.1 we find

$$\mathrm{E}_{\psi\to\omega}(\lambda)u(\psi) = \mathrm{M}_{|\xi|\to in/2-\lambda}\mathrm{F}_{x\to\xi}\mathrm{M}^{-1}_{\mu+in/2\to|x|}\tilde{f}(\mu+in/2)u(\psi), \tag{6.2}$$

$$\mathrm{F}'_{x'\to\xi'} = \mathrm{M}^{-1}_{i(n-1)/2-\nu\to|\xi'|}\mathrm{E}'_{\psi\to\xi'/|\xi'|}(\nu)\mathrm{M}_{|x'|\to\nu+i(n-1)/2}.$$

Putting $(\xi',\xi_n) = |\xi|(1+s^2)^{-1/2}(\mathring{\omega},s)$ we rewrite the equation $\mathrm{F} = \mathfrak{F}\mathrm{F}'$ in the form

$$\mathrm{F}_{x\to\xi} = \mathfrak{F}_{x_n\to|\xi|s(1+s^2)^{-1/2}}\mathrm{M}^{-1}_{i(n-1)/2-\nu\to|\xi|(1+s^2)^{-1/2}}\ \times$$

$$\times\ \mathrm{E}'_{\psi\to\mathring{\omega}}(\nu)\mathrm{M}_{|x'|\to\nu+i(n-1)/2}.$$

Substitute this in (6.2). Then

$$(\mathrm{E}_{\psi\to\omega}(\lambda)u)\circ\kappa(\mathring{\omega},s) = \mathrm{M}_{|\xi|\to in/2-\lambda}\mathfrak{F}_{x_n\to|\xi|s(1+s^2)^{-1/2}}\ \times$$

$$\times\ \mathrm{M}^{-1}_{i(n-1)/2-\nu\to|\xi|(1+s^2)^{-1/2}}\mathrm{E}'_{\psi\to\mathring{\omega}}(\nu)\mathrm{M}_{|x'|\to\nu+i(n-1)/2}\ \times$$

$$\times\ \mathrm{M}^{-1}_{\mu+in/2\to|x|}\tilde{f}(\mu+in/2)u(\kappa(\mathring{\omega},x_n/|x'|)).$$

By the change of variables $\rho = |\xi|(1+s^2)^{-1/2}$ we are led to the relation

$$(E_{\psi\to\omega}(\lambda)u)\circ\kappa(\mathring{\omega},s) =$$
$$= p_\lambda(s)M_{\rho\to in/2-\lambda}\mathfrak{F}_{x_n\to\rho s}M^{-1}_{i(n-1)/2-\nu\to\rho}\times$$
$$\times E'_{\psi\to\mathring{\omega}}(\nu)M_{|x'|\to\nu+i(n-1)/2}\times$$
$$\times M^{-1}_{\mu+in/2\to|x|}\tilde{f}(\mu+in/2)u(\kappa(\psi,x_n/|x'|)).$$

As the result of the new change $x = \rho^{-1}y$ this equation takes the form

$$(E_{\psi\to\omega}(\lambda)u)\circ\kappa(\mathring{\omega},s) = \tag{6.3}$$
$$= p_\lambda(s)M_{\rho\to i(in/2-\lambda)}\rho^{-1}\mathfrak{F}_{y_n\to s}M^{-1}_{i(n-1)/2-\nu\to\rho}\times$$
$$\times E'_{\psi\to\mathring{\omega}}(\nu)M_{|y'|\rho^{-1}\to\nu+i(n-1)/2}M^{-1}_{\mu+in/2\to y\rho^{-1}}\times$$
$$\times\tilde{f}(\mu+in/2)u(\kappa(\psi,y_n/|y'|)) =$$
$$= p_\lambda(s)M_{\rho\to(-\lambda)}\mathfrak{F}_{y_n\to s}M^{-1}_{-\nu\to\rho}\rho^{i\nu}E'_{\psi\to\mathring{\omega}}(\nu)M_{|y'|\to i(n-1)/2+\nu}\times$$
$$\times M^{-1}_{\mu+in/2\to|y|}\rho^{-i\mu}\tilde{f}(\mu+in/2)u(\kappa(\psi,y_n/|y'|)).$$

Recall the inversion formula (1.2.6). We see that composition of the operators $M^{-1}_{-\nu\to\rho}\rho^{i\nu}$ is integration with respect to ν and multiplication by $(2\pi)^{-1/2}$. Further,

$$M^{-1}_{\mu+in/2\to|y|}\rho^{-i\mu} = M^{-1}_{-\mu\to\rho}|y|^{i(\mu+in/2)}.$$

The transforms $M_{\rho\to-\lambda}$ and $M^{-1}_{-\mu\to\rho}$ are inverses of each other. Represent $M_{|y'|\to\nu+i(n-1)/2}$ as the limit, as $\epsilon\to+0$, of the operator $M_{|y'|\to\nu}|y'|^{(n-1)/2}e^{-\epsilon|y'|}$. For a fixed $\epsilon>0$ the operators $M_{|y'|\to\nu}e^{-\epsilon|y'|}$ and $M^{-1}_{-\mu\to\rho}$ can be interchanged. Hence (6.3) implies

$$(E_{\psi\to\omega}(\lambda)u)\circ\kappa(\mathring{\omega},s) =$$
$$= \frac{1}{(2\pi)^{3/2}}p_\lambda(s)\int_{-\infty}^{+\infty}e^{-isy_n}dy_n\int_{-\infty}^{+\infty}E'_{\psi\to\mathring{\omega}}(\nu)d\nu\times$$
$$\times\lim_{\epsilon\to+0}\int_0^{+\infty}|y'|^{-i\nu+(n-3)/2}e^{-\epsilon|y'|}|y|^{i\lambda-n/2}\times$$

$$\times\, u(\kappa(\psi, y_n / |y'|))d\,|y'|,$$

which takes, after the change of variable $y_n = t\,|y'|$, the form

$$(\mathrm{E}(\lambda)u)\circ\kappa(\mathring{\omega},s) = \tag{6.4}$$

$$= \lim_{\epsilon\to+0}\frac{1}{(2\pi)^{3/2}}p_\lambda(s)\int_{-\infty}^{+\infty}\mathrm{E}'(v)dv\int_{-\infty}^{+\infty}p_{-\lambda}^{-1}(t)u(\kappa(\psi,t))dt\,\times$$

$$\times\int_0^{+\infty}e^{-it|y'|s-\epsilon|y'|}\,|y'|^{i(\lambda-v)-1/2}d\,|y'|.$$

We have (compare with (2.10))

$$\lim_{\epsilon\to+0}\int_0^{+\infty}e^{i\eta r-\epsilon r}r^{ia-1/2}dr = \tag{6.5}$$

$$= \Gamma(ia+1/2)\,|\eta|^{-ia-1/2}e^{(-a+i/2)\pi\,\mathrm{sgn}\,\eta/2}.$$

Applying this formula for $a = \lambda - v$, $\eta = -ts$, (6.1) derives from (6.4). ■

Denote by $\chi_\pm$ the characteristic function of the halfaxis $\mathbb{R}_\pm$. For functions given on $\mathbb{R}$ (on S^{n-1}), put $(P^\pm v)(t) = \chi_\pm(t)v(t)$ (resp. $(\mathrm{P}^\pm u) = \chi_\pm(\omega_n)u(\omega)$, where $\omega = (\omega', \omega_n)$, $|\omega| = 1$). For functions $u \in C^\infty(S^{n-1})$ we define $(\mathrm{T}(\lambda)u)(\mathring{\omega},\sigma) = \mathcal{F}^{-1}_{s\to\sigma}p_\lambda(s)(\mathrm{E}_{\psi\to\omega}(\lambda)u)(\kappa(\mathring{\omega},s))$, $(\mathring{\omega},\sigma) \in S^{n-2}\times\mathbb{R}$.

L e m m a 6.2. *For* $\lambda \in \mathbb{R}$, $u \in C^\infty(S^{n-1})$ *the equations*

$$(\mathrm{P}^\pm\mathrm{T}(\lambda)u)(\mathring{\omega},\sigma) = \mathrm{T}(\lambda)(\mathrm{P}^\pm u)(\mathring{\omega},\sigma) = \tag{6.6}$$

$$= \pm\frac{1}{2\pi}\int_{\mathbb{R}}|\sigma|^{i(\lambda-\mu)-1/2}\mathrm{E}'_{\psi\to\mathring{\omega}}(\mu)d\mu\int_{\mathbb{R}_\pm}|t|^{i(\mu-\lambda)-1/2}p_{-\lambda}^{-1}(u\circ\kappa)(\psi,t)dt$$

hold.

P r o o f. Using (6.1) we write the operator $\mathrm{T}(\lambda)$ in the form

$$(\mathrm{T}(\lambda)u)(\mathring{\omega},\sigma) = \tag{6.7}$$

$$= \lim_{\epsilon\to+0}\frac{1}{(2\pi)^{1/2}}\int_{-\infty}^{+\infty}e^{i\sigma s-\epsilon|s|}(\mathrm{E}(\lambda)u)\circ\kappa(\mathring{\omega},s)ds =$$

$$= \frac{1}{(2\pi)^2} \int_{-\infty}^{+\infty} \mathrm{E}'(\mu)\Gamma(i(\lambda-\mu)+1/2)d\mu \times$$

$$\times \int_{-\infty}^{+\infty} p_{-\lambda}^{-1}(t)|t|^{i(\mu-\lambda)-1/2}(u\circ\kappa)(\psi,t)dt \times$$

$$\times \lim_{\epsilon\to+0} \int_{-\infty}^{+\infty} e^{i\sigma s-\epsilon|s|+\operatorname{sgn}(st)(\lambda-\mu-i/2)\pi/2}|s|^{i(\mu-\lambda)-1/2}ds.$$

By (6.5),

$$\lim_{\epsilon\to+0} \int_{-\infty}^{+\infty} e^{i\sigma s-\epsilon|s|+\operatorname{sgn}(st)(\lambda-\mu-i/2)\pi/2}|s|^{i(\mu-\lambda)-1/2}ds = \tag{6.8}$$

$$= |\sigma|^{i(\lambda-\mu)-1/2}\Gamma(i(\mu-\lambda)+1/2) \times$$

$$\times (e^{(\lambda-\mu)\pi(\operatorname{sgn}\sigma+\operatorname{sgn}t)/2+i\pi(\operatorname{sgn}\sigma-\operatorname{sgn}t)/4}+$$

$$+e^{-(\lambda-\mu)\pi(\operatorname{sgn}\sigma+\operatorname{sgn}t)/2-i\pi(\operatorname{sgn}\sigma-\operatorname{sgn}t)/4}).$$

For $\operatorname{sgn}\sigma+\operatorname{sgn}t = 0$ the expression between brackets (on the right) vanishes, while for $\operatorname{sgn}\sigma = \operatorname{sgn}t$ it equals $(e^{\pi(\lambda+\mu)}+e^{-\pi(\lambda-\mu)})\operatorname{sgn}\sigma$. Since $\Gamma(i(\lambda-\mu)+1/2)\Gamma(i(\mu-\lambda)+1/2) = 2\pi(e^{\pi(\lambda-\mu)}+e^{-\mu(\lambda-\mu)})^{-1}$, formulas (6.7) and (6.8) imply (6.6). ■

We introduce the space $H^s_\pm(\lambda,S^{n-1})$ as the closure of the set $C_0^\infty(S^{n-1}_\pm)$ with respect to the norm of $H^s(\lambda,S^{n-1})$. Denote by $H^s_\pm(\mathbb{R})$ the completion of the set $C_0^\infty(\mathbb{R}_\pm)$ with respect to the norm of $H^s(\mathbb{R})$. Let, further, $L_2(S^{n-2},B)$ be the space of functions given on $S^{n-2} = \{\omega=(\omega',\omega_n)\in S^{n-1}: \omega_n=0\}$ with values in the normed space B, and let, moreover,

$$\|u;L_2(S^{n-2},B)\| = \left[\int_{S^{n-2}} \|u(\omega');B\|^2 d\omega'\right]^{1/2}.$$

P r o p o s i t i o n 6.3. *For* $\lambda \neq i(k+n/2)$, $k = 0,1,\cdots$, *the operator*

$$\mathrm{T}(\lambda): H_\pm^{-\operatorname{Im}\lambda}(\lambda,S^{n-1}) \to L_2(S^{n-2},H_\pm^{-\operatorname{Im}\lambda}(\mathbb{R})) \tag{6.9}$$

realizes an isomorphism (algebraically and topologically).

P r o o f. Recall that by proposition 5.5 the map

$E(\lambda): H_{\pm}^{-\mathrm{Im}\lambda}(\lambda,S^{n-1}) \to H^0(S^{n-1})$ is continuous. The volume element $d\omega$ on S^{n-1} can be written in the form $d\omega = (1+t^2)^{-n/2}dtd\mathring{\omega}$, where, as before, $\omega = (\omega',\omega_n) = (1+t^2)^{-1/2}(\mathring{\omega},t)$. This implies the inequality

$$\|T(\lambda)u;L_2(S^{n-2},H^{-\mathrm{Im}\lambda}(\mathbb{R}))\| \leqslant c\|u;H_{\pm}^{-\mathrm{Im}\lambda}(\lambda,S^{n-1})\|. \tag{6.10}$$

By lemma 6.2, for $u \in C_0^\infty(S_{\pm}^{n-1})$ and real λ,

$$P^{\pm}T(\lambda)u = T(\lambda)P^{\pm}u = T(\lambda)u. \tag{6.11}$$

Using analytic extension, relation (6.11) can be extended to all complex $\lambda = i(k+n/2)$, $k = 0,1,\cdots$. Together with (6.10) this means that $T(\lambda)$ realizes a continuous map (6.9), which clearly is a monomorphism.

We will show that the operator (6.9) is epimorphic. Applying proposition 5.5 we obtain that the operator $T(\lambda)^{-1} = E(\lambda)^{-1}p_\lambda\mathfrak{F}$: $L_2(S^{n-2},H^{-\mathrm{Im}\lambda}(\mathbb{R})) \to H^{-\mathrm{Im}\lambda}(\lambda,S^{n-2})$ is continuous. Put $u = T(\lambda)^{-1}f$ for a given $f \in L_2(S^{n-2},H_{\pm}^{-\mathrm{Im}\lambda}(\mathbb{R}))$. If $\mathrm{Im}\lambda \leqslant 0$, then $u \in H^0(S^{n-1})$, and we can apply the projection $P^{\pm}$ to u. From lemma 6.2 we obtain that $P^{\pm}u = P^{\pm}T(\lambda)^{-1}f = T(\lambda)^{-1}P^{\pm}f = T(\lambda)^{-1}f = u$, i.e. $u \in H_{\pm}^{-\mathrm{Im}\lambda}(\lambda,S^{n-1})$. Thus it has been proved that $T(\lambda)$ is epimorphic if $\mathrm{Im}\lambda \leqslant 0$. If, however $\mathrm{Im}\lambda > 0$, we choose a sequence $\{f_k\}$ converging to f in $L_2(S^{n-2},H_{\pm}^{-\mathrm{Im}\lambda}(\mathbb{R}))$ such that $T(\lambda)^{-1}f_k \in H^0(S^{n-1})$. Since $P^{\pm}T(\lambda)^{-1}f_k = T(\lambda)^{-1}P^{\pm}f_k = T(\lambda)^{-1}f_k$, we have $u = T(\lambda)^{-1}f = \lim T(\lambda)^{-1}f_k \in H_{\pm}^{-\mathrm{Im}\lambda}(\lambda,S^{n-1})$. ∎

In essence, the next assertion is well-known (the Paley-Wiener theorem; cf., e.g., [74]).

P r o p o s i t i o n 6.4. *Let* $f \in L_2(S^{n-2},H_+^\nu(\mathbb{R}))$. *Then* $\tau \mapsto \mathfrak{F}_{\sigma\to s}(f(\mathring{\omega},\sigma)e^{\sigma\tau}) = (\mathfrak{F}f)(\mathring{\omega},s+i\tau)$ *is, for* $\tau \leqslant 0$, *a continuous function with values in* $L_2(S^{n-2},\mathfrak{F}H_+^\nu(\mathbb{R}))$. *For almost all* $\mathring{\omega} \in S^{n-2}$ *the function* $s+i\tau \mapsto (\mathfrak{F}f)(\mathring{\omega},s+i\tau)$ *is analytic in the halfplane* $\tau < 0$, *and*

$$\int_{-\infty}^{+\infty} |(\mathfrak{F}f)(\mathring{\omega},s+i\tau)|^2(1+s^2+\tau^2)^\nu ds \leqslant c(\mathring{\omega}) < \infty,$$

where $c(\mathring{\omega})$ *is independent of* $\tau \leqslant 0$. *The inequality*

$$\int_{S^{n-2}} d\mathring{\omega} \int_{-\infty}^{+\infty} |(\mathfrak{F}f)(\mathring{\omega},s+i\tau)|^2(1+s^2\tau^2)^\nu ds \leqslant c \tag{6.12}$$

holds, in which the constant $c < \infty$ *is also independent of* $\tau \leqslant 0$. *Conversely, if for almost all* $\mathring{\omega} \in S^{n-2}$ *the function* $(\mathcal{F}f)(\mathring{\omega},\cdot)$ *is analytic in the halfplane indicated and satisfies* (6.12), *then* $f \in L_2(S^{n-2}, H^{\nu}_{+}(\mathbb{R}))$. *The proposition remains valid if* H_+ *is replaced by* H_- *and* $\tau < 0$ *by* $\tau > 0$, *everywhere.*

Let $\mathcal{H}_+(\mu)$ (resp. $\mathcal{H}_-(\mu)$) be the subspace of $H^0(S^{n-1})$ formed by the elements ν satisfying the conditions: a) for almost all $\xi'/|\xi'| \in S^{n-2}$ the homogeneous function $\xi = (\xi',\xi_n) \mapsto |\xi|^{\mu}\nu(\xi/|\xi|)$, $\xi' \neq 0$, can be analytically extended to the halfplane $\operatorname{Im}\xi_n > 0$ (resp. $\operatorname{Im}\xi_n < 0$) (this is equivalent to the possibility of analytically extending the function $s \mapsto w(\mathring{\omega},s) = (1+s^2)^{\mu/2}\nu \circ \kappa(\mathring{\omega},s)$ to the corresponding halfplane); b) the inequality

$$\int_{S^{n-2}} d\mathring{\omega} \int_{-\infty}^{+\infty} |w(\mathring{\omega},s+i\tau)|^2(1+s^2+\tau^2)^{-\operatorname{Re}\mu - n/2} ds \leqslant c < \infty$$

is fulfilled with a constant c which is independent of $\tau \leqslant 0$ (resp. $\tau \geqslant 0$).

T h e o r e m 6.5. *For* $\lambda \neq i(k+n/2)$, $k = 0,1,\cdots$, *the operator* E(λ) *maps* $H_{\pm}^{-\operatorname{Im}\lambda}(\lambda, S^{n-1})$ *onto* $\mathcal{H}_{\mp}(-i\lambda - n/2)$. *This map is an isomorphism (algebraically and topologically).*

P r o o f. By proposition 5.5 the operator $\mathrm{E}(\lambda)\colon H_{\pm}^{-\operatorname{Im}\lambda}(\lambda,S^{n-1}) \to H^0(S^{n-1})$ is continuous. By proposition 6.3 the membership $\nu \in H_{\pm}^{-\operatorname{Im}\lambda}(\lambda,S^{n-1})$ implies that $\mathrm{T}(\lambda)\nu \in L_2(S^{n-2}, H_{\pm}^{-\operatorname{Im}\lambda}(\mathbb{R}))$. By applying proposition 6.4 we find that the function $s \mapsto (\mathcal{F}\mathrm{T}(\lambda)\nu)(\mathring{\omega},s)$ can be analytically extended to the halfplane $\operatorname{Im} s \lesseqgtr 0$ for almost all $\mathring{\omega}$. This means that the function $\omega \mapsto (p_{\lambda}\mathcal{F}\mathrm{T}(\lambda)\nu)\circ\kappa^{-1}(\omega) = (\mathrm{E}(\lambda)\nu)(\omega)$, $\omega = \xi/|\xi|$, can be analytically extended as a homogeneous function of degree $-i\lambda - n/2$ to the halfplane $\operatorname{Im}\xi_n \lesseqgtr 0$. Thus we have a monomorphism $\mathrm{E}(\lambda)\colon H_{\pm}^{-\operatorname{Im}\lambda}(\lambda,S^{n-1}) \to \mathcal{H}_{\mp}(-i\lambda - n/2)$. The fact that the map is epimorphic follows from propositions 6.3 and 6.4. ∎

Chapter 2

The Fourier transform and convolution operators on spaces with weighted norms

In this Chapter we consider operators of the form $A = F^{-1}_{\xi\to x}\Phi(\xi)F_{y\to\xi}$, where F is the Fourier transform on $\mathbb{R}^m$, and Φ is a positively homogeneous function of complex degree a, i.e. $\Phi(t\xi) = t^a\Phi(\xi)$ for $\xi \in \mathbb{R}^m \setminus 0$, $t > 0$. We study the problem of continuity of the maps $A: H^s(\mathbb{R}^m) \to H_\beta^{s-\mathrm{Re}\,a}(\mathbb{R}^m)$ and $A: H_\beta^s(\mathbb{R}^m, \mathbb{R}^{m-n}) \to H_\beta^{s-\mathrm{Re}\,a}(\mathbb{R}^m, \mathbb{R}^{m-n})$.

The space $H_\beta^s(\mathbb{R}^m)$, for nonnegative integers $s, \beta \in \mathbb{R}$, is endowed with the norm

$$\left[\int_{\mathbb{R}^m} \sum_{|\alpha|=0}^{s} |x|^{2(\beta-s+|\alpha|)} |D_x^\alpha u(x)|^2 dx\right]^{1/2},$$

while in the space $H_\beta^s(\mathbb{R}^m, \mathbb{R}^{m-n})$ we have the norm

$$\left[\int_{\mathbb{R}^m} \sum_{|\alpha|=0}^{s} |x^{(1)}|^{2(\beta-s+|\alpha|)} |D_x^\alpha u(x)|^2 dx\right]^{1/2},$$

where $n \leqslant m$, $x = (x^{(1)}, x^{(2)})$, $x^{(1)} = (x_1, \cdots, x_n)$, $x^{(2)} = (x_{n+1}, \cdots, x_m)$. For $n = m$ the space $H_\beta^s(\mathbb{R}^m, \mathbb{R}^{m-n})$ coincides with $H_\beta^s(\mathbb{R}^n)$.

In §1 we introduce, for arbitrary real numbers s and β, the space $H_\beta^s(\mathbb{R}^n)$. In §2 it will turn out that the Fourier transform F, at first defined on a suitable dense subset of $H_\beta^s(\mathbb{R}^n)$, can be extended to the whole space $H_\beta^s(\mathbb{R}^n)$, and realizes a continuous map $F: H_\beta^s(\mathbb{R}^n) \to H_\beta^s(\mathbb{R}^n)$ (if $\beta - s \neq k + n/2$, $k = 0, 1, \cdots$). For $k + n/2 < \beta - s < k + 1 + n/2$ the operator F is at first defined on the set

$$\left\{v \in C_0^\infty(\mathbb{R}^n \setminus 0): \int_{\mathbb{R}^n} v(x) x^\gamma dx = 0, \quad |\gamma| = 0, 1, \cdots, k\right\},$$

while in case $\beta-s<n/2$ we may assume that F is given on $C_0^\infty(\mathbb{R}^n\setminus 0)$. §3 is devoted to the map $A:H_\beta^s(\mathbb{R}^n)\to H_\beta^{s-\mathrm{Re}\,a}(\mathbb{R}^n)$. If $\beta-s\neq k+n/2$, $\beta-s\neq -k-\mathrm{Re}\,a-n/2$ $(k=0,1,\cdots)$, then the operator A (at first given on some dense subset of $H_\beta^s(\mathbb{R}^n)$) is continuous, hence its closure $\overline{A}$ is an operator defined on all of $H_\beta^s(\mathbb{R}^n)$. The representation for $\overline{A}$ obtained in §3 is the starting point for the study of the algebras of pseudodifferential operators with discontinuous symbols in the next Chapters. In §4 - §6 analogous problems for the map $A:H_\beta^s(\mathbb{R}^m,\mathbb{R}^{m-n})\to H_\beta^{s-\mathrm{Re}\,a}(\mathbb{R}^m,\mathbb{R}^{m-n})$ are considered.

§1. The spaces $H_\beta^s(\mathbb{R}^n)$

Let, initially, s be a nonnegative integer and $\beta\in\mathbb{R}$. We denote by $H_\beta^s(\mathbb{R}^n)$ the completion of the set $C_0^\infty(\mathbb{R}^n\setminus 0)$ with respect to the norm

$$\left[\sum_{|\alpha|=0}^{s}\int_{\mathbb{R}^n}|x|^{2(\beta-s+|\alpha|)}|D_x^\alpha u(x)|^2dx\right]^{1/2}. \tag{1.1}$$

This norm is equivalent to the norm

$$\left[\int_0^\infty r^{2(\beta-s)+n-1}\sum_{j=0}^{s}\|(rD_r)^j u(r,\cdot);H^{s-j}(S^{n-1})\|^2dr\right]^{1/2},$$

where, as before, $r=|x|$. Using the Parseval equality (1.2.7) for the Mellin transform, we obtain yet another equivalent norm:

$$\left[\int_{\mathrm{Im}\lambda=\beta-s}\sum_{j=0}^{s}|\lambda|^{2j}\|\tilde{u}(\lambda+in/2,\cdot);H^{s-j}(S^{n-1})\|^2d\lambda\right]^{1/2}.$$

Proposition 1.5.1 and the obvious truths

$$|\lambda|^{2j}(1+m^2)^{s-j}\leqslant(1+|\lambda|^2+m^2)^s\leqslant c\sum_{j=0}^{s}|\lambda|^{2j}(1+m^2)^{s-j}$$

imply that the norm (1.1) is equivalent to the norm

$$\|u;H_\beta^s(\mathbb{R}^n)\| = \left[\int_{\mathrm{Im}\lambda=\beta-s}\|\tilde{u}(\lambda+in/2,\cdot);H^s(\lambda,S^{n-1})\|^2d\lambda\right]^{1/2}. \tag{1.2}$$

The righthand side of (1.2) makes sense for arbitrary real s, cf. (1.5.20). For all

$s,\beta \in \mathbb{R}$ the space $H^s_\beta(\mathbb{R}^n)$ is defined as the completion of $C_0^\infty(\mathbb{R}^n \setminus 0)$ with respect to the norm (1.2). The following assertions are readily derived from the definitions of the norms.

P r o p o s i t i o n 1.1. *The operator of multiplication by a homogeneous function of degree a,* $\Phi(\xi) = \rho^a f(\psi)$*, realizes a continuous map* $H^s_\beta(\mathbb{R}^n) \to H^s_{\beta-\mathrm{Re}\,a}(\mathbb{R}^n)$ *if and only if f is a multiplier in* $H^s(S^{n-1})$ *(i.e. if the operator of multiplication by f is continuous on* $H^s(S^{n-1})$*).*

It is well-known that for $2s > n-1$ the space $H^s(S^{n-1})$ is a ring under ordinary multiplication. Hence for $2s > n-1$ the operator $\Phi: H^s_\beta(\mathbb{R}^n) \to H^s_{\beta-\mathrm{Re}\,a}(\mathbb{R}^n)$ is continuous only if $f \in H^s(S^{n-1})$. (A description of the multipliers in the Sobolev-Slobodetskii spaces can be found in, e.g., [37].)

P r o p o s i t i o n 1.2. *For arbitrary positive* δ *the inclusion* $H^{s+\delta}_{\beta+\delta}(\mathbb{R}^n) \subset H^s_\beta(\mathbb{R}^n)$ *holds; the inequality* $\|u;H^s_\beta(\mathbb{R}^n)\| \leqslant \|u;H^{s+\delta}_{\beta+\delta}(\mathbb{R}^n)\|$ *holds also.*

P r o p o s i t i o n 1.3. *The spaces* $H^s_\beta(\mathbb{R}^n)$ *and* $H^{-s}_{-\beta}(\mathbb{R}^n)$ *are the duals of each other under the extension of the scalar product in* $L_2(\mathbb{R}^n) = H^0_0(\mathbb{R}^n)$.

§2. The Fourier transform on the spaces $H^s_\beta(\mathbb{R}^n)$

For $u \in C_0^\infty(\mathbb{R}^n \setminus 0)$ we put

$$(\mathrm{F}_\sigma^{\pm 1}u)(r,\phi) = \frac{1}{\sqrt{2\pi}} \int_{\mathrm{Im}\,\lambda=\sigma} r^{i(in/2\mp\lambda)} \mathrm{E}_{\psi\to\phi}(\lambda)^{\pm 1} \tilde{u}(in/2\pm\lambda,\psi)\,d\lambda; \quad (2.1)$$

the operator F_σ (resp. F_σ^{-1}) is defined under the condition that $\sigma \neq k+n/2$ (resp. $\sigma \neq -k-n/2$), where $k = 0,1,\cdots$. It is obvious that for $\sigma \neq \pm(k+n/2)$ the operators F_σ and F_σ^{-1} are the inverses of each other, and F_0 coincides with the Fourier transform (proposition 1.2.1).

P r o p o s i t i o n 2.1. *Let* σ *and* σ' *be real numbers subject to the inequalities* $\sigma < n/2$, $p+n/2 < \sigma' < p+1+n/2$*, where p is a certain nonnegative integer. Then the formulas*

$$(\mathrm{F}_{\pm\sigma}^{\pm 1}u)(x) = \tag{2.2}$$

$$= \frac{1}{(2\pi)^{(n-1)/2}} \sum_{|\gamma|=0}^{p} \frac{(\mp i)^{|\gamma|+2}}{\gamma!} x^\gamma \int_{S^{n-1}} y^\gamma u(y)dy + (\mathrm{F}_{\pm\sigma'}^{\pm 1}u)(x)$$

hold, in which the upper (lower) sign must be chosen everywhere. Moreover, for arbitrary $h<0$,

$$(\mathrm{F}_{\pm\sigma}^{\pm 1}u)(x) = O(|x|^h) \quad \text{as } x \to \infty, \tag{2.3}$$

$$(\mathrm{F}_{\pm\sigma'}^{\pm 1}u)(x) = O(|x|^{p+1}) \quad \text{as } x \to 0. \tag{2.4}$$

Thus, the formulas (2.2) *give for $x \to 0$ the asymptotic behavior of $\mathrm{F}_{\pm\sigma}^{\pm 1}u$, and for $x \to \infty$ those of $\mathrm{F}_{\pm\sigma'}^{\pm 1}u$.*

P r o o f. We verify the assertion for, e.g., $\mathrm{F}_\sigma u$. Since $u \in C_0^\infty(\mathbb{R}^n \setminus 0)$, the function $\lambda \mapsto \tilde{u}(\lambda + in/2, \cdot)$ is analytic in the whole λ-plane and decreases faster than any power of $|\lambda|$ in every strip $|\mathrm{Im}\lambda| < N$. This and inequality (1.5.21) imply (2.3) and the estimate

$$(\mathrm{F}_{\sigma'}u)(x) = O(|x|^{\sigma'}) \quad \text{as } x \to 0. \tag{2.5}$$

Further,

$$\mathrm{res}\, \mathrm{E}_{\psi\to\phi}(\lambda)\tilde{u}(\lambda + in/2, \psi)|_{\lambda = i(k+n/2)} = \tag{2.6}$$

$$= \frac{(-i)^{k+1}}{(2\pi)^{(n+1)/2}} \sum_{|\gamma|=k} \frac{\phi^\gamma}{\gamma!} \int_{\mathbb{R}^n} y^\gamma u(y)dy.$$

Therefore, replacing in (2.1) the integration curve $\mathrm{Im}\lambda = \sigma$ by $\mathrm{Im}\lambda = \sigma'$ we are led to (2.2). If $\sigma'' \in (p+1+n/2, p+2+n/2)$, then the difference $\mathrm{F}_{\sigma'}u - \mathrm{F}_{\sigma''}u$ is equal to $2\pi i |x|^{p+1}$ times the residue (2.6) for $k = p+1$. Together with (2.5) (in which σ' must be replaced by σ'') this leads to (2.4). ■

P r o p o s i t i o n 2.2.
1) *Let $\beta - s \neq k + n/2$, $k = 0, 1, \cdots$. Then the maps*

$$\mathrm{F}_{\beta - s}: H_\beta^s(\mathbb{R}^n) \to H_s^\beta(\mathbb{R}^n), \; \mathrm{F}_{s-\beta}^{-1}: H_\beta^s(\mathbb{R}^n) \to H_s^\beta(\mathbb{R}^n) \tag{2.7}$$

are continuous. Under the condition $\beta - s \neq -k - n/2$, $k = 0, 1, \cdots$, each of the maps (2.7) *is an isomorphism.*

2) *Let* $\beta - s = k + n/2$ *and let* $H^s_{\beta,k}(\mathbb{R}^n)$ *(resp.* $H^\beta_{s,k}(\mathbb{R}^n)$*) be the subspaces of* $H^s_\beta(\mathbb{R}^n)$ *(resp.* $H^\beta_s(\mathbb{R}^n)$*) obtained by closure of the set of functions from* $C_0^\infty(\mathbb{R}^n \setminus 0)$ *satisfying the conditions*

$$\int_{S^{n-1}} v(r,\phi) Y_{m,q}(\phi) d\phi \equiv 0, \quad m \leqslant k, \quad m \equiv k (\mathrm{mod}\, 2). \tag{2.8}$$

Then every map

$$\mathrm{F}_{\beta - s}: H^s_{\beta,k}(\mathbb{R}^n) \to H^\beta_{s,k}(\mathbb{R}^n), \ \mathrm{F}^{-1}_{s-\beta}: H^s_{\beta,k}(\mathbb{R}^n) \to H^\beta_{s,k}(\mathbb{R}^n)$$

is an isomorphism.

P r o o f. 1) It suffices to convince ourselves of the continuity of the operators (2.7). By (2.1), $\widetilde{(\mathrm{F}_{\beta-s} u)}(-\lambda + in/2), \cdot) = \mathrm{E}(\lambda)\ \tilde{u}(\lambda + in/2, \cdot)$. Using proposition 1.5.5 and the definition of the norm (1.2), we are led to the inequality $\|\mathrm{F}_{\beta-s} u; H^\beta_s(\mathbb{R}^n)\| \leqslant c \|u; H^s_\beta(\mathbb{R}^n)\|$. Continuity of the first map (2.7) has been proved; continuity of the second is verified similarly. 2) The restriction of $\mathrm{E}(\lambda)$ (resp. $\mathrm{E}(\lambda)^{-1}$) to the lineal $\mathfrak{L}_k^\perp$ does not have poles on the line $\mathrm{Im}\,\lambda = k + n/2$ (resp. $\mathrm{Im}\,\lambda = -k - n/2$); moreover, $\mathfrak{L}_k^\perp$ is invariant under these operators (see proposition 1.4.3). Therefore we can reason similarly to the proof of assertion 1). ■

Let M_p be the set of functions v from $C_0^\infty(\mathbb{R}^n \setminus 0)$ subject to the condition $\tilde{v}(i(n+q), \phi) \equiv 0$, $q = 0, \cdots, p$.

P r o p o s i t i o n 2.3. *The set* M_p *is dense in* $H^s_\beta(\mathbb{R}^n)$ *for any nonnegative integer* p, *and for* $\beta - s \neq k + n/2$, $k = 0, 1, \cdots$.

P r o o f. We first verify that M_0 is dense. For an arbitrary function $v \in C_0^\infty(\mathbb{R}^n \setminus 0)$ we introduce the sequence $\{v_k\}_{k=1}^\infty$ for which $v_k(r,\phi) = k^n v(kr, \phi)$ if $\beta - s > n/2$ and $v_k(r,\phi) = k^{-n} v(k^{-1} r, \phi)$ if $\beta - s < n/2$. Clearly $v - v_k \in M_0$. Further, $\tilde{v}_k(\lambda + in/2, \cdot) = k^{\pm(i\lambda + n/2)} \tilde{v}(\lambda + in/2, \cdot)$ if $\beta - s \gtrless n/2$. This and (1.2) imply that $v_k \to 0$ in $H^s_\beta(\mathbb{R}^n)$. We have proved that M_0 is dense. We now assume that M_{p-1} is dense also. It suffices to convince ourselves that every function in M_{p-1} is the limit of a sequence of functions from M_p. Let $w \in M_{p-1}$, $w_k(r,\phi) = k^{\pm(n+p)} w(k^{\pm 1} r, \phi)$ for $\beta - s \gtrless p + n/2$. It remains to note that $w - w_k \in M_p$ and that $\|w_k; H^s_\beta(\mathbb{R}^n)\| \to 0$. ■

Note that proposition 2.3 remains valid for $\beta - s = k + n/2$ too, we only need modify the definitions of the sequences $\{v_k\}, \{w_k\}$ (cf. [44]).

We denote by $\mathfrak{M}_p$ the set of functions from $C_0^\infty(\mathbb{R}^n \setminus 0)$ satisfying the conditions

$$\int v(x) x^\gamma dx = 0, \quad |\gamma| = 0, 1, \cdots, p. \tag{2.9}$$

Theorem 2.4.

1) *Let* $\beta - s \neq k + n/2$, $k = 0, 1, \cdots$, *and let* F *be the Fourier transform defined on* $\mathfrak{M}_p$ *for* $\beta - s \in (p + n/2, p + 1 + n/2)$ *and on* $C_0^\infty(\mathbb{R}^n \setminus 0)$ *for* $\beta - s < n/2$. *Then the domain of definition,* $\mathfrak{D}(\mathrm{F})$, *of* F *is dense in* $H^s_\beta(\mathbb{R}^n)$. *For all* $u \in \mathfrak{D}(\mathrm{F})$ *the estimate*

$$\|\mathrm{F}u; H^\beta_s(\mathbb{R}^n)\| \leqslant c \|u; H^s_\beta(\mathbb{R}^n)\| \tag{2.10}$$

holds. The operator $\overline{\mathrm{F}}$ *(the closure of* F*), defined on all of* $H^s_\beta(\mathbb{R}^n)$, *realizes a continuous map* $\overline{\mathrm{F}}: H^s_\beta(\mathbb{R}^n) \to H^\beta_s(\mathbb{R}^n)$. *All assertions made remain valid if* β *and* s *are interchanged and if* F^{-1} *is written instead of* F. *The formulas* $\overline{\mathrm{F}} = \mathrm{F}_{\beta - s}$, $\overline{\mathrm{F}^{-1}} = \mathrm{F}^{-1}_{\beta - s}$ *(cf. (2.1)) hold for the operators* $\overline{\mathrm{F}}$ *and* $\overline{\mathrm{F}^{-1}}$, *which are defined on* $H^s_\beta(\mathbb{R}^n)$ *and* $H^\beta_s(\mathbb{R}^n)$, *respectively. Under the additional restriction* $\beta - s \neq -(k + n/2)$ *the operators* $\overline{\mathrm{F}}$ *and* $\overline{\mathrm{F}^{-1}}$ *are inverse to each other.*

2) *Let* $\beta - s = k + n/2$ *and let* F *be the Fourier transform defined on the set of functions from* $C_0^\infty(\mathbb{R}^n \setminus 0)$ *satisfying* (2.9) *for* $|\gamma| = 0, \cdots, k-1$ *and* (2.8) *(if* $k = 0$, *condition* (2.9) *can be omitted). The closure of this set of functions coincides with the subspace* $H^s_{\beta,k}(\mathbb{R}^n)$, *and the map* $\overline{\mathrm{F}}: H^s_{\beta,k}(\mathbb{R}^n) \to H^\beta_{s,k}(\mathbb{R}^n)$ *is an isomorphism. The operators* $\overline{\mathrm{F}}$ *and* $\mathrm{F}_{\beta - s}$ *coincide on* $H^s_{\beta,k}(\mathbb{R}^n)$. *All assertions remain valid if* β *and* s *are interchanged and* F^{-1} *is written instead of* F.

Proof. 1) That $\mathfrak{D}(\mathrm{F}^{\pm 1})$ is dense follows from proposition (2.3). On $\mathfrak{D}(\mathrm{F})$ (resp. $\mathfrak{D}(\mathrm{F}^{-1})$) the operator F (resp. F^{-1}) coincides with $\mathrm{F}_{\beta - s}$ (resp. $\mathrm{F}^{-1}_{\beta - s}$) (proposition 2.1). The estimate (2.10) follows from proposition 2.2.

2) Note that the denseness of the set of functions on which the Fourier transform F is initially defined follows from the proof of proposition 2.3. The remaining assertions are verified as is done in the first part of the theorem. ■

R e m a r k 2.5. If the numbers $\beta-s$, $\beta'-s'$ belong to distinct intervals $(-\infty,n/2)$, $(q+n/2,q+1+n/2)$, where $q=0,1,\cdots$, then the operators $\overline{\mathrm{F}}$ and $\overline{\mathrm{F}}'$ defined as the closure of the Fourier transform on $H^s_\beta(\mathbb{R}^n)$ and $H^{s'}_{\beta'}(\mathbb{R}^n)$ are, generally speaking, distinct on functions u from $C_0^\infty(\mathbb{R}^n\setminus 0)$. The equality $\mathrm{F}u=\mathrm{F}'u$ holds if $u\in\mathfrak{M}_p$ where $p=\max\{[\beta'-s'-n/2],[\beta-s-n/2]\}$.

§3. The convolution operator on the spaces $H^s_\beta(\mathbb{R}^n)$

Let Φ be a homogeneous function of (complex) degree a on $\mathbb{R}^n\setminus 0$ which is infinitely differentiable on the sphere S^{n-1}. We consider the operator

$$\mathrm{A}=\mathrm{F}^{-1}\Phi(\xi)\mathrm{F}. \tag{3.1}$$

The aim of the present paragraph lies in clarifying the problem of the boundedness of the map $\mathrm{A}:H^s_\beta(\mathbb{R}^n)\to H^{s-\mathrm{Re}\,a}_\beta(\mathbb{R}^n)$.

P r o p o s i t i o n 3.1. *Let* $\mathrm{Re}\,a>-n/2$. *Then the representation*

$$(\mathrm{A}u)(r,\phi)=\frac{1}{\sqrt{2\pi}}\int_{-\infty}^{+\infty}r^{i(\lambda+ia+in/2)}\mathrm{E}_{\omega\to\phi}(\lambda+ia)^{-1}\times \tag{3.2}$$

$$\times\Phi(\omega)\mathrm{E}_{\theta\to\omega}(\lambda)\tilde{u}(\lambda+in/2,\theta)d\lambda$$

holds for the operator (3.1), *defined on* $C_0^\infty(\mathbb{R}^n\setminus 0)$.

P r o o f. By (1.28),

$$\rho^a\Phi(\omega)(\mathrm{F}u)(\rho,\omega)=\frac{1}{\sqrt{2\pi}}\int_{\mathrm{Im}\lambda=\mathrm{Re}\,a}\rho^{i(in/2-\lambda)}\Phi(\omega)\times \tag{3.3}$$

$$\times\mathrm{E}(\lambda-ia)\tilde{u}(\lambda-ia+in/2,\cdot)d\lambda.$$

Now note that in the strip between the lines $\mathrm{Im}\lambda=0$, $\mathrm{Im}\lambda=\mathrm{Re}\,a$ the integrand is holomorphic, since $\mathrm{Re}\,a>-n/2$, while the poles of the function $\mu\mapsto\mathrm{E}(\mu)$ are located at $\mu=i(k+n/2)$, $k=0,1,\cdots$. The membership $u\in C_0^\infty(\mathbb{R}^n\setminus 0)$ implies that the function $\lambda\mapsto\tilde{u}(\lambda+in/2,\cdot)$ is rapidly decreasing as $\lambda\to\infty$ in any strip $|\mathrm{Im}\lambda|<h$. Together with (1.5.21) this allows us to replace in (3.3) the integration line $\mathrm{Im}\lambda=\mathrm{Re}\,a$ by the line $\mathrm{Im}\lambda=0$. By now applying the inverse Fourier transform $\mathrm{F}^{-1}=\mathrm{F}_0^{-1}$ (cf. (2.1)) to (3.3) we obtain

$$(Au)(r,\phi) = \frac{1}{\sqrt{2\pi}} \int_{-\infty}^{+\infty} r^{i(in/2+\lambda)} E_{\omega\to\phi}(\lambda)^{-1}\Phi(\omega) \times$$

$$\times E_{\theta\to\omega}(\lambda - ia)\tilde{u}(\lambda - ia + in/2, \theta) d\lambda.$$

Put $\mu = \lambda - ia$ and then replace the line $\text{Im}\,\mu = -\text{Re}\,a$ by the line $\text{Im}\,\mu = 0$. As the result we obtain (3.2). ■

P r o p o s i t i o n 3.2. *Let* $\tau \neq k + n/2$, $\tau \neq -k - \text{Re}\,a - n/2$, *where* $k = 0, 1, \cdots$, *and let* A *be the operator defined on* $C_0^\infty(\mathbf{R}^n \setminus 0)$ *by* (3.1) *for* $\text{Re}\,a > -n/2$. *The following representation holds for this operator (or for its analytic extension in a) when* $a \neq -l - n$, *where* $l = 0, 1, \cdots$:

$$(Au)(r,\phi) = \frac{1}{\sqrt{2\pi}} \int_{\text{Im}\lambda=\tau} r^{i(\lambda + ia + in/2)} \times \tag{3.4}$$

$$\times E_{\omega\to\phi}(\lambda + ia)^{-1}\Phi(\omega) E_{\theta\to\omega}(\lambda)\tilde{u}(\lambda + in/2, \theta) d\lambda +$$

$$+ \frac{1}{(2\pi)^n} e^{i\pi(n+1+a)/2} \Bigg\{ \sum_{k=0}^{[\tau - n/2]} \Gamma(n + a + k) \times$$

$$\times \int_{S^{n-1}} (x\omega + i0)^{-n-a-k} \Phi(\omega) \sum_{|\gamma|=k} \frac{\omega^\gamma}{\gamma!} d\omega \int_{\mathbf{R}^n} y^\gamma u(y) dy +$$

$$- \sum_{k=0}^{[-\tau - n/2 - \text{Re}\,a]} (-1)^k \Gamma(n + a + k) \times$$

$$\times \sum_{|\gamma|=k} \frac{x^\gamma}{\gamma!} \int_{\mathbf{R}^n} u(y) dy \int_{S^{n-1}} (-y\omega + i0)^{-n-a-k} \Phi(\omega)\omega^\gamma d\omega \Bigg\},$$

where there is no summation if the corresponding upper limit is negative. If a is a nonnegative integer and Φ *is a homogeneous polynomial of degree a, then* A *is a homogeneous differential operator of order a. For such an operator a representation* (3.4) *in which only the first righthand term occurs holds.*

P r o o f. The integrand in (3.2) has singularities at $\lambda = i(k + n/2)$, $\lambda = -i(k + a + n/2)$, $k = 0, 1, \cdots$, only, and these singular points are simple poles. Changing in (3.2) the integration line and writing out explicitly the residues leads to (3.4). If a is a nonnegative integer and Φ is a homogeneous polynomial of degree a, the residues are annihilated (see proposition 1.4.3). ■

Now we consider the case $a = -l-n$. In general, for such an a formula (3.4) cannot be written down since certain terms between the brackets have a pole at a. If we additionally require Φ to satisfy

$$\int_{S^{n-1}} \Phi(\omega)\omega^\gamma d\omega = 0 \tag{3.5}$$

for all multi-indices γ such that $|\gamma| = 0, \cdots ,l$, then the residues of the terms in (3.4) are annihilated for $a = -l-n$. As the result we obtain a representation for A also in case $a = -l-n$. It coincides with (3.4), in which we must substitute $a = -l-n$ and transform the terms for $0 \leqslant k \leqslant l$ taking into account (1.2.5) and (1.4.3).

Assume that $\beta - s \neq k + n/2$, $\beta - s \neq -k - \mathrm{Re}\, a - n/2$, $k = 0,1, \cdots$, and denote by $\mathfrak{N}(\beta,s,a)$ the set of functions u from $C_0^\infty(\mathbb{R}^n \setminus 0)$ such that $\tilde{u}(i(q+n),\phi) \equiv 0$ for $q = 0, \cdots ,[\beta - s - n/2]$ and $\tilde{u}(-i(k+a),\phi) \equiv 0$ for $q = 0, \cdots ,[s - \beta - \mathrm{Re}\, a - n/2]$ (if one of $[\beta - s - n/2]$, $[s - \beta - \mathrm{Re}\, a - n/2]$ is negative, the corresponding condition is omitted). In the following theorem, A denotes the operator (3.1) or the operator obtained from (3.1) by analytic extension in a.

Theorem 3.3. *Let $\beta - s \neq k + n/2$, $\beta - s \neq -k - \mathrm{Re}\, a - n/2$, where $k = 0,1, \cdots$. Then on the set $\mathfrak{N}(\beta,s,a)$ the following representation is valid for* A:

$$(\mathrm{A}u)(x) = \frac{1}{\sqrt{2\pi}} \int_{\mathrm{Im}\lambda = \beta - s} r^{i(\lambda + ia + in/2)} \mathrm{E}_{\omega \to \phi}(\lambda + ia)^{-1} \times \tag{3.6}$$

$$\times\, \Phi(\omega) \mathrm{E}_{\theta \to \omega}(\lambda) \tilde{u}(\lambda + in/2, \theta) d\lambda.$$

This representation can be written as

$$\mathrm{A} = \mathrm{F}^{-1}_{\beta - s + \mathrm{Re}\, a} \Phi(\xi) \mathrm{F}_{\beta - s}; \tag{3.7}$$

here $\mathrm{F}_{\beta - s}$ (resp. $\mathrm{F}^{-1}_{\beta - s + \mathrm{Re}\, a}$) is the closure of the Fourier transform (resp. the inverse Fourier transform) on $H^s_\beta(\mathbb{R}^n)$ (resp. $H^\beta_{s - \mathrm{Re}\, a}(\mathbb{R}^n)$) described in theorem 2.4. On $\mathfrak{N}(\beta,s,a)$ the estimate

$$\|\mathrm{A}u; H_\beta^{s - \mathrm{Re}\, a}(\mathbb{R}^n)\| \leqslant c \|u; H^s_\beta(\mathbb{R}^n)\| \tag{3.8}$$

holds. The set $\mathfrak{N}(\beta,s,a)$ is dense in $H^s_\beta(\mathbb{R}^n)$, hence $\overline{\mathrm{A}}$ (the closure of A*) is defined on all of $H^s_\beta(\mathbb{R}^n)$ and realizes a continuous map $\overline{\mathrm{A}}: H^s_\beta(\mathbb{R}^n) \to H_\beta^{s - \mathrm{Re}\, a}(\mathbb{R}^n)$. The* *representations* (3.6) *and* (3.7) *hold for* $\overline{\mathrm{A}}$.

P r o o f. Formula (3.6) follows immediately from (3.4) and the definition of $\mathfrak{N}(\beta,s,a)$. We show that (3.6) can be rewritten as (3.7). By (3.6), (2.1) and $\overline{\mathrm{F}^{-1}} = \mathrm{F}^{-1}_{\beta-s+\mathrm{Re}a}$, which holds on $H^{\beta}_{s-\mathrm{Re}a}(\mathbb{R}^n)$ (theorem 2.4), we have $\mathrm{A}u = \mathrm{F}^{-1}_{\beta-s+\mathrm{Re}a}v$, where $\tilde{v}(-\lambda-ia+in/2,\cdot) = \Phi\mathrm{E}(\lambda)\tilde{u}(\lambda+in/2,\cdot)$, $\mathrm{Im}\lambda = \beta-s$. Inverting the Mellin transform (cf. (1.2.6)) and taking into account (2.1) we obtain $v(\rho,\omega) = \rho^a\Phi(\omega)(\mathrm{F}_{\beta-s}u)(\rho,\omega)$. This implies (3.7). In order to obtain (3.8) it suffices to use (3.7) and to recall that each map $\mathrm{F}_{\beta-s}:H^s_\beta(\mathbb{R}^n)\to H^\beta_s(\mathbb{R}^n)$, $\Phi:H^\beta_s(\mathbb{R}^n)\to H^\beta_{s-\mathrm{Re}a}(\mathbb{R}^n)$, $\mathrm{F}^{-1}_{\beta-s+\mathrm{Re}a}: H^\beta_{s-\mathrm{Re}a}(\mathbb{R}^n)\to H^{s-\mathrm{Re}a}_\beta(\mathbb{R}^n)$ is continuous (theorem 2.4, proposition 1.1). The denseness of $\mathfrak{N}(\beta,s,a)$ is verified similarly as in the proof of proposition 2.3. The representations (3.6), (3.7) for $\overline{\mathrm{A}}$ follow from the corresponding formulas for A and the continuity of the three maps just listed. ■

R e m a r k 3.4. The relation between the operators (3.1) and $\overline{\mathrm{A}}$ for $a = -l-n$ can be established under the additional conditions (3.5) only. However, $\overline{\mathrm{A}}$ is itself defined without these conditions. Unless stated otherwise, $\overline{\mathrm{A}}$ is considered for arbitrary functions Φ.

R e m a r k 3.5. If $\beta-s\neq k+n/2$, $\beta-s\neq -k-\mathrm{Re}a-n/2$, the operator (3.6) realizes a continuous map $H^s_\beta(\mathbb{R}^n)\to H^{s-\mathrm{Re}a}_\beta(\mathbb{R}^n)$ if and only if the function $S^{n-1}\ni\theta\mapsto\Phi(\theta)$ is a multiplier in the Sobolev-Slobodetskii space $H^\beta(S^{n-1})$ (cf. proposition 1.1). In particular, for $|\beta|>(n-1)/2$ the condition $\Phi\in H^{|\beta|}(S^{n-1})$ is necessary and sufficient for the operator (3.6) to be continuous.

R e m a r k 3.6. The operators (3.6) and $\mathrm{F}^{-1}\Phi(\xi)\mathrm{F}$ coincide also on a set larger than $\mathfrak{N}(\beta,s,a)$; it is only necessary that the terms between the brackets in (3.4) cancel each other.

§4. The spaces $H^s_\beta(\mathbb{R}^m,\mathbb{R}^{m-n})$*)

Let $1 \leqslant n \leqslant m$, $x^{(1)} = (x_1,\cdots,x_n) \in \mathbb{R}^n$, $x^{(2)} = (x_{n+1},\cdots,x_m) \in \mathbb{R}^{m-n}$, $x = (x^{(1)},x^{(2)}) \in \mathbb{R}^m$. Put $C_0^\infty(\mathbb{R}^m,\mathbb{R}^{m-n}) = C_0^\infty(\mathbb{R}^m \setminus \{x=(x^{(1)},x^{(2)}): x^{(1)}=0\})$. We first assume that s is a nonnegative integer and that $\beta \in \mathbb{R}$, and we introduce the space $H^s_\beta(\mathbb{R}^m,\mathbb{R}^{m-n})$ of functions on $\mathbb{R}^m$ as the completion of the set $C_0^\infty(\mathbb{R}^m,\mathbb{R}^{m-n})$ with respect to the norm

$$\|u;H^s_\beta(\mathbb{R}^m,\mathbb{R}^{m-n})\| = \tag{4.1}$$

$$= \left[\sum_{|\alpha|=0}^{s} \int_{\mathbb{R}^m} |x^{(1)}|^{2(\beta-s+|\alpha|)} |D_x^\alpha u(x)|^2 dx\right]^{1/2}.$$

For $n = m$ the space $H^s_\beta(\mathbb{R}^m,\mathbb{R}^{m-n})$ coincides with $H^s_\beta(\mathbb{R}^m)$.

We denote by $\hat{u} = \mathcal{F}u$ the Fourier transform of a function u with respect to the $x^{(2)} = (x_{n+1},\cdots,x_m)$ variable, and we put $z = |x^{(1)}|\eta$, $v(z,\eta) = |\eta|^{s-\beta-n/2}\hat{u}(z/|\eta|,\eta)$. The norm (4.1) is equivalent to the norm

$$\left[\int_{\mathbb{R}^{m-n}} \sum_{j=0}^{s} \|v(\cdot,\eta);H^j_\beta(\mathbb{R}^n)\|^2 d\eta\right]^{1/2}.$$

Unless otherwise said, all spaces occurring in the sequel are defined as completions of $C_0^\infty(\mathbb{R}^n \setminus 0)$ or $C_0^\infty(\mathbb{R}^m,\mathbb{R}^{m-n})$ in corresponding norms.

We introduce the space $\mathcal{E}^s_\beta(\mathbb{R}^n)$ of functions on $\mathbb{R}^n$, by taking

$$\|w;\mathcal{E}^s_\beta(\mathbb{R}^n)\| = \left[\sum_{j=0}^{s} \|w;H^j_\beta(\mathbb{R}^n)\|^2\right]^{1/2}.$$

Taking into consideration the definition of the norm in $H^s_\rho(\mathbb{R}^n)$ (see (1.1)) we find that an equivalent norm in $\mathcal{E}^s_\beta(\mathbb{R}^n)$ is given by

$$\left[\int_{\mathbb{R}^n} |z|^{2\beta} \sum_{|\kappa|=0}^{s} (|z|^{2(|\kappa|-s)}+1)|D_z^\kappa w(z)|^2 dz\right]^{1/2}.$$

We now define the space $H^s_\beta(\mathbb{R}^m,\mathbb{R}^{m-n})$ for an arbitrary nonnegative index s. For functions u with support in the set $\{x=(x^{(1)},x^{(2)}) \in \mathbb{R}^m; 2^{-1} < |x^{(1)}| < 2\}$ the norm in $H^s_\beta(\mathbb{R}^m,\mathbb{R}^{m-n})$ will be taken to be equivalent to the norm in the

*) §4 - §6 are not used in the sequel.

Sobolev-Slobodetskii space $H^s(\mathbb{R}^m)$. As is well-known [28], for such functions u the norm in $H^s(\mathbb{R}^m)$ has the form

$$\Bigg[\int_{\mathbb{R}^m}\sum_{|\alpha|=0}^{[s]}|D_x^\alpha u(x)|^2dx+\sum_{|\alpha|=[s]}\int_{\mathbb{R}^{m-n}}dx^{(2)}\int_{\mathbb{R}^n}\int_{\mathbb{R}^n}|D_y^\alpha u(y,x^{(2)})+ \tag{4.2}$$

$$-D_z^\alpha u(z,x^{(2)})|^2|y-z|^{-n+2([s]-s)}dydz+$$

$$+\int_{\mathbb{R}^n}dx^{(1)}\int_{\mathbb{R}^{m-n}}\int_{\mathbb{R}^{m-n}}\sum_{|\alpha|=[s]}|D_\xi^\alpha u(x^{(1)},\xi)+$$

$$-D_\eta^\alpha u(x^{(1)},\eta)|^2|\xi-\eta|^{n-m+2([s]-s)}d\xi d\eta\Bigg]^{1/2}.$$

We assume that also for arbitrary s the norm in $H_\beta^s(\mathbb{R}^m,\mathbb{R}^{m-n})$ is homogeneous of degree $\beta-s+m/2$. This means that the norm is multiplied by $t^{\beta-s+m/2}$ as the result of a similitude transformation $x\mapsto tx$ in $\mathbb{R}^m$. Using (4.2) and homogeneity of the norm, we define the norm in $H_\beta^s(\mathbb{R}^m,\mathbb{R}^{m-n})$ for functions u from $C_0^\infty(\mathbb{R}^m,\mathbb{R}^{m-n})$ with arbitrary support by

$$\|u;H_\beta^s(\mathbb{R}^n,\mathbb{R}^{m-n})\|= \tag{4.3}$$

$$=\Bigg[\int_{\mathbb{R}^m}|x^{(1)}|^{2(\beta-s)}\sum_{|\alpha|=0}^{[s]}|x^{(1)}|^{2|\alpha|}|D_x^\alpha u(x)|^2dx+$$

$$+\int_{\mathbb{R}^{m-n}}dx^{(2)}\int_{\mathbb{R}^n}\int_{\mathbb{R}^n}\sum_{|\alpha|=[s]}\big||y|^\beta D_y^\alpha u(y,x^{(2)})-|z|^\beta D_z^\alpha u(z,x^{(2)})\big|^2\times$$

$$\times|y-z|^{-n+2([s]-s)}dydz+$$

$$+\sum_{|\alpha|=[s]}\int_{\mathbb{R}^n}|x^{(1)}|^{2\beta}dx^{(1)}\int_{\mathbb{R}^{m-n}}\int_{\mathbb{R}^{m-n}}|D_\xi^\alpha u(x^{(1)},\xi)-D_\eta^\alpha u(x^{(1)},\eta)|^2\times$$

$$\times|\xi-\eta|^{n-m+2([s]-s)}d\xi d\eta\Bigg]^{1/2}.$$

Let $\zeta\in C_0^\infty(\mathbb{R}^{m-n})$, $\zeta\not\equiv 0$. We will assume that the norm of a function $u(x^{(1)},x^{(2)})=\zeta(x^{(2)})w(x^{(1)})$ in $H_\beta^s(\mathbb{R}^m,\mathbb{R}^{m-n})$ is equivalent to the norm of w in $\mathfrak{S}_\beta^s(\mathbb{R}^n)$. Substituting the expression for u in (4.3) we are led to the following expression for the norm in $\mathfrak{S}_\beta^s(\mathbb{R}^n)$:

$$\|w;\mathfrak{S}_\beta^s(\mathbb{R}^n)\|= \tag{4.4}$$

$$=\Bigg[\int_{\mathbb{R}^n}|x|^{2\beta}\sum_{|\alpha|=0}^{[s]}(|x|^{2(|\alpha|-s)}+1)|D_x^\alpha w(x)|^2dx+$$

$$+\int_{\mathbb{R}^n}\int_{\mathbb{R}^n}\sum_{|\alpha|=s}\frac{\big|\,|x|^\beta D^\alpha_x w(x)-|y|^\beta D^\alpha_y w(y)\big|^2}{|x-y|^{n+2(s-[s])}}dxdy\Bigg]^{1/2}.$$

We rewrite the expression for $\|\cdot;\mathcal{E}^s_\beta(\mathbb{R}^n)\|$ in terms of the Mellin transform. Replace (4.4) by the equivalent norm

$$\left[\|w;H^s_\beta(\mathbb{R}^n)\|^2+\sum_{j=0}^{[s]}\|w;H^j_\beta(\mathbb{R}^n)\|^2\right]^{1/2}.$$

Recalling (1.2) we obtain

$$\|w;\mathcal{E}^s_\beta(\mathbb{R}^n)\|^2 = \tag{4.5}$$

$$=\int_{\mathrm{Im}\lambda=\beta-s}\|\tilde{w}(\lambda+in/2,\cdot);H^s(\lambda,S^{n-1})\|^2d\lambda+$$

$$+\sum_{j=0}^{[s]}\int_{\mathrm{Im}\lambda=\beta-j}\|\tilde{w}(\lambda+in/2,\cdot);H^j(\lambda,S^{n-1})\|^2d\lambda.$$

Note that for $\beta\in\mathbb{R}$, $s\geqslant 0$,

$$\|u;H^s_\beta(\mathbb{R}^m,\mathbb{R}^{m-n})\| = \tag{4.6}$$

$$=\left\{\int\|\,|\eta|^{s-\beta-n/2}\hat{u}(\cdot/|\eta|,\eta);\mathcal{E}^s_\beta(\mathbb{R}^n)\|^2d\eta\right]^{1/2}.$$

P r o p o s i t i o n 4.1. *Let* $v\in C_0^\infty(\mathbb{R}^n\setminus 0)$, $s,\beta\in\mathbb{R}$, $s\geqslant 0$. *For arbitrary* $l\in[0,s]$ *the inequality*

$$\|v;H^s_\beta(\mathbb{R}^n)\|^2\leqslant c(\|v;H^0_\beta(\mathbb{R}^n)\|^2+\|v;H^s_\beta(\mathbb{R}^n)\|^2) \tag{4.7}$$

holds, with a constant c independent of v.

P r o o f. Let p be an arbitrary number subject to the condition $p>\max\{\beta^2,(\beta-s)^2\}$. Then the inequality $\mathrm{Re}(\lambda^2+p)>0$ holds for $\beta-s\leqslant\mathrm{Im}\lambda\leqslant\beta$. Put $\mathrm{U}(\lambda)=(1-\delta)^{-(i\lambda+\beta)/2}(1+\lambda^2+p-\delta)^{(i\lambda+\beta)/2}$, where $\beta-s\leqslant\mathrm{Im}\lambda\leqslant\beta$ and δ is the Laplace-Beltrami operator. For $\mathrm{Im}\lambda=\beta$ the operator $\mathrm{U}(\lambda)$ is unitary on $L_2(S^{n-1})$, while for $\mathrm{Im}\lambda=\beta-s$ the map $\mathrm{U}(\lambda)$: $H^s(\lambda,S^{n-1})\to H^s(S^{n-1})$ is an isomorphism; moreover, $\|\mathrm{U}(\lambda)\|+\|\mathrm{U}(\lambda)^{-1}\|\leqslant$ const. This implies that the righthand side of (4.7) is equivalent to

$$\int\limits_{\mathrm{Im}\lambda=\beta-s} \|\mathrm{U}(\lambda)\tilde{v}(\lambda+in/2,\cdot);H^s(S^{n-1})\|^2 d\lambda + \tag{4.8}$$

$$+ \int\limits_{\mathrm{Im}\lambda=\beta} \|\mathrm{U}(\lambda)\tilde{v}(\lambda+in/2,\cdot);L_2(S^{n-1})\|^2 d\lambda.$$

Let $\lambda \mapsto f(\lambda) \in C^\infty(S^{n-1})$ be a function which is analytic in the strip $\beta - s < \mathrm{Im}\lambda < \beta$ and continuous in the closure of this strip. Assume that the inequality

$$\int\limits_{\mathrm{Im}\lambda=\beta-s} \|f(\lambda);H^s(S^{n-1})\|^2 d\lambda + \int\limits_{\mathrm{Im}\lambda=\beta} \|f(\lambda);L_2(S^{n-1})\|^2 d\lambda < \infty$$

holds. Putting $z = x + i(\beta-(1-\theta)s)$, $0<\theta<1$, we can write $f(z)$ as a Poisson integral:

$$f(z) = \frac{\sin\pi\theta}{2}\int\limits_{\mathbb{R}}\left[\frac{f(\sigma+i(\beta-s))}{\cosh\pi(x-s)-\cos\pi\theta} + \frac{f(\sigma+i\beta)}{\cosh\pi(x-\sigma)+\cos\pi\theta}\right]d\sigma \equiv$$

$$\equiv \mu_0(\theta,\cdot)\star f_0 + \mu_1(\theta,\cdot)\star f_1,$$

where f_0 and f_1 are the traces of f on the lower and the upper edge of the strip, respectively.

As is well-known, the space $H^l(S^{n-1})$ is interpolating with respect to the pair $(H^s(S^{n-1}),L_2(S^{n-1}))$, $H^l(S^{n-1}) = (H^s(S^{n-1}),L_2(S^{n-1}))_{\theta,2}$ for $l = (1-\theta)s$, and the inequality

$$\|f(z);H^l(S^{n-1})\|^2 \leqslant \tag{4.9}$$

$$\leqslant c(\|\mu_0(\theta,\cdot)\star f_0;H^s(S^{n-1})\|^2 + \|\mu_1(\theta,\cdot)\star f_1;L_2(S^{n-1})\|^2)$$

holds.

Substitute in (4.9) $f(z) = \mathrm{U}(z)\tilde{v}(z+in/2,\cdot)$ and subsequently integrate over the line $\mathrm{Im} z = \beta - l = \beta-(1-\theta)s$. Since the kernels $\mu_j(\theta,\cdot)$, $j = 1,2$, are smooth and rapidly decreasing, the quantity

$$\int\limits_{\mathrm{Im}\lambda=\beta-l} \|\mathrm{U}(z)\tilde{v}(z+in/2,\cdot);H^l(S^{n-1})\|^2 dz$$

can be bounded from above by (4.8). Taking into account the properties of the operator-function U thus leads to (4.7). ■

This proposition allows us to define an equivalent norm in $\mathfrak{S}^s_\beta(\mathbb{R}^n)$ by

$$\|w;\mathfrak{S}^s_\beta(\mathbb{R}^n)\| = \Bigg[\int_{\operatorname{Im}\lambda=\beta-s} \|\tilde{w}(\lambda+in/2,\cdot);H^s(\lambda,S^{n-1})\|^2 d\lambda +$$

$$+ \int_{\operatorname{Im}\lambda=\beta} \|\tilde{w}(\lambda+in/2,\cdot);H^0(S^{n-1})\|^2 d\lambda\Bigg]^{1/2}.$$

§5. Transversal operators and their special representations

1. A transversal operator. Let Φ be positively homogeneous function of degree a on $\mathbb{R}^m \setminus 0$ that is infinitely differentiable on S^{m-1}. We denote the operator of multiplication by Φ by the same letter Φ. Consider the operator

$$\mathrm{A}u = \mathrm{F}^{-1}\Phi\mathrm{F}u, \tag{5.1}$$

where F is the Fourier transform on $\mathbb{R}^m$ and $u \in C_0^\infty(\mathbb{R}^m, \mathbb{R}^{m-n})$. We denote, as before, by $\hat{u} = \mathfrak{F}u$ the partial Fourier transform with respect to the variables $x_{n+1}, \cdots, x_m$. We put $\theta = \xi^{(2)}/|\xi^{(2)}|$, $X = x^{(1)}|\xi^{(2)}|$, $Y = Y^{(1)}|\xi^{(2)}|$. By using the fact that Φ is homogeneous we obtain that the quantity $|\xi^{(2)}|^{-a}\widehat{(\mathrm{A}u)}(X|\xi^{(2)}|^{-1}, \xi^{(2)})$ is equal to

$$\frac{1}{(2\pi)^n}\int_{\mathbb{R}^n} e^{iX\eta}\Phi(\eta,\theta)d\eta \int_{\mathbb{R}^n} e^{-iY\eta}\hat{u}(Y|\xi^{(2)}|^{-1}, \xi^{(2)})dY.$$

In the sequel we will denote X and Y by x and y ($\in \mathbb{R}^n$), and $\hat{u}(Y|\xi^{(2)}|^{-1}, \xi^{(2)})$ and $|\xi^{(2)}|^{-a}\widehat{(\mathrm{A}u)}(X|\xi^{(2)}|^{-1}, \xi^{(2)})$ by $v(y)$ and $(\mathrm{A}(\theta)v)(x)$. Thus,

$$(\mathrm{A}(\theta)v)(x) = \mathfrak{F}^{-1}_{\eta\to x}\Phi(\eta,\theta)\mathfrak{F}_{y\to\eta}v(y). \tag{5.2}$$

For every $\theta \in S^{m-n-1}$ we call the operator $\mathrm{A}(\theta)$ a *transversal operator* (corresponding to the operator (5.1)).

2. Special representations of a transversal operator. In the investigation in §3 of the boundedness of a convolution operator in the scale $H^s_\beta(\mathbb{R}^n)$ we needed a special representation of such an operator (cf. (3.4)). Our aim in the present section is to derive corresponding representations for $\mathrm{A}(\theta)$. We will need them in order to give estimates in the scale $\mathfrak{S}^s_\beta(\mathbb{R}^n)$.

Let $v \in C_0^\infty(\mathbb{R}^n \setminus 0)$ and $\tau \neq k + n/2$, $k = 0, 1, \cdots$. Put

$$J_\tau(t,\omega) = \int_{\operatorname{Im}\mu=\tau} e^{-i\mu t}\mathrm{E}_{\psi\to\omega}(\mu)\tilde{v}(\mu+in/2,\psi)d\mu. \tag{5.3}$$

L e m m a 5.1. *Let* $\eta_- \in C^\infty(\mathbf{R})$ *with, moreover,* $\eta_-(t) = 0$ *for* $t > 1/2$ *and* $\eta_-(t) = 1$ *for* $t < -1/2$. *The integral*

$$\int_{-\infty}^{+\infty} e^{i\lambda t}\eta_-(t)\Phi(e^t\omega,\theta)J_\tau(t,\omega)dt, \quad |\omega| = |\theta| = 1, \tag{5.4}$$

defined for $\operatorname{Im}\lambda < n/2$, *has an analytic extension to the whole* λ*-plane as a meromorphic function; only the points* $\lambda = i(p+n/2)$, *where* $p = p_\tau, p_\tau+1, \cdots$, *and* $p_\tau = \max\{[\tau-n/2]+1,0\}$, *can be singular. At these points the integral* (5.4) *can have poles of order one only. The residue at* $\lambda = i(p+n/2)$ *is equal to*

$$\frac{1}{(2\pi)^{(n-1)/2}} \sum_{|\gamma|+|\kappa|=p} \omega^{\gamma+\kappa}\frac{(-i)^{|\gamma+1|}}{\gamma!\kappa!}\partial^\kappa\Phi(0,\theta)\int_{\mathbf{R}^n} y^\gamma v(y)dy, \tag{5.5}$$

where $(\partial^\kappa\Phi)(x,z) \equiv (\partial_x^\kappa\Phi)(x,z)$. *As* λ *tends to* ∞ *while remaining in some strip* $\{\lambda: |\operatorname{Im}\lambda| < h\}$, *then the integral* (5.4) *(its analytic extension) decreases faster than any power of* $|\lambda|$.

P r o o f. Note that $J_\tau(t,\omega) = \sqrt{2\pi}\rho^{n/2}(\mathrm{F}_\tau v)(\rho,\omega)$, where $\rho = e^t$ and F_τ is defined by (2.1). By (2.2) we have

$$J_\tau(t,\omega) = \frac{1}{(2\pi)^{(n-1)/2}}\sum_{|\gamma|=p_\tau}^{M} (-i)^{|\gamma|} e^{(n/2+|\gamma|)t}\frac{\omega^\gamma}{\gamma!}\times \tag{5.6}$$

$$\times\int_{\mathbf{R}^n} y^\gamma v(y)dy + J_{\tau'}(t,\omega),$$

with $\tau' \in (M+n/2, M+1+n/2)$. Expand Φ by Taylor's formula and rewrite (5.4) as

$$\frac{1}{(2\pi)^{(n-1)/2}}\sum_{p=p_\tau}^{M+N}\sum_{|\kappa|+|\gamma|=p}\frac{(-i)^{|\gamma|}\omega^{\gamma+\kappa}}{\gamma!\kappa!}\times \tag{5.7}$$

$$\times\int_{\mathbf{R}^n} y^\gamma v(y)dy\cdot\partial^\kappa\Phi(0,\theta)\int_{-\infty}^{+\infty} e^{(i\lambda+n/2+|\kappa|+|\gamma|)t}\eta_-(t)dt +$$

$$+\int_{-\infty}^{+\infty} e^{i\lambda t}\eta_-(t)\Phi_{N+1}(t,\omega,\theta)J_\tau(t,\omega)dt + \int_{-\infty}^{+\infty} e^{i\lambda t}\eta_-(t)\Phi(e^t\omega,\theta)J_{\tau'}(t,\omega)dt.$$

In view of the estimates $\Phi_{N+1}(t,\omega,\theta) = O(e^{(N+1)t})$, $J_{\tau'}(t,\omega) = O(e^{(N+1+n/2)t})$, $J_\tau(t,\omega) = O(e^{(p_\tau+n/2)t})$ as $t\to-\infty$ (cf. (2.4)), the last two terms in (5.7) are functions analytic in the halfplane $\operatorname{Im}\lambda < h_0 = \min\{N+p_\tau, M\}+1+n/2$ and rapidly decreasing as $|\lambda|\to\infty$, $|\operatorname{Im}\lambda| \leqslant h < h_0$. In order to convince ourselves of the required behavior of the remaining terms of (5.7), it suffices to apply the formula for integration by parts to the integrals with respect to t. ■

P r o p o s i t i o n 5.2. *Let $-n/2<\tau+\operatorname{Re} a$, $q+n/2<\tau<q+1+n/2$, where q is a nonnegative integer. Then we have for the operator $\mathrm{A}(\theta)$, defined by (5.2), and $v\in C_0^\infty(\mathbf{R}^n\setminus 0)$ the representation*

$$(\mathrm{A}(\theta)v)(x) = \tag{5.8}$$

$$= \frac{1}{\sqrt{2\pi}}\int_{\operatorname{Im}\lambda=\tau} r^{i(in/2+\lambda+ia)}\mathrm{E}_{\omega\to\phi}(\lambda+ia)^{-1}\Phi_{\mu\to\lambda}(\omega,\theta)\times$$

$$\times\,\mathrm{E}_{\psi\to\omega}(\mu)\tilde{v}(\mu+in/2,\psi)d\mu + \sum_{|\gamma|=0}^{q} Z_\gamma(x,\theta)\int_{\mathbf{R}^n} y^\gamma v(y)dy.$$

Here $r=|x|$, $\phi = x/|x|$,

$$\Phi_{\mu\to\lambda}(\omega,\theta)w(\mu,\omega) = \tag{5.9}$$

$$= \frac{1}{2\pi}\int_{-\infty}^{+\infty} e^{i\lambda t}\Phi(\omega,e^{-t}\theta)dt\int_{\operatorname{Im}\mu=\tau} e^{-i\mu t}w(\mu,\omega)d\mu,$$

$$Z_\gamma(x,\theta) = \frac{(-i)^{|\gamma|}}{(2\pi)^{(n+1)/2}\gamma!}\left\{\int_{\operatorname{Im}\gamma=h} r^{i(in/2+\lambda)}\mathrm{E}_{\omega\to\phi}(\lambda)^{-1}\omega^\gamma d\lambda\times\right.$$

$$\times\int_{-\infty}^{+\infty}\eta_+(t)e^{(i\lambda+n/2+|\gamma|)t}\Phi(e^t\omega,\theta)dt + \int_{-\infty}^{+\infty} r^{i(in/2+\lambda)}\mathrm{E}(\lambda)^{-1}\omega^\gamma d\lambda\times$$

$$\left.\times\int_{-\infty}^{+\infty}\eta_-(t)\,e^{(i\lambda+n/2+|\gamma|)t}\Phi(e^t\omega,\theta)dt\right\},$$

the $\eta_\pm$ are nonnegative functions from $C^\infty(\mathbf{R})$, $\eta_-(t)=0$ as $t>1/2$, $\eta_-(t)=1$ as $t<-1/2$, $\eta_+(t)=1-\eta_-(t)$; and h is an arbitrary number satisfying $h>\max\{-n/2, |\gamma|+\operatorname{Re} a+n/2\}$. Under the conditions $-n/2<\tau+\operatorname{Re} a$, $\tau<n/2$ the formula obtained from (5.8) by discarding the terms containing Z_γ

holds.

P r o o f. Applying proposition 1.2.1 we rewrite (5.2) in the form

$$(A(\theta)v)(x) = \frac{1}{\sqrt{2\pi}} \int_{-\infty}^{+\infty} r^{i(in/2+\lambda)} E_{\omega\to\phi}(\lambda)^{-1} d\lambda \times \tag{5.10}$$

$$\times \int_{-\infty}^{+\infty} e^{i\lambda t} \Phi(e^t\omega,\theta) J_0(t,\omega) dt.$$

Our aim is to replace J_0 by J_τ (cf. (5.3)) and the line $\text{Im}\lambda = 0$ by $\text{Im}\lambda = \tau$. Represent (5.10) as the sum $I_+(x)+I_-(x)$, where the $I_\pm$ are equal to the right-hand side of (5.10) with Φ replaced by $\eta_\pm(t)\Phi$. By the equation $J_\tau(t,\omega) = \sqrt{2\pi}\rho^{n/2}(F_\tau v)(\rho,\omega)$ and estimate (2.3), the inner integral in I_+, containing η_+, is an analytic function on the whole λ-plane and decreases faster than any power of $|\lambda|$ in every strip $|\text{Im}\lambda| < N$. Hence for $\tau + \text{Re}a > -n/2$,

$$I_+(x) = \frac{1}{(2\pi)^{3/2}} \int_{\text{Im}\lambda=\tau} r^{i(in/2+\lambda+ia)} E(\lambda+ia)^{-1} d\lambda \times \tag{5.11}$$

$$\times \int_{-\infty}^{+\infty} e^{i\lambda t} \eta_+(t)\Phi(\omega,e^{-t}\theta) J_0(t,\omega) dt.$$

Note that (2.4) implies that the integral (5.4) is analytic in the halfplane $\text{Im}\lambda < q+1+n/2$ (under the condition $\tau > q+n/2$). For $-n/2 < \tau + \text{Re}a < q+1+n/2$, by using (5.6) (in which we must put $\tau = 0$, $\tau' = \tau$) we obtain

$$I_-(x) = \frac{1}{(2\pi)^{3/2}} \int_{\text{Im}\lambda=\tau+\text{Re}a} r^{i(in/2+\lambda+ia)} E(\lambda)^{-1} d\lambda \times \tag{5.12}$$

$$\times \int_{-\infty}^{+\infty} e^{i\lambda-ia)t} \eta_-(t)\Phi(\omega,e^{-t}\theta) J_\tau(t,\omega) dt +$$

$$+ \frac{1}{(2\pi)^{(n+1)/2}} \int_{\text{Im}\lambda=0} r^{i(in/2+\lambda)} E(\lambda)^{-1} d\lambda \times$$

$$\times \int_{-\infty}^{+\infty} e^{i\lambda t} \eta_-(t)\Phi(e^t\omega,\theta) \sum_{k=0}^{q} e^{(k+n/2)t} dt\, (-i)^k \sum_{|\gamma|=k} \frac{\omega^\gamma}{\gamma!} \int_{\mathbf{R}^n} y^\gamma v(y) dy.$$

Substituting in (5.11) the expression for $J_0(t,\omega)$ from (5.6) (where, again, we put

$\tau = 0, \tau' = \tau$) and using (5.12) leads to (5.8).

For $\tau + \mathrm{Re}\, a > q+1+n/2$ the poles of the integral (5.4) fall within the strip $0 < \mathrm{Im}\lambda < \tau + \mathrm{Re}\, a$. However, proposition 1.4.3 and formula (5.5) imply that the poles disappear after an application of $\mathrm{E}(\lambda)^{-1}$ to this integral. Hence (5.12) remains valid also under the single restriction $-n/2 < \tau + \mathrm{Re}\, a$. The proof is finished by combining (5.11) and (5.12); in advance it is necessary to replace $J_0(t,\omega)$ by (5.6), in which $\tau = 0$, $\tau' = \tau$. ■

Proposition 5.3. *Let* $\tau < n/2$, $-p-1-n/2 < \tau + \mathrm{Re}\, a < -p-n/2$, *p a nonnegative integer. Then we have for the operator* (5.2) *and* $v \in C_0^\infty(\mathbb{R}^n \setminus 0)$ *the representation* (5.8) *in which the sum of the terms containing* Z_γ *must be replaced by*

$$\sum_{|\gamma|=0}^{p} x^\gamma (v, \zeta_\gamma(\cdot,\theta)).$$

Here (f,g) *is the scalar product in* $L_2(\mathbb{R}^n)$ *and* $\zeta_\gamma(x,\theta)$ *coincides with the righthand side of* (5.9) *in which* Φ *must be replaced by* $\overline{\Phi}$.

Proof. Lemma 5.1 implies that the integral

$$\int_{-\infty}^{+\infty} e^{i\lambda t}\Phi(e^t\omega,\theta)J_0(t,\omega)dt$$

is an analytic function in the halfplane $\mathrm{Im}\lambda < n/2$. Recall that $J_0(t,\omega) = J_\tau(t,\omega)$ for $\tau < n/2$. We will first assume that a is such that $-n/2 < \tau + \mathrm{Re}\, a < \delta - n/2$, with δ a sufficiently small positive number. Then the equation obtained from (5.8) by discarding the terms containing Z_γ, $|\gamma| = 0, \cdots, q$, holds. This and (1.4.2) imply that

$$(\mathrm{A}(\theta)v)(x) = \frac{1}{(2\pi)^{3/2}} \int_{\mathrm{Im}\lambda=\tau-\delta} r^{i(in/2+\lambda)}\mathrm{E}(\lambda+ia)^{-1}d\lambda \times$$

$$\times \int_{-\infty}^{+\infty} e^{i\lambda t}\Phi(\omega, e^{-t}\theta)J_\tau(t,\omega)dt +$$

$$-\frac{1}{(2\pi)^{(n+1)/2}}\, i \int_{S^{n-1}} d\omega \int_{-\infty}^{+\infty} e^{tn/2}\Phi(e^t\omega,\theta)J_\tau(t,\omega)dt.$$

Clearly, equality also holds under the condition $\delta - 1 - n/2 < \tau + \mathrm{Re}\, a < -n/2$. If this condition is satisfied, we may again replace the line $\mathrm{Im}\lambda = \tau - \delta$ by $\mathrm{Im}\lambda = \tau$. Repeating in a suitable manner these considerations we are led to the formula (for $\tau < n/2$, $-p-1-n/2 < \tau + \mathrm{Re}\, a < -p-n/2$):

$$(A(\theta)v)(x) = \frac{1}{(2\pi)^{3/2}} \int_{\mathrm{Im}\lambda=\tau} r^{i(in/2+\lambda+ia)} E(\lambda+ia)^{-1} d\lambda \times$$

$$\times \int_{-\infty}^{+\infty} e^{i\lambda t}\Phi(\omega, e^{-t}\theta) J_\tau(t,\omega)dt - \frac{1}{(2\pi)^{(n+1)/2}} \times$$

$$\times \sum_{|\gamma|=0}^{p} \frac{i^{|\gamma|+1}}{\gamma!} x^\gamma \int_{S^{n-1}} \omega^\gamma d\omega \int_{-\infty}^{+\infty} e^{(n/2+|\gamma|)t}\Phi(e^t\omega, \theta) J_0(t,\omega)dt.$$

Consider the integral

$$\int \omega^\gamma d\omega \int e^{(n/2+|\gamma|)t}\eta_+(t)\Phi(e^t\omega,\theta)J_0(t,\omega)dt =$$

$$= \int \omega^\gamma d\omega \int e^{(n/2+|\gamma|)t}\eta_+(t)\Phi(e^t\omega,\theta)dt \times$$

$$\times \int_{\mathrm{Im}\mu=-h} e^{-i\mu t}E(\mu)\tilde{v}(\mu+in/2,\cdot)d\mu,$$

where h is chosen to satisfy the inequality $h > \max\{-n/2, |\gamma| + \mathrm{Re}\, a + n/2\}$. Changing the order of integration, we rewrite this expression in the form

$$\int_{\mathbf{R}^n} v(x)dx \int_{\mathrm{Im}\mu=-h} r^{i(in/2-\mu)} E_{\omega\to\psi}(\mu) \times$$

$$\times \left[\omega^\gamma \int_{-\infty}^{+\infty} \eta_+(t) e^{(-i\mu+n/2+|\gamma|)t}\Phi(e^t\omega,\theta)dt \right] d\mu.$$

The expression with η_- instead of η_+ can be written similarly; we must only put $h = 0$ in this case. Recalling (1.2.1) and (1.4.1) we obtain

$$\overline{E_{\omega\to\psi}(\mu)f(\omega)} = E_{\omega\to\psi}(\overline{\mu})^{-1}\overline{f(\omega)}.$$

It remains to combine the equations given. ■

Using a reasoning close to the above proves

P r o p o s i t i o n 5.4. *Let* $-p-1-n/2 < \tau + \mathrm{Re}\, a < -p-n/2$, $q+n/2 < \tau < q+1+n/2$, *p and q nonnegative integers. Then we have for the* *operator* (5.2) *and* $v \in C_0^\infty(\mathbf{R}^n \setminus 0)$ *the representation* (5.8) *in which at the righthand side we must add the sum*

$$\sum_{|\gamma|=0}^{p} x^\gamma(v, \eta_{\gamma,q}(\cdot,\theta)),$$

where

$$\eta_{\gamma,q}(x,\theta) = \frac{(-i)^{|\gamma|}}{(2\pi)^{(n+1)/2}\gamma!} \int_{\operatorname{Im}\mu=-\tau} r^{i(in/2+\mu)} \mathrm{E}(\mu)^{-1} \omega^\gamma d\mu \times \tag{5.13}$$

$$\times \int_{-\infty}^{+\infty} e^{i\mu t+(n/2+|\gamma|)t} \overline{\Phi(e^t\omega,\theta)}dt.$$

3. Some properties of the special representations for A(θ).

P r o p o s i t i o n 5.5. *The functional*

$$C_0^\infty(\mathbf{R}^n \setminus 0) \ni v \mapsto \int_{\mathbf{R}^n} v(y) y^\gamma dy \tag{5.14}$$

is continuous on $\mathfrak{S}_\beta^s(\mathbf{R}^n)$ *if and only if* $|\gamma| + n/2 < \beta < s + |\gamma| + n/2$.

P r o o f. The definition implies that $\mathfrak{S}_\beta^s(\mathbf{R}^n)$ is the intersection of the spaces $H_\beta^j(\mathbf{R}^n)$, $j = 0, \cdots, [s]$, and $H_\beta^s(\mathbf{R}^n)$. Since the inclusion $H_{\beta+\delta}^{l+\delta}(\mathbf{R}^n) \subset H_\beta^l(\mathbf{R}^n)$ holds for all l, β and arbitrary $\delta > 0$, the functional is continuous on $\mathfrak{S}_\beta^s(\mathbf{R}^n)$ if it is continuous on the intersection of the spaces $H_{\beta-j}^0(\mathbf{R}^n)$, $j = 0, \cdots, [s]$, and $H_{\beta-s}^0(\mathbf{R}^n)$. The latter property is guaranteed by the relations $\chi y^\gamma \in (H_{\beta-s}^0(\mathbf{R}^n))^* = H_{s-\beta}^0(\mathbf{R}^n)$ and $(1-\chi)y^\gamma \in (H_\beta^0(\mathbf{R}^n))^* = H_{-\beta}^0(\mathbf{R}^n)$, where $\chi \in C_0^\infty(\mathbf{R}^n)$ with $\chi = 1$ in a neighborhood of the origin. It is obvious that these relations follow from the inequalities to which β, γ and s are subjected. It has been established that the conditions of the proposition are sufficient: necessity, in particular, is contained in the following proposition.

P r o p o s i t i o n 5.6. *Let* p_1, p_2 *be integers,* $0 \leq p_1 \leq p_2$, *and let* $\mathfrak{M}(p_1,p_2) = \{v \in C_0^\infty(\mathbf{R}^n \setminus 0) \colon \tilde{v}(i(q+n),\phi) \equiv 0,\ q = p_1, \cdots, p_2\}$. *In* $\mathfrak{S}_\beta^s(\mathbf{R}^n)$ *each of the following sets is dense:* $\mathfrak{M}(0,p_2)$ *if* $\beta > s + p_2$, $\mathfrak{M}(p_1,p_2)$ *if* $\beta < p_1 + n/2$.

P r o o f. The proof is obtained from the proof of proposition 2.3 by making some obvious changes in it. ■

We now turn to the functionals (v, ζ_γ) and $(v, \eta_{\gamma,q})$ from propositions 5.3 and 5.4.

P r o p o s i t i o n 5.7. *The functions $x \to \zeta_\gamma(x,\theta)$, defined in proposition 5.3, are infinitely differentiable outside the origin. As $x \to \infty$ they decrease faster than any power of $|x|$, uniformly in $\theta \in S^{m-n-1}$. As $x \to 0$ the estimates*

$$\zeta_\gamma(x,\theta) = \begin{cases} O(1) \text{ for } \mathrm{Re}\, a \leqslant -n-\gamma,\ a+n+|\gamma| \neq 0, \\ O(\ln|x|) \text{ for } a+n+|\gamma| = 0, \\ O(|x|^{-n-\mathrm{Re}\, a-|\gamma|}) \text{ for } \mathrm{Re}\, a > -n-|\gamma|, \end{cases} \tag{5.15}$$

hold. Everything said remains valid for the functions $Z_\gamma(\cdot,\theta)$ also.

P r o o f. The function $\zeta_\gamma(\cdot,\theta)$ coincides with the righthand side of (5.9) in which Φ must be replaced by $\overline{\Phi}$. Since in the first term of (5.9) the number h can be taken arbitrary large, using proposition 1.5.5 we find that this term decreases faster than any power of $|x|$ as $x \to \infty$.

We now turn to the second term in (5.9). The inner integral has a meromorphic extension onto the whole λ-plane. The points $\lambda = i(|\gamma|+|\kappa|+n/2)$, $|\kappa| = 0,1,\cdots$, turn out to be (first order) poles, and the corresponding residues are equal to $i(\kappa!)^{-1}\partial^\kappa\overline{\Phi(0,\theta)}\omega^\kappa$. As $\lambda \to \infty$ such that $|\mathrm{Im}\lambda| < N$ for some N, then this extension decreases faster than any power of $|\lambda|$. After multiplication of the inner integral by ω^γ and an application of the operator $\mathrm{E}(\lambda)^{-1}$ the poles disappear (by proposition 1.4.3). This allows us to replace in the second term of (5.9) the line of integration $\mathrm{Im}\lambda = 0$ by the line $\mathrm{Im}\lambda = h$. From this we obtain that the second term also decreases rapidly as $x \to \infty$.

We now verify formula (5.15). Using the fact that the poles of $\mathrm{E}(\lambda)^{-1}$ are located at the points $\lambda = -i(k+n/2)$, $k = 0,1,\cdots$, and by moving downwards the line of integration of the second term in (5.9) we find that this term is $O(1)$. Consider the first term. The inner integral extends as a meromorphic function onto the whole λ-plane, and only the points $\lambda = i(|\gamma|+a-|\alpha|+n/2)$ (are first order) poles (in order to see this we must expand $\Phi(\omega, e^{-t}\theta)$ by Taylor's formula at $(\omega,0)$). By moving the line of integration downwards, we find that the first term in (5.9) is $O(|x|^{-a-|\gamma|-n})$ as $x\to 0$ (if $a \neq -|\gamma|-n$) or $O(\ln|x|)$ (if $a=-|\gamma|-n$; in this case the integrand has a second order pole at $\lambda=-in/2$). ■

A similar reasoning proves

Proposition 5.8. *Let* $q+n/2<\tau<q+1+n/2$ *and* $-p-1-n/2<\tau+\mathrm{Re}\,a<-p-n/2$, *where* p,q *are nonnegative integers. The functions* $\eta_{\gamma,q}(\cdot,\theta)$, *defined by* (5.13), *are infinitely differentiable outside the origin. As* $x\to\infty$ *we have, for any* N,

$$\eta_{\gamma,q}(x)+\frac{(-i)^{|\gamma|-1}}{(2\pi)^{n-1/2}\gamma!}\sum_{|\kappa|=0}^{q}\frac{(ix)^{\kappa}}{\kappa!}\int_{S^{n-1}}\omega^{\gamma+\kappa}d\omega\times$$

$$\times\int_{-\infty}^{+\infty}e^{(n+|\kappa|+|\gamma|)t}\overline{\Phi(e^{t}\omega,\theta)}dt=O(|x|^{N}).$$

As $x\to 0$ *the estimate*

$$\eta_{\gamma,q}(x,\theta)=$$

$$=\begin{cases} O(|x|^{q+1}) \text{ for } q+1\leqslant -\mathrm{Re}\,a-n/2-|\gamma|, \\ \qquad\qquad q+1\neq -a-n/2-|\gamma|, \\ O(|x|^{q+1}\ln|x|) \text{ for } q+1=-a-n/2-|\gamma|, \\ O(|x|^{-\mathrm{Re}\,a-n/2-|\gamma|}) \text{ for } q+1>-\mathrm{Re}\,a-n/2-|\gamma| \end{cases}$$

holds.

In order to verify the following two assertions it is necessary to use propositions 5.7 and 5.8 (compare with the proof of proposition 5.5).

Proposition 5.9. *If* $\beta-s<\min\{n/2,-\mathrm{Re}\,a-|\gamma|-n/2\}$, *then the functionals* $v\mapsto(v,\zeta_\gamma)$ *are continuous on* $\mathcal{E}^s_\beta(\mathbb{R}^n)$.

Proposition 5.10. *Let* $q+n/2<\beta-s<q+1+n/2$, $-p-1-n/2<\beta-s+\mathrm{Re}\,a<-p-n/2$, *where* p,q *are nonnegative integers. Then the functionals* $v\to(v,\eta_{\gamma,q})$, *where* $|\gamma|=0,\cdots,p$, *are continuous on* $\mathcal{E}^s_\beta(\mathbb{R}^n)$. *The closure* $\overline{\mathfrak{N}}$ *in* $\mathcal{E}^s_\beta(\mathbb{R}^n)$ *of the set* $\mathfrak{N}=\{v\in C_0^\infty(\mathbb{R}^n\setminus 0):(v,\eta_{\gamma,q})=0,$ $|\gamma|=0,\cdots,p,\ (v,y^\kappa)=0,\ |\kappa|=0,\cdots,q\}$ *coincides with the subspace of* *common zeros of the functionals* $(v, \eta_{\gamma,q})$.

§6. Estimates for the convolution operator on the spaces $H^s_\beta(\mathbb{R}^m, \mathbb{R}^{m-n})$

We first give a boundedness theorem for the transversal operators $A(\theta)$ (cf. (5.2)) in the scale $\mathcal{E}^s_\beta(\mathbb{R}^n)$. As a corollary we subsequently obtain a statement concerning continuity of the original operator (5.1) in the scale $H^s_\beta(\mathbb{R}^m, \mathbb{R}^{m-n})$.

Theorem 6.1. *Let a,b and s be such that* $\beta - s + \mathrm{Re}\, a > -n/2$, $\mathrm{Re}\, a \leqslant s$, $s \geqslant 0$, *and* $\beta - s \neq k + n/2$, $\beta \neq k + n/2$, $k = 0, 1, \cdots$. *For* $\beta - s > n/2$ *we put*

$$\mathfrak{L} = \left\{ v \in C_0^\infty(\mathbb{R}^n \setminus 0) : \int_{\mathbb{R}^n} v(y) y^\gamma dy = 0, \quad |\gamma| = 0, \cdots, [\beta - s - n/2] \right\},$$

and for $\beta - s < n/2$ *we will assume* $\mathfrak{L} = C_0^\infty(\mathbb{R}^n \setminus 0)$.

Then the operator $A(\theta)$, *given on* $\mathfrak{L}$ *by* (5.3), *allows the estimate*

$$\|A(\theta)v; \mathcal{E}^{s-\mathrm{Re}\, a}_\beta(\mathbb{R}^n)\| \leqslant c \|v; \mathcal{E}^s_\beta(\mathbb{R}^n)\| \tag{6.1}$$

with a constant c which is independent of $\theta \in S^{m-n-1}$. *The set* $\mathfrak{L}$ *is dense in* $\mathcal{E}^s_\beta(\mathbb{R}^n)$, *and for* $\overline{A(\theta)}$ *(the closure of* $A(\theta)$*), which is defined on all of* $\mathcal{E}^s_\beta(\mathbb{R}^n)$, *the representation*

$$(\overline{A(\theta)}v)(x) = \frac{1}{\sqrt{2\pi}} \int_{\mathrm{Im}\lambda = \beta - s} r^{i(in/2 + \lambda + ia)} E_{\omega \to \phi}(\lambda + ia)^{-1} \times \tag{6.2}$$

$$\times \Phi_{\mu \to \lambda}(\omega, \theta) E_{\psi \to \omega}(\mu) \tilde{v}(\mu + in/2, \psi) d\mu$$

holds.

Proof. By proposition 5.2, for $v \in \mathfrak{L}$ the expression $A(\theta)v$ coincides with the righthand side of (6.2). Proposition 5.6 implies that $\mathfrak{L}$ is dense. Hence it suffices to verify the estimate (6.1).

Put $w = A(\theta)v$. After application of the Mellin transform this equation takes the form (for $\mathrm{Im}\lambda = \mathrm{Im}\mu = \beta - s$)

$$\tilde{w}(\lambda + ia + in/2, \phi) = \tag{6.3}$$

$$= E_{\omega \to \phi}(\lambda + ia)^{-1} \Phi_{\mu \to \lambda}(\omega, \theta) E_{\psi \to \omega}(\mu) \tilde{v}(\mu + in/2, \psi),$$

where the expression $\Phi_{\mu\to\lambda}(\omega,\theta)\mathrm{E}(\mu)\tilde{v}(\mu+in/2,\cdot)$ is to be understood as in proposition 5.2. We show that $\lambda\mapsto\tilde{w}(\lambda+ia+n/2,\cdot)$, defined for $\mathrm{Im}\,\lambda=\beta-s$ by (6.3), has an analytic extension to the halfplane $\mathrm{Im}\,\lambda\geqslant\beta-s$.

We have

$$\Phi_{\mu\to\lambda}(\omega,\theta)\mathrm{E}_{\psi\to\omega}(\mu)\tilde{v}(\mu+in/2,\psi)= \tag{6.4}$$

$$=\int_{-\infty}^{+\infty}\eta_-(t)e^{i\lambda t}\Phi(\omega,e^{-t}\theta)J_{\beta-s}(t,\omega)dt+$$

$$+\int_{-\infty}^{+\infty}\eta_+(t)e^{i\lambda t}\Phi(\omega,e^{-t}\theta)J_{\beta-s}(t,\omega)dt$$

(the notations are those of §5.2). Formula (5.6) and proposition 2.1 imply that $J_{\beta-s}(t,\omega)=O(e^{ht})$ as $t\to+\infty$ and $h=[\beta-s-n/2]+n/2$. Recall that the operator function $\lambda\mapsto\mathrm{E}(\lambda+ia)^{-1}$ depends analytically on λ in the halfplane $\mathrm{Im}\,\lambda+\mathrm{Re}\,a>-n/2$. Hence the result of applying $\mathrm{E}(\lambda+ia)^{-1}$ to the second term on the right in (6.4) is also an analytic function on the halfplane $\mathrm{Im}\,\lambda\geqslant\beta-s$. The first term has a meromorphic extension to the whole λ-plane (cf. lemma 5.1). It was shown in the proof of proposition 5.2 that the value of $\mathrm{E}(\lambda+ia)^{-1}$ on the first term is also an analytic function in the halfplane $\mathrm{Im}\,\lambda+\mathrm{Re}\,a>-n/2$.

Let's turn to estimating the norm

$$\|w;\mathcal{E}_\beta^{s-\mathrm{Re}\,a}(\mathbb{R}^n)\|=$$

$$=\left[\sum_{j=0}^{[s-\mathrm{Re}\,a]}\int_{\mathrm{Im}\,\lambda=\beta-j-\mathrm{Re}\,a}\|\tilde{w}(\lambda+ia+in/2,\cdot);H^j(\lambda,S^{n-1})\|^2d\lambda+\right.$$

$$\left.+\int_{\mathrm{Im}\,\lambda=\beta-s}\|\tilde{w}(\lambda+ia+in/2,\cdot);H^{s-\mathrm{Re}\,a}(\lambda,S^{n-1})\|^2d\lambda\right]^{1/2}.$$

We denote the first and the second term on the right in (6.4) by $\tilde{g}_-(\lambda,\omega,\theta)$ and $\tilde{g}_+(\lambda,\omega,\theta)$, respectively. We have

$$\tilde{w}(\lambda+ia+in/2,\phi)= \tag{6.5}$$

$$=\mathrm{E}_{\omega\to\phi}(\lambda+ia)^{-1}\tilde{g}_-(\lambda,\omega,\theta)+\mathrm{E}_{\omega\to\phi}(\lambda+ia)^{-1}\tilde{g}_+(\lambda,\omega,\theta).$$

We first consider the expression $\mathrm{E}(\lambda+ia)^{-1}\tilde{g}_+(\lambda,\omega,\theta)$. By assumption $\beta-s+\mathrm{Re}\,a>-n/2$, hence for $\tau=\beta-s$ or $\tau=\beta-\mathrm{Re}\,a$ the line $\mathrm{Im}\,\lambda=\tau$ does not contain poles of the operator-function $\lambda \mapsto E(\lambda+ia)^{-1}$. This and proposition 1.5.5 imply that

$$\int_{\operatorname{Im}\lambda=\tau} \|E(\lambda+ia)^{-1}\tilde{g}_+(\lambda,\cdot,\theta);\, H^l(\lambda,S^{n-1})\|^2 d\lambda \leqslant \tag{6.6}$$

$$\leqslant c \int_{\operatorname{Im}\lambda=\tau} \|\tilde{g}_+(\lambda,\cdot,\theta);\, H^\beta(\lambda,S^{n-1})\|^2 d\lambda,$$

where $l = \beta-\tau-\operatorname{Re} a$. By putting $\tilde{f}(\mu,\omega) = E_{\psi\to\omega}(\mu)\tilde{v}(\mu+in/2,\psi)$ we obtain

$$\tilde{g}_+(\xi+i\tau,\omega,\theta) = \tag{6.7}$$

$$= \int_{-\infty}^{+\infty} e^{i\xi t}\eta_+(t)e^{(\beta-s-\tau)t}\Phi(\omega,e^{-t}\theta)dt \int_{-\infty}^{+\infty} e^{-i\sigma t}\tilde{f}(\sigma+i(\beta-s),\omega)d\sigma.$$

Obviously, $\beta-s-\tau\leqslant 0$. Assume for the moment that $\beta \geqslant 0$. Then the norm

$$\left[\int_{\operatorname{Im}\lambda=\tau} \|\tilde{g}_+(\lambda,\cdot,\theta); H^\beta(\lambda,S^{n-1})\|^2 d\lambda\right]^{1/2}$$

is equivalent to the norm

$$\left\{\int_{-\infty}^{+\infty} [(1+\xi^2)^\beta \|\tilde{g}_+(\xi+i\tau,\cdot,\theta);\, L_2(S^{n-1})\|^2 + \right. \tag{6.8}$$

$$\left. + \|\tilde{g}_+(\xi+i\tau,\cdot,\theta);\, H^\beta(S^{n-1})\|^2] d\xi\right\}^{1/2},$$

where, as usual, $H^\beta(S^{n-1})$ is a Sobolev-Slobodetskii space. Further, the operator of multiplication by the function $t \mapsto \eta_+(t)e^{(\beta-s-\tau)t}\Phi(\omega,e^{-t}\theta)$ is continuous on $H^\beta(\mathbb{R})$ and, moreover, its norm is uniformly bounded in ω and θ, $|\theta| = |\omega| = 1$. Hence the integral of the first term in (6.8) is at most

$$c\int_{-\infty}^{+\infty} (1+\sigma^2)^\beta \|\tilde{f}(\sigma+i(\beta-s),\cdot);L_2(S^{n-1})\|^2 d\sigma.$$

The same operator of multiplication is continuous also on the space $L_2(\mathbb{R};H^\beta(S^{n-1}))$ of square-integrable functions with values in $H^\beta(S^{n-1})$. Thus, the integral of the second term in (6.8) is at most

$$c\int_{-\infty}^{+\infty} \|\tilde{f}(\sigma+i(\beta-s),\cdot);H^\beta(S^{n-1})\|^2 d\sigma.$$

Thus,

$$\int_{\operatorname{Im}\lambda=\tau} \|\tilde{g}_+(\lambda,\cdot,\theta);H^\beta(\lambda,S^{n-1})\|^2 d\lambda \leqslant \tag{6.9}$$

$$\leqslant c \int_{\operatorname{Im}\lambda=\beta-s} \|\tilde{f}(\mu,\cdot);H^\beta(\mu,S^{n-1})\|^2 d\mu.$$

The operator adjoint to the operator $\tilde{f}\mapsto\tilde{g}_+$ with respect to the duality

$$<u,v> = \int_{\mathbf{R}}\int_{S^{n-1}} u(\xi,\omega)\overline{v(\xi,\omega)}d\xi d\omega$$

is obtained from (6.7) by replacing Φ by $\overline{\Phi}$. The estimates proved for $\beta\geqslant 0$ for the adjoint operator imply estimates for the operator (6.7) for $\beta<0$ also.

By the requirements in the theorem, there are no poles of $E(\mu)$ on the line $\operatorname{Im}\mu = \beta-s$. By applying proposition 1.5.5 we derive from (6.9) that

$$\int_{\operatorname{Im}\lambda=\tau} \|\tilde{g}_+(\lambda,\cdot,\theta);H^\beta(\lambda,S^{n-1})\|^2 d\lambda \leqslant \tag{6.10}$$

$$\leqslant \int_{\operatorname{Im}\mu=\beta-s} \|\tilde{v}(\mu+in/2,\cdot);H^s(\mu,S^{n-1})\|^2 d\mu.$$

We will now consider the expression $E_{\omega\to\phi}(\lambda+ia)^{-1}\tilde{g}_-(\lambda,\omega,\theta)$. By (5.6), on the line $\operatorname{Im}\lambda = \tau$ ($\tau = \beta-\operatorname{Re}a$ or $\tau = \beta-s$) we have

$$E_{\omega\to\phi}(\lambda+ia)^{-1}\tilde{g}_-(\lambda,\omega,\theta) = \tag{6.11}$$

$$= E_{\omega\to\phi}(\lambda+ia)^{-1}\int_{-\infty}^{+\infty} e^{i(\lambda+ia)t}\eta_-(t)\Phi(e^t\omega,\theta)J_\beta(t,\omega)dt +$$

$$+\frac{1}{(2\pi)^{(n-1)/2}}\sum_{|\gamma|=\kappa_0}^{k_1}\frac{(-i)^{|\gamma|}}{\gamma!}\int_{\mathbf{R}^n} v(y)y^\gamma dy\, E_{\omega\to\phi}(\lambda+ia)^{-1}\omega^\gamma \times$$

$$\times\int_{-\infty}^{+\infty}\eta_-(t)e^{(i(\lambda+ia)+n/2+|\gamma|)t}\Phi(e^t\omega,\theta)dt,$$

where $k_0 = \max\{[\tau-n/2]+1,0\}$, $k_1 = [\beta-n/2]$. By proposition 2.1, as $t\to-\infty$ we have $J_\beta(t,\omega) = O(e^{([\beta]+1)t})$ (if $\beta>n/2$) or $J_\beta(t,\omega) = O(e^{tn/2})$ (if $\beta<n/2$). Hence the integral in the first term on the right in (6.11) converges. The inner integral in the remaining terms can be regarded as the analytic extension of the convergent integral defined by

$$\int_{-\infty}^{+\infty} \eta_-(t)e^{(i\lambda-a+n/2+|\gamma|)t}\Phi(e^t,\omega,\theta)dt =$$

$$= \int_{-\infty}^{+\infty} \eta_-(t)e^{(i\lambda-a+n/2+|\gamma|)t}\Phi_{N+1}(t,\omega,\theta)dt +$$

$$-\sum_{|\kappa|=0}^{N} \frac{\partial^\kappa\Phi(0,\theta)\omega^\kappa}{\kappa!(i\lambda-a+n/2+|\gamma|+|\kappa|)} \times$$

$$\times \int_{-\infty}^{+\infty} e^{(i\lambda-a+n/2+|\gamma|+|\kappa|)t}\eta_-'(t)dt.$$

Here Φ_{N+1} is the remainder in the Taylor expansion of Φ. If $|\mathrm{Im}\lambda| <$ const and $\lambda\to\infty$, then this extension decreases faster than any power of $|\lambda|$. An application of $E(\lambda+ia)^{-1}$ preserves the character of decrease (proposition 1.5.5), while the poles disappear (proposition 1.4.3).

The functionals (5.14) occur in (6.11) only if $|\gamma|+n/2<\beta$. Thus, these functionals are continuous on $\mathfrak{S}_\beta^s(\mathbb{R}^n)$. Taking into account that the terms in (6.11) containing them are rapidly decreasing as $\lambda\to\infty$, we find that the norm in $\mathfrak{S}_\beta^{s-\mathrm{Re}\,a}(\mathbb{R}^n)$ of each such term is at most $c\|v;\mathfrak{S}_\beta^s(\mathbb{R}^n)\|$.

Denote the first term on the right in (6.11) by $E_{\omega\to\phi}(\lambda+ia)^{-1}\tilde{h}_-(\lambda,\omega,\theta)$. The inequality

$$\int_{\mathrm{Im}\lambda=\tau} \|E(\lambda+ia)^{-1}\tilde{h}_-(\lambda,\cdot,\theta);H^l(\lambda,S^{n-1})\|^2 d\lambda \leqslant$$

$$\leqslant c \int_{\mathrm{Im}\lambda=\tau} \|\tilde{h}_-(\lambda,\cdot,\theta);H^\beta(\lambda,S^{n-1})\|^2 d\lambda$$

holds. Putting, as before, $\tilde{f}(\mu,\omega) = E_{\psi\to\omega}(\mu)\tilde{v}(\mu+in/2,\psi)$ we are led to the formula

$$\tilde{h}_-(\xi+i(\tau+\mathrm{Re}\,a),\omega,\theta) =$$

$$= \int_{-\infty}^{+\infty} e^{i\xi t+(\beta-\tau-\mathrm{Re}\,a)t}\eta_-(t)\Phi(e^t\omega,\theta)dt \int_{-\infty}^{+\infty} e^{i\sigma t}\tilde{f}(\sigma+i\beta,\omega)d\sigma.$$

In order to estimate $\tilde{h}_-$ we repeat the considerations by which we derived (6.10). We also learn that $E(\mu)$ does not have poles on the line $\mathrm{Im}\,\mu = \beta$. As the result we obtain

$$\int_{\operatorname{Im}\lambda=\tau} \|\mathrm{E}(\lambda+ia)^{-1}\tilde{h}_-(\lambda,\cdot,\theta);H^l(\lambda,S^{n-1})\|^2 d\lambda \leqslant \tag{6.12}$$

$$\leqslant c \int_{\operatorname{Im}\mu=\beta} \|\tilde{f}(\mu,\cdot);H^\beta(\mu,S^{n-1})\|^2 d\mu \leqslant$$

$$\leqslant c \int_{\operatorname{Im}\mu=\beta} \|\tilde{v}(\mu+in/2,\cdot);H^0(\mu,S^{n-1})\|^2 d\mu.$$

By combining (6.5), (6.6), (6.10) - (6.12) we thus have

$$\|w;\mathfrak{S}_\beta^{s-\operatorname{Re}a}(\mathbf{R}^n)\| \leqslant$$

$$\leqslant c\left\{\|v;\mathfrak{S}_\beta^s(\mathbf{R}^n)\|^2 + \int_{\operatorname{Im}\mu=\beta-s} \|\tilde{v}(\mu+in/2,\cdot);H^s(\mu,S^{n-1})\|^2 d\mu + \right.$$

$$\left. + \int_{\operatorname{Im}\mu=\beta} \|\tilde{v}(\mu+in/2,\cdot);H^0(\mu,S^{n-1})\|^2 d\mu\right\}^{1/2}.$$

Since both integrals on the right occur in the expression for the norm $\|v,\mathfrak{S}_\beta^s(\mathbf{R}^n)\|$ (cf. (4.5)), the last inequality immediately leads to (6.1). ■

If $\beta-s+\operatorname{Re}a<-n/2$, the operator $\mathrm{A}(\theta)$, generally speaking, 'gets out of hand'. In order to make the map $\mathrm{A}(\theta):\mathfrak{S}_\beta^s(\mathbf{R}^n)\to\mathfrak{S}_\beta^{s-\operatorname{Re}a}(\mathbf{R}^n)$ bounded, we must restrict it to a subspace with finite-dimensional deficiency and (or) add to it a finite-dimensional term. We give two theorems of this kind.

Let p,q be nonnegative integers, $p\leqslant q$, let s be an arbitrary nonnegative number, and let α,β be such that

$$\operatorname{Re}a\leqslant s,\quad -n/2-q-1<\beta-s+\operatorname{Re}a<-n/2-q. \tag{6.13}$$

Let, also, $\zeta_\gamma(\cdot,\theta)$ be the function defined in proposition 5.3. If $p<q$ we put

$$\mathfrak{C} = \left\{v\in C_0^\infty(\mathbf{R}^n\setminus 0):\ \int_{\mathbf{R}^n} v(y)\overline{\zeta_\gamma(y,\theta)}dy = 0,\quad |\gamma| = p+1,\cdots,q\right\}, \tag{6.14}$$

while if $p=q$ we will assume $\mathfrak{C}=C_0^\infty(\mathbf{R}^n\setminus 0)$.

We introduce on $\mathfrak{C}$ the operator $\mathfrak{A}(\theta)$, defined for $\beta\in(-p-1-n/2,-p-n/2)$ by

$$(\mathfrak{A}(\theta)v)(x) = (\mathrm{A}(\theta)v)(x) - \sum_{|\gamma|=0}^{p} x^\gamma(v,\zeta_\gamma(\cdot,\theta)),$$

while for $\eta > -n/2$ we put $\mathcal{A}(\theta) = \mathrm{A}(\theta)$, where $\mathrm{A}(\theta)$ is the operator (5.2).

Theorem 6.2. *Assume $\beta \neq k+n/2$, $\beta-[s-\mathrm{Re}\,a] \neq -k-n/2$, where $k = 0,1,\cdots$, and $\beta-\mathrm{Re}\,a < n/2$. Let, moreover, conditions (6.13) be fulfilled. Then the functionals (v,ζ_γ) occurring in the definition of $\mathfrak{C}$ are continuous on $\mathfrak{S}^s_\beta(\mathbb{R}^n)$. For $v \in \mathfrak{C}$ the estimate*

$$\|\mathcal{A}(\theta)v\,;\mathfrak{S}^{s-\mathrm{Re}\,a}_\beta(\mathbb{R}^n)\| \leqslant c\|v\,;\mathfrak{S}^s_\beta(\mathbb{R}^n)\| \tag{6.15}$$

holds.

For the operator $\overline{\mathcal{A}(\theta)}$ (the closure of $\mathcal{A}(\theta)$), given on the subspace $\overline{\mathfrak{C}} \subset \mathfrak{S}^s_\beta(\mathbb{R}^n)$ (which coincides for $\beta \in (-q-1-n/2, -q-n/2)$ with the whole space $\mathfrak{S}^s_\beta(\mathbb{R}^n)$), the representation

$$(\overline{\mathcal{A}(\theta)}v)(x) = \frac{1}{\sqrt{2\pi}} \int_{\mathrm{Im}\lambda=\beta-s} r^{i(in/2+\lambda+ia)} \times \tag{6.16}$$

$$\times\, \mathrm{E}_{\omega\to\phi}(\lambda+ia)^{-1}\Phi_{\mu\to\lambda}(\omega,\theta)\mathrm{E}_{\psi\to\omega}(\mu)\tilde{v}(\mu+in/2,\psi)d\lambda$$

is valid.

Proof. The assumption $\beta-\mathrm{Re}\,a < n/2$ and conditions (6.13) imply the inequality $\beta-s < \min\{n/2, -q-\mathrm{Re}\,a-n/2\}$. Continuity of the functionals (v,ζ_γ) from (6.14) is guaranteed by proposition 5.9. For a function $v \in C_0^\infty(\mathbb{R}^n\setminus 0)$ the representation (6.16) (with the replacement of $\mathrm{Im}\lambda = \beta-s$ by $\mathrm{Im}\lambda = \beta-\mathrm{Re}\,a$) follows from proposition 5.3. Put $w = \mathcal{A}(\theta)v$. After Mellin transformation we obtain on the line $\mathrm{lim}\lambda = \beta-\mathrm{Re}\,a$ equation (6.3), in which $\mathrm{Im}\,\mu = \beta-\mathrm{Re}\,a$. We show that for $v \in \mathfrak{C}$ the function $\lambda \mapsto \tilde{w}(\lambda+ia+in/2,\cdot)$ is analytic in the strip $\beta-s \leqslant \mathrm{Im}\lambda \leqslant \beta-\mathrm{Re}\,a$. By lemma 5.1 the integral

$$\int_{-\infty}^{+\infty} e^{i\lambda t}\Phi(\omega,e^{-t}\theta)J_{\beta-\mathrm{Re}\,a}(t,\omega)dt \tag{6.17}$$

is analytic in the halfplane $\mathrm{Im}\lambda < -\mathrm{Re}\,a+n/2$. Hence only the poles $\lambda = -i(a+k+n/2)$, $k = 0,1,\cdots$, of the operator-function $\mathrm{E}(\lambda+ia)^{-1}$ can be singular points for $\tilde{w}(\lambda+ia+in/2,\cdot)$ if $\beta < n/2$; moreover, their order is one. The residue of $\tilde{w}(\lambda+ia+in/2,\cdot)$ at $\lambda = -i(a+k+n/2)$ is zero if $(v,\zeta_\gamma) = 0$ for all multi-indices γ such that $|\gamma| = k$. Hence $v \in \mathfrak{C}$ implies that $\tilde{w}$ is analytic in the strip $\beta - s \leqslant \operatorname{Im}\lambda \leqslant \beta - \operatorname{Re} a$ if $\beta \leqslant n/2$. For $\beta > n/2$ there do not occur new singularities of $\mathrm{E}(\lambda + ia)^{-1}$ in the given strip. By lemma 5.1, the integral (6.17) has a meromorphic extension onto the λ-plane. The poles of this extension disappear after an application of $\mathrm{E}(\lambda + ia)^{-1}$. So, if $v \in \mathfrak{C}$, then $\lambda \to \tilde{w}(\lambda + ia + in/2, \cdot)$ is an analytic function in the strip mentioned above. This leads to the representation (6.16).

We now turn to the estimation of the norm $\|w; \mathcal{E}_\beta^{s - \operatorname{Re} a}(\mathbb{R}^n)\|$. Put

$$\tilde{g}_\pm(\lambda, \omega, \theta) = \int_{-\infty}^{+\infty} \eta_\pm(t) e^{i\lambda t} \Phi(\omega, e^{-t}\theta) J_{\beta - \operatorname{Re} a}(t, \omega) dt;$$

then (6.5) holds for $\tilde{w}(\lambda + ia + in/2, \cdot)$. In view of the conditions of the theorem, there are no poles of $\mathrm{E}(\lambda + ia)^{-1}$ on the line $\operatorname{Im}\lambda = \tau$ for $\tau = \beta - \operatorname{Re} a$ or $\tau = \beta - s$. This implies inequality (6.6). By putting $f(\mu, \cdot) = \mathrm{E}(\mu)\tilde{v}(\mu + in/2, \cdot)$ and using the fact that $J_{\beta - \operatorname{Re} a}(t, \omega) = J_{\beta - s}(t, \omega)$ (since $\beta - \operatorname{Re} a < n/2$) we obtain (6.7), in which, according to the above said, $\beta - s - \tau \leqslant 0$. Thus, (6.10) holds. The remaining considerations coincide literally with the corresponding part of the proof of theorem 6.1. ■

R e m a r k 6.3. We view the operator $\mathcal{Q}(\theta)$ from a different perspective. Let $s = a = 0, p + n/2 < \beta < p + 1 + n/2$, and let $\overline{\mathrm{A}(\theta)}: \mathcal{E}_\beta^0(\mathbb{R}^n) \to \mathcal{E}_\beta^0(\mathbb{R}^n)$ be the continuous map described in theorem 6.1. Clearly, $\mathcal{E}_\beta^0(\mathbb{R}^n) = H_\beta^0(\mathbb{R}^n)$ and $(\mathcal{E}_\beta^0(\mathbb{R}^n))^* = \mathcal{E}_{-\beta}^0(\mathbb{R}^n)$. The operator $(\overline{\mathrm{A}(\theta)})^*: \mathcal{E}_{-\beta}^0(\mathbb{R}^n) \to \mathcal{E}_{-\beta}^0(\mathbb{R}^n)$, adjoint to $\overline{\mathrm{A}(\theta)}$ with respect to the scalar product in $L_2(\mathbb{R}^n)$, coincides with the righthand side of (6.16) in which β is replaced by $-\beta$, Φ by $\overline{\Phi}$ and in which we have put $a = s = 0$. By theorem 6.2, $(\overline{\mathrm{A}(\theta)})^*$ is the closure of $\mathcal{Q}(\theta)$ (in which also Φ is replaced by $\overline{\Phi}$). The latter operator is 'formally adjoint' to the operator

$$v \mapsto \mathrm{A}(\theta)v - \sum_{|\gamma| = 0}^{p} Z_\gamma(\cdot, \theta) \int_{\mathbb{R}^n} y^\gamma v(y) dy,$$

which coincides with $\mathrm{A}(\theta)$ on a dense set in $\mathcal{E}_\beta^0(\mathbb{R}^n)$ (see proposition 5.2 and theorem 6.1).

T h e o r e m 6.4. *Let* $q + n/2 < \beta - s < q + 1 + n/2$ *and* $-p - 1 - n/2 < \beta - s + \operatorname{Re} a < -p - n/2$, *where* p, q *are nonnegative integers. Assume that* $\beta - [s - \operatorname{Re} a] \neq -k - n/2$, $\beta \neq k + n/2$ $(k = 0, 1, \cdots)$. *Define on the set*

$$\mathfrak{N} = \left\{ v \in C_0^\infty(\mathbb{R}^n \setminus 0) : \int v(y)\overline{\eta_{k,q}(y,\Theta)}dy = 0, \right.$$

$$\left. |\kappa| = 0, \cdots, p;\ \int v(y) y^\gamma dy = 0,\ |\gamma| = 0, \cdots, q \right\},$$

where the functions $\eta_{\kappa,q}$ *are given by* (5.25) *for* $\tau = \beta - s$, *the operator* $\mathrm{A}(\theta)$ *by formula* (5.3).

Then for $v \in \mathfrak{C}$ *the estimate* (6.2) *holds. Representation* (6.2) *is valid for the operator* $\overline{\mathrm{A}(\theta)}$ *(the closure of* $\mathrm{A}(\theta)$*), given on the subspace*

$$\overline{\mathfrak{N}} = \left\{ v \in \mathfrak{S}_\beta^s(\mathbb{R}^n) : \int v(y)\overline{\eta_{\kappa,q}(y,\theta)}dy = 0,\ |\kappa| = 0, \cdots, p \right\}. \tag{6.18}$$

P r o o f. Boundedness of the functionals from (6.18), and equation (6.18) itself, follow from proposition 5.10. By applying proposition 5.4 we obtain representation (6.2) for the operator $\mathrm{A}(\theta)$ on the set $\mathfrak{N}$. The remaining considerations are completely analogous to those given in the proof of theorems 6.1 and 6.2. ■

Let's turn to the initial operator (5.1). Boundedness results for this operator in the scale of spaces $H_\beta^s(\mathbb{R}^m, \mathbb{R}^{m-n})$ will be based on boundedness theorems for the operator $\mathrm{A}(\theta)$ in the scale $\mathfrak{S}_\beta^s(\mathbb{R}^n)$.

T h e o r e m 6.5. *Let the numbers* $a \in \mathbb{C}$, $\beta \in \mathbb{R}$, *and* $s \geqslant 0$ *be such that* $\mathrm{Re}\, a \leqslant s$, $\beta - s + \mathrm{Re}\, a > -n/2$, *and* $\beta - s \neq k + n/2$, $\beta \neq k + n/2$, $k = 0, 1, \cdots$. *For* $\beta - s > n/2$ *put*

$$\mathfrak{L} = \left\{ u \in C_0^\infty(\mathbb{R}^m, \mathbb{R}^{m-n}) : \int (\mathfrak{F}u)(y,\xi) y^\gamma dy = 0,\ |\gamma| = 0, \cdots, [\beta - s - n/2] \right\},$$

and for $\beta - s < n/2$ *assume that* $\mathfrak{L} = C_0^\infty(\mathbb{R}^m, \mathbb{R}^{m-n})$.

Then the estimate

$$\|\mathrm{A}u; H_\beta^{s - \mathrm{Re}\, a}(\mathbb{R}^m, \mathbb{R}^{m-n})\| \leqslant c\|u; H_\beta^s(\mathbb{R}^m, \mathbb{R}^{m-n})\| \tag{6.19}$$

holds for the operator A *given on* $\mathfrak{L}$ *by* (5.1). *The set* $\mathfrak{L}$ *is dense in* $H_\beta^s(\mathbb{R}^m, \mathbb{R}^{m-n})$, *and for the operator* $\overline{\mathrm{A}}$ *(the closure of* A*), defined on the whole space*

$H^s_\beta(\mathbf{R}^m, \mathbf{R}^{m-n})$, the representation

$$(Au)(x^{(1)}, x^{(2)}) = \tag{6.20}$$

$$= \mathfrak{F}^{-1}_{\xi^{(2)} \to x^{(2)}} \int_{\operatorname{Im}\lambda = \beta - s} r^{i(in/2+\lambda+ia)} \mathrm{E}_{\omega\to\phi}(\lambda+ia)^{-1} \times$$

$$\times \Phi_{\mu\to\lambda}(\omega,\theta) |\xi^{(2)}|^{i(\lambda-\mu)} \mathrm{E}_{\psi\to\omega}(\mu) \widetilde{(\mathfrak{F}_{y^{(2)}\to\xi^{(2)}} u)}(\mu+in/2, \psi; \xi^{(2)}) d\lambda$$

holds. Here, as before, $\mathfrak{F}$ is the Fourier transform of $x^{(2)} \mapsto u(x^{(1)}, x^{(2)})$, $\widetilde{(\mathfrak{F}u)}(\mu+in/2, \psi; \xi^{(2)})$ is the Mellin transform of $r \mapsto (\mathfrak{F}u)\ (r, \psi; \xi^{(2)})$, $r = |x^{(1)}|$, $\phi = x^{(1)}/|x^{(1)}|$, $\theta = \xi^{(2)}/|\xi^{(2)}|$, $\psi = y^{(1)}/|y^{(1)}|$, and the operator $\Phi_{\mu\to\lambda}(\omega,\theta)$ is the same as in proposition 5.2.

P r o o f. Put $w = \mathrm{A}u$. Instead of $\xi^{(2)}$ we simply write ξ. By (4.6) it suffices to verify the inequality

$$\| |\xi|^{-\alpha} (\mathfrak{F}_{y\to\xi} w)(\cdot/|\xi|, \xi); \mathcal{E}^{s-\operatorname{Re}\alpha}_\beta(\mathbf{R}^n)\| \leqslant \tag{6.21}$$

$$\leqslant c \|(\mathfrak{F}_{y\to\xi} u)(\cdot/|\xi|, \xi); \mathcal{E}^s_\beta(\mathbf{R}^n)\|.$$

The estimate (6.19) is obtained by multiplication of (6.21) by $|\xi|^{s-\beta-n/2}$ and subsequent integration with respect to ξ. Recalling the definition of $\mathrm{A}(\theta)$ we derive (6.21) immediately from (6.1). It must here also be taken into account that $v(y) \equiv (\mathfrak{F}u)\ (y/|\xi|, \xi)$, $(\mathrm{A}(\theta), v)(x) \equiv |\xi|^{-\alpha} (\mathfrak{F}(\mathrm{A}u))(x/|\xi|, \xi)$. The representation (6.20) clearly follows from (6.2).

We turn to the proof of the fact that $\mathfrak{L}$ is dense. Passing to the variables $r = |y|$, $\psi = y/|y|$, we write $(\mathfrak{F}u)(y, \xi)$ in the form $(\mathfrak{F}u)(r, \psi; \xi)$. We first verify that the set of functions u from $C_0^\infty(\mathbf{R}^m, \mathbf{R}^{m-n})$ subject to the condition

$$\int_0^\infty (\mathfrak{F}u)(r, \psi; \xi) r^{n-1} dr = 0 \tag{6.22}$$

is dense in $H^s_\beta(\mathbf{R}^m, \mathbf{R}^{m-n})$.

For an arbitrary function $u \in C_0^\infty(\mathbf{R}^m, \mathbf{R}^{m-n})$ we introduce the sequence $\{u_k\}$ for which $(\mathfrak{F}u_k)(y, \xi) = k^n (\mathfrak{F}u)(ky, \xi)$. It is obvious that the difference $u - u_k$ satisfies (6.22). Furthermore, $\widetilde{(\mathfrak{F}u_k)}(\lambda+in/2, \psi; \xi) = k^{i\lambda+n/2} \widetilde{(\mathfrak{F}u)}(\lambda+in/2, \psi; \xi)$, and by (4.5),

$$\|(\mathfrak{F}u_k)(\cdot/|\xi|, \xi); \mathcal{E}^s_\beta(\mathbf{R}^n)\| = k^{n-2(\beta-s)} \|(\mathfrak{F}u)(\cdot/|\xi|, \xi); \mathcal{E}^s_\beta(\mathbf{R}^n)\|,$$

where the dot indicates the y variables. This and (4.6) imply

$$\|u_k;H^s_\beta(\mathbb{R}^m,\mathbb{R}^{m-n})\| \leqslant k^{n-2(\beta-s)}\|u;H^s_\beta(\mathbb{R}^m,\mathbb{R}^{m-n})\|.$$

Hence $u_k \to 0$ in $H^s_\beta(\mathbb{R}^m,\mathbb{R}^{m-n})$ as $k\to\infty$, and it has been proved that the set of functions satisfying (6.22) is dense. The remaining reasoning is obvious (cf. the proof of proposition 2.3). ∎

We now give a statement concerning boundedness of the operator A; this statement follows from theorem 6.2 concerning the operator A(θ). Let p,q be nonnegative integers, $p\leqslant q$. If $p<q$ we denote by $\mathcal{S}$ the set of functions from $C_0^\infty(\mathbb{R}^m,\mathbb{R}^{m-n})$ satisfying for all $\xi\in\mathbb{R}^{m-n}$ the conditions

$$\int(\mathcal{F}u)(y,\xi)\zeta_\gamma(y\,|\xi|,\xi/|\xi|)dy = 0,\ |\gamma|=p+1,\cdots,q;$$

here $y\equiv y^{(1)}$, $\xi\equiv\xi^{(2)}$. If $p = q$ we assume $\mathcal{S} = C_0^\infty(\mathbb{R}^m,\mathbb{R}^{m-n})$. It is easy to prove that

$$\zeta_\gamma(y\,|\xi|,\xi/|\xi|) =$$

$$= |\xi|^{-\alpha-|\gamma|}\frac{(-i)^{|\gamma|}}{(2\pi)^{(n+1)/2}\gamma!}\int_{\operatorname{Im}\mu=h} r^{i(in/2+\mu)}\mathrm{E}(\mu)^{-1}\omega^\gamma d\mu\,\times$$

$$\times\int_0^\infty \rho^{i\mu+n/2+|\gamma|-1}\overline{\Phi(\rho\omega,\xi)}d\rho,$$

where $r=|y|$, $\phi=y/|y|$, h is an arbitrary number satisfying the inequality $h>\max\{-n/2,|\gamma|+\operatorname{Re}a+n/2\}$, and the inner integral is understood as the analytic extension (with respect to μ) of the corresponding convergent integral.

Put

$$\mathcal{K}_\gamma(y^{(1)},y^{(2)}) = \mathcal{F}^{-1}_{\xi\to y^{(2)}}(|\xi|^{a+|\gamma|}\zeta_\gamma(y^{(1)}|\xi|,\xi/|\xi|))$$

(the Fourier transform is understood in the sense of the theory of generalized functions). We introduce on $\mathcal{S}$ the operator $\mathcal{A}$, defined for $\beta\in(-p-1-n/2,-p-n/2)$ by

$$(\mathcal{A}u)(x^{(1)},x^{(2)}) = (\mathrm{A}u)(x^{(1)},x^{(2)})+$$

$$-\sum_{|\gamma|=0}^{p}(x^{(1)})^\gamma\int\mathcal{K}_\gamma(y^{(1)},x^{(2)}-y^{(2)})u(y^{(1)},y^{(2)})dy^{(1)}dy^{(2)},$$

while for $\beta > -n/2$ we put $\mathcal{A} = \mathrm{A}$, with A the operator (5.1).

T h e o r e m 6.6. *Assume that* $\beta \neq k + n/2$, $\beta - [s - \mathrm{Re}\, a] \neq -k - n/2$, *where* $k = 0, 1, \cdots$, *and* $\beta - \mathrm{Re}\, a < n/2$. *Assume also that conditions* (6.13) *are satisfied. Then for* $u \in \mathbb{S}$ *the estimate*

$$\|\mathcal{A}u; H_\beta^{s - \mathrm{Re}\, a}(\mathbf{R}^m, \mathbf{R}^{m-n})\| \leqslant c \|u; H_\beta^s(\mathbf{R}^m, \mathbf{R}^{m-n})\|$$

holds.

A representation of the form (6.20) *holds for the operator* $\overline{\mathcal{A}}$ *(the closure of* $\mathcal{A}$*), given on the subspace* $\mathbb{S}$.

We will not halt in order to formulate a boundedness theorem for A following from theorem 6.4 on $\mathrm{A}(\theta)$. We finally note that the representations for $\overline{\mathrm{A}(\theta)}$ and $\overline{\mathrm{A}}$ obtained in this paragraph can be written using the 'generalized' Fourier transform from theorem 2.4 (compare with theorem 3.5).

Chapter 3

Meromorphic pseudodifferential operators

Meromorphic pseudodifferential operators arise, e.g., after applying the Mellin transform on the right and on the left to the operator of convolution with a homogeneous function. A 'canonical' meromorphic pseudodifferential operator of order a has the form $\mathrm{E}_{\theta\to\phi}(\lambda+ia)^{-1}\Phi(\phi,\theta)\mathrm{E}_{\psi\to\theta}(\lambda)$. Thus, the operator $\mathrm{E}(\lambda)$ plays the same role for meromorphic pseudodifferential operators as does the Fourier transform for ordinary pseudodifferential operators.

In §1 we define canonical meromorphic pseudodifferential operators, and show some simplest properties of them. In particular, a meromorphic pseudodifferential operator is represented as an integral operator, whose kernel is described. In §2 we study actions on canonical operators: composition, transition to the adjoint operator, shift, and differentiation with respect to a parameter. These operations lead, in general, outside the set of canonical meromorphic pseudodifferential operators. A larger class of meromorphic pseudodifferential operators that is invariant under these actions is discussed in §3. In §4 we define for a meromorphic pseudodifferential operator of arbitrary order the trace and prove formulas expressing the trace of an operator in terms of its symbol. The asymptotic behavior of the values of a meromorphic pseudodifferential operator on strongly oscillating functions is clarified in §5. §6 is devoted to the generalization to meromorphic pseudodifferential operators of certain inequalities which are well-known for classical pseudodifferential operators (Gårding's inequality, etc.). In §7 we study 'periodic' meromorphic pseudodifferential operators, and the final §8 contains rules for change of variables in meromorphic pseudodifferential operators.

§1. Canonical meromorphic pseudodifferential operators

1. Definitions of operators. Continuity. Let a be an arbitrary complex number. For $\lambda \neq i(k+n/2)$, $\lambda \neq -i(k+a+n/2)$, $k = 0,1,\cdots$, we introduce the operator

$$\mathfrak{U}_{\psi\to\phi}(\lambda) = \mathrm{E}_{\theta\to\phi}(\lambda+ia)^{-1}\Phi(\theta)\mathrm{E}_{\psi\to\theta}(\lambda). \tag{1.1}$$

By proposition 1.5.5 the maps $\mathrm{E}(\lambda){:}H^s(\lambda,S^{n-1})\to H^{s+\mathrm{Im}\lambda}(\lambda,S^{n-1})$ and $\mathrm{E}(\lambda+ia)^{-1}{:}H^{s+\mathrm{Im}\lambda}(\lambda,S^{n-1})\to H^{s-\mathrm{Re}a}(\lambda,S^{n-1})$ are continuous. Hence the estimate

$$\|\mathfrak{U}(\lambda)\| \leqslant c\|\Phi;H^{s+\mathrm{Im}\lambda}(\lambda,S^{n-1})\to H^{s+\mathrm{Im}\lambda}(\lambda,S^{n-1})\|$$

holds for the norm of the operator $\mathfrak{U}(\lambda)$: $H^s(\lambda,S^{n-1})\to H^{s-\mathrm{Re}a}(\lambda,S^{n-1})$, where on the right we have the norm of the operator of multiplication by the function Φ on $H^{s+\mathrm{Im}\lambda}(\lambda,S^{n-1})$. Using the inequalities (1.5.21) and (1.5.22) we obtain the following

P r o p o s i t i o n 1.1. *Let $\mathfrak{F}$ be an arbitrary closed set located in a strip $|\mathrm{Im}\lambda| < h$ and not containing poles of the meromorphic operator-function $\lambda \mapsto \mathfrak{U}(\lambda)$ (which can be located only at the points $\lambda = i(k+n/2)$, $\lambda = -i(k+a+n/2)$, $k = 0,1,\cdots$). Then the estimate*

$$\|\mathfrak{U}(\lambda);H^s(\lambda,S^{n-1})\to H^{s-\mathrm{Re}a}(\lambda,S^{n-1})\| \leqslant c(\mathfrak{F})\|\Phi;C^q(S^{n-1})\| \tag{1.2}$$

holds, where $\lambda \in \mathfrak{F}$ and q is a number depending on h only.

In the sequel it is assumed that $\Phi \in C^\infty(S^{n-1})$, and hence the choice of a concrete value of q does not play a role.

The operator (1.1) made its appearance already in the previous Chapter (cf. formula (2.3.6)); it is also the simplest meromorphic pseudodifferential operator of order a. An operator of the form

$$\mathfrak{U}_{\psi\to\phi}(\lambda) = \mathrm{E}_{\theta\to\phi}(\lambda+ia)^{-1}\Phi(\phi,\theta)\mathrm{E}_{\psi\to\theta}(\lambda) \tag{1.3}$$

will be called a canonical *meromorphic pseudodifferential operator of order a.*

Proposition 1.5.1 and formula (1.5.4) imply the following assertion

P r o p o s i t i o n 1.2. *Let $\Phi \in C^\infty(S^{n-1}\times S^{n-1})$ and let*

$$\Phi(\phi,\theta) = \sum_{m=0}^{\infty} \sum_{k=1}^{k_m} a_{mk}(\phi) Y_{mk}(\theta). \tag{1.4}$$

Then for arbitrary $\gamma,\delta = 0,1,\cdots$, *the estimate*

$$\|a_{mk}; C^\gamma(S^{n-1})\| \leqslant c_{\gamma\delta}(1+m)^{-\delta}$$

holds.

Proposition 1.3. *Let* $\Phi \in C^\infty(S^{n-1}\times S^{n-1})$ *and let* $\mathfrak{U}(\lambda)$ *be the operator defined by* (1.3) *for* $\lambda \neq i(k+n/2)$, $\lambda \neq -i(k+a+n/2)$, $k = 0,1,\cdots$. *Then the map* $\mathfrak{U}(\lambda)$: $H^s(\lambda,S^{n-1}) \to H^{s-\mathrm{Re}a}(\lambda,S^{n-1})$ *is continuous. On every closed set* $\mathfrak{F}$ *located in a strip* $|\mathrm{Im}\lambda| < h$ *and not containing poles of the operator-function* $\lambda \mapsto \mathfrak{U}(\lambda)$ *the estimate*

$$\|\mathfrak{U}(\lambda); H^s(\lambda,S^{n-1}) \to H^{s-\mathrm{Re}a}(\lambda,S^{n-1})\| \leqslant c(s,\mathfrak{F}) \tag{1.5}$$

holds.

Proof. Expand Φ in a series (1.4), and denote by $\mathfrak{U}_{mk}(\lambda)$ the operator $E_{\theta\to\phi}(\lambda+ia)^{-1} Y_{mk}(\theta) E_{\psi\to\theta}(\theta)$. By proposition 1.1, inequality (1.2) with Φ replaced by Y_{mk} holds for the operator $\mathfrak{U}_{mk}$. In view of the estimate $\|Y_{mk}; C^q(S^{n-1})\| \leqslant O(m^{-1+q+n/2})$ (cf. [29] or [40]), this and proposition 1.2 imply that the series

$$\sum_{m=0}^{\infty} \sum_{k=1}^{k_m} \sup_{\lambda\in\mathfrak{F}} \|\mathfrak{U}_{mk}(\lambda); H^s(\lambda,S^{n-1}) \to H^{s-\mathrm{Re}a}(\lambda,S^{n-1})\|$$

converges, and thus imply the relation (1.5). ■

The operator-function $\lambda \to \mathfrak{U}(\lambda)$, given by (1.3), is meromorphic in the whole λ-plane. If $a \neq -l-n$ ($l=0,1,\cdots$), then this function can have poles at $\lambda = i(k+n/2)$, $\lambda = -i(k+a+n/2)$ of the first order only, and there are no other poles. The residue of the function $\lambda \mapsto \mathfrak{U}(\lambda)u$ at $\lambda = i(k+n/2)$ is

$$\frac{1}{(2\pi)^n} e^{i\pi(n+a)/2} \Gamma(n+k+a) \times \tag{1.6}$$

$$\times \int (\phi\theta + i0)^{-k-n-a} \Phi(\phi,\theta) \sum_{|\gamma|=k} \frac{1}{\gamma!} \theta^\gamma d\theta \int \psi^\gamma u(\psi) d\psi,$$

and the residue at $\lambda = -i(k+a+n/2)$ is

$$\frac{1}{(2\pi)^n} e^{i\pi(n+a)/2}\Gamma(n+k+a) \sum_{|\gamma|=k} \frac{(-1)^k}{\gamma!}\phi^\gamma \times \tag{1.7}$$

$$\times \int \theta^\gamma \Phi(\phi,\theta)d\theta \int (-\theta\psi + i0)^{-k-n-a} u(\psi)d\psi.$$

If a is a nonnegative integer and if the function $\theta \mapsto \Phi(\phi,\theta)$ coincides with the values on S^{n-1} of a homogeneous polynomial $x \mapsto P(\phi,x)$ of degree κ ($\theta = x/|x|$, the coefficients of the polynomial depend on ϕ), with moreover $\kappa \leqslant a$, $\kappa \equiv a (\text{mod}\, 2)$, then these residues are equal to zero; the operator-function $\mathfrak{U}(\lambda)$ is then holomorphic in the whole plane (cf. proposition 1.4.3).

In case $a = -l-n$, the points $\lambda = i(k+n/2)$ for $k = 0, \cdots, l$ are poles of order two, and the points $\lambda = i(k+n/2)$, $\lambda = i(l-k+n/2)$ for $k = l+1, l+2, \cdots$ are poles of order one. The residues at poles of order one can be computed by formulas (1.6), (1.7). The coefficient in front of $(i\lambda + k + n/2)^{-2}$ in the Laurent expansion of the function $\lambda \mapsto \mathfrak{U}(\lambda)u$ is equal to

$$\frac{(-1)^{k+1} i^l}{k!(l-k)!(2\pi)^n} \int (\phi\theta)^{l-k} \Phi(\phi,\theta)d\theta \int (\theta\psi)^k u(\psi)d\psi,$$

and the coefficient in front of $(i\lambda + k + n/2)^{-1}$ coincides with

$$\frac{(-1)^{k+1} i^l}{k!(l-k)!(2\pi)^n} \Big[\{\psi(l-k+1) - \psi(k+1)\} \times$$

$$\times \int (\phi\theta)^{l-k} \Phi(\phi,\theta)d\theta \int (\theta\psi)^k u(\psi)d\psi +$$

$$+ \int (\phi\theta)^{l-k} \Phi(\phi,\theta)d\theta \int (\theta\psi)^k \ln(-\theta\psi + i0) u(\psi)d\psi +$$

$$- \int (\phi\theta)^{l-k} \Phi(\phi,\theta) \ln(\phi\theta + i0)d\theta \int (\theta\psi)^k\, u(\psi)d\psi \Big],$$

where $\psi(z) = \Gamma'(z)/\Gamma(z)$. These formulas can be obtained by a direct computation; we only have to use the representation ([5], 1.17 (11)) $\Gamma(z) = (-1)^k \{(z+k)^{-1} + \psi(k+1) + O(z+k)\}/k!$ and the expansion

$$(\pm\phi\theta + i0)^{-z} = (\pm\phi\theta + i0)^k - (z+k)(\pm\phi\theta + i0)^k \ln(\pm\phi\theta + i0) + \cdots .$$

Thus, at the points $\lambda = -i(k+a+n/2)$ the function $\mathfrak{U}$ is, in general, not finite-meromorphic, while at the other poles it is finite-meromorphic. (A meromorphic operator-function is called *finite-meromorphic* at a pole $\lambda = \lambda_0$ if the coefficients in the principal part of its Laurent expansion are finite-dimensional operators.) At the poles $\lambda = -i(k+a+n/2)$ the coefficients in the principal part of the expansion of the function (1.3) are integral operators with $C^\infty(S^{n-1}\times S^{n-1})$ kernels.

2. One-dimensional meromorphic pseudodifferential operators. In the one-dimensional case ϕ, θ, ψ take the two values ± 1, and the integral of a function u over the zero-dimensional sphere is the sum $u(1)+u(-1)$. In this case the operators $\mathrm{E}(\lambda)^{\pm 1}$ are defined by (1.4.10) and (1.4.11), and the vector $v(\phi) = \mathrm{E}_{\theta\to\phi}(\lambda+ia)^{-1}\Phi(\phi,\theta)\mathrm{E}_{\psi\to\theta}(\lambda)u(\psi)$ is given by

$$\begin{bmatrix} v(1) \\ v(-1) \end{bmatrix} = \frac{\Gamma(1/2+i\lambda)\Gamma(1/2-i\lambda+a)}{2\pi}\mathfrak{B}(\lambda)\begin{bmatrix} u(1) \\ u(-1) \end{bmatrix},$$

where the entries b_{jk} of the matrix $\mathfrak{B}(\lambda)$ are as follows:

$$b_{11} = e^{i\pi(a-2i\lambda)/2}\Phi(1,1)+e^{-i\pi(a-2i\lambda)/2}\Phi(1,-1),$$

$$b_{12} = e^{i\pi(1+a)/2}\Phi(1,1)+e^{-i\pi(1+a)/2}\Phi(1,-1),$$

$$b_{21} = e^{-i\pi(a+1)/2}\Phi(-1,1)+e^{i\pi(a+1)/2}\Phi(-1,-1),$$

$$b_{22} = e^{-i\pi(a-2i\lambda)}\Phi(-1,1)+e^{i\pi(a-2i\lambda)}\Phi(-1,-1).$$

Clearly, $\Phi(\phi,\theta) = c(\phi)+d(\phi)\operatorname{sgn}\theta$, where $c(\phi) = (\Phi(\phi,1)+\Phi(\phi,-1))/2$, $d(\phi) = (\Phi(\phi,1)-\Phi(\phi,-1))/2$. Hence a canonical meromorphic pseudodifferential operator of order a has the form

$$\frac{\Gamma(1/2+i\lambda)\Gamma(1/2-i\lambda+a)}{\pi}\times$$

$$\times\left\{\begin{bmatrix} c(1)\cos\frac{\pi}{2}(a-2i\lambda) & -c(1)\sin\frac{\pi}{2}a \\ -c(-1)\sin\frac{\pi}{2}a & c(-1)\cos\frac{\pi}{2}(a-2i\lambda)\end{bmatrix}+\right.$$

$$\left.+i\begin{bmatrix} d(1)\sin\frac{\pi}{2}(a-2i\lambda) & d(1)\cos\frac{\pi}{2}a \\ -d(-1)\cos\frac{\pi}{2}a & -d(-1)\sin\frac{\pi}{2}(a-2i\lambda)\end{bmatrix}\right\}.$$

Since $\Gamma(i\lambda+1/2)\Gamma(-i\lambda+1/2) = \pi/\cos(\pi i\lambda)$, for $a = 0$ we are led to the following expression for $\mathfrak{U}(\lambda)$:

$$\begin{bmatrix} c(1)+d(1)\tanh \pi\lambda & id(1)/\cosh \pi\lambda \\ -id(-1)/\cosh \pi\lambda & c(-1)-d(-1)\tanh \pi\lambda \end{bmatrix}.$$

3. The operator $\mathfrak{U}(\lambda)$ as an integral operator, and its kernel. Consider the operator (1.1). We first assume that the number a is subject to the inequalities $-n/2<\mathrm{Re}\, a<0$. We put $f = \mathrm{E}(ia+in/2)^{-1}\Phi$, and introduce the function G, which is homogeneous of degree $-a-n$, by

$$G(x) = r^{-n-a}f(\phi), \ r = |x|, \ \phi = x/|x|.$$

For a function $u \in C_0^\infty(\mathbb{R}^n \setminus 0)$ we define

$$(G\star u)(x) = \int G(x-y)u(y)dy.$$

Since $(\mathrm{F}G)(\xi) = |\xi|^a \mathrm{E}_{\phi\to\theta}(ia+in/2)f(\phi) = |\xi|^a\Phi(\theta)$ (cf. (1.4.7)), we have $G\star u = \mathrm{F}^{-1}|\xi|^a\Phi(\theta)\mathrm{F}u$. By proposition 2.3.1 we have for the operator $G\star u$ the representation

$$(G\star u)(x) = \frac{1}{\sqrt{2\pi}}\int_{-\infty}^{+\infty} r^{i(\lambda+ia+in/2)}\mathfrak{U}(\lambda)\tilde{u}(\lambda+in/2,\cdot)d\lambda, \tag{1.8}$$

where $\mathfrak{U}(\lambda)$ is the operator (1.1).

We apply to the convolution $G\star u$ the Mellin transform $\mathrm{M}_{r\to\lambda+ia+in/2}$ for $\mathrm{Im}\lambda = 0$. Putting $x = r\phi$, $y = \rho\psi$ we have

$$\frac{1}{\sqrt{2\pi}}\int_0^{+\infty} r^{-i(\lambda+ia+in/2)-1}(G\star u)(x)dr = \tag{1.9}$$

$$= \frac{1}{\sqrt{2\pi}}\int_0^{+\infty} r^{-i(\lambda+ia+in/2)-1}dr\int_{S^{n-1}}d\psi\int_0^{+\infty} G(r\phi-\rho\psi)u(\rho,\psi)\rho^{n-1}d\rho =$$

$$= \int_{S^{n-1}}d\psi\int_0^{+\infty} r^{-i(\lambda+ia+in/2)-1}G(t\phi-\psi)\tilde{u}(\lambda+in/2,\psi)dt.$$

Now (1.8) and (1.9) imply

$$\mathfrak{U}_{\psi\to\phi}(\lambda)\tilde{u}(\lambda+in/2,\psi) = \int G(\phi,\psi;\lambda)\tilde{u}(\lambda+in/2,\psi)d\psi,$$

where

$$G(\phi,\psi;\lambda) = \int_0^{+\infty} t^{-i(\lambda+ia+in/2)-1}G(t\phi-\psi)dt. \tag{1.10}$$

Thus, for an arbitrary function $v \in C^\infty(S^{n-1})$,

$$\mathfrak{U}_{\psi\to\phi}(\lambda)v(\psi) = \int G(\phi,\psi;\lambda)v(\psi)d\psi. \tag{1.11}$$

Equation (1.11) has been proved for real λ and under the restriction $-n/2 < \mathrm{Re}\, a < 0$. By analytic extension it can be proved for all complex a and λ, except for the poles.

We clarify a manner for analytically extending the integral (1.10). Let χ_1,χ_2 be nonnegative functions of class $C^\infty[0,\infty)$, $\chi_1(t)+\chi_2(t) \equiv 1$, where moreover χ_1 has compact support and equals one near $t = 0$. The integral

$$G_1(\phi,\psi;\lambda) \equiv \int_0^\infty t^{-i(\lambda+ia+in/2)-1}G(t\phi-\psi)\chi_1(t)dt$$

is, for $\phi \neq \psi$, an analytic function in the halfplane $\mathrm{Im}\lambda > -\mathrm{Re}\, a - n/2$. Analytic extension of $G_1(\phi,\psi;\lambda)$ to the domain $\mathrm{Im}\lambda > -\mathrm{Re}\, a - p - n/2$ (p a natural number) is realized using the formula

$$G_1(\phi,\psi;\lambda) = \frac{(-1)^p}{(a+n/2-i\lambda)\cdots(a+n/2+p-1-i\lambda)} \times \tag{1.12}$$

$$\times \int_0^\infty t^{-i(\lambda+ia+in/2)+p-1}\frac{d^{p-1}}{dt^{p-1}}(\chi_1(t)G(t\phi-\psi))dt$$

(the points $\lambda = -i(a+n/2), \cdots, -i(a+p-1+n/2)$ turn out to be poles).

The function

$$G_2(\phi,\psi;\lambda) = \int_0^\infty t^{-i(\lambda+ia+in/2)-1}G(t\phi-\psi)\chi_2(t)dt$$

is analytic in the halfplane $\mathrm{Im}\lambda < n/2$. After the change of variable $t = \tau^{-1}$ it takes the form

$$G_2(\phi,\psi;\lambda) = \int_0^\infty \tau^{-i\lambda+n/2-1}G(\phi-\tau\psi)\chi_2(1/\tau)d\tau.$$

The analytic extension of $G_2(\phi,\psi;\lambda)$ to the domain $\mathrm{Im}\lambda < p+n/2$ is constructed according to the formula

$$G_2(\phi,\psi;\lambda) = \frac{(-1)^p}{(i\lambda+n/2)\cdots(i\lambda+n/2+p-1)} \times \tag{1.13}$$

$$\times \int_0^\infty \tau^{i\lambda+n/2+p-1}\frac{d^p}{d\tau^p}(G(\phi-\tau\psi)\chi_2(\tau^{-1}))d\tau.$$

By $G(\phi,\psi;\lambda)$ we will understand the function defined by

$$G(\phi,\psi;\lambda) = G_1(\phi,\psi;\lambda)+G_2(\phi,\psi;\lambda).$$

Note that for $a = 0,1,\cdots$, the operator $E(ia+in/2)^{-1}$ is not an isomorphism (it annihilates the subspace $\mathfrak{L}_a$; cf. proposition 1.4.3), while for $a = -l-n$, $l = 0,1,\cdots$, the operator $E(ia+in/2)^{-1}$ is simply not defined. Summarizing all this leads to the following assertion.

P r o p o s i t i o n 1.4. *Let $a \neq 0,1,\cdots$ and $a \neq -l-n$, where $l = 0,1,\cdots$. Let also $\lambda \neq i(k+n/2)$, $\lambda \neq -i(k+a+n/2)$, $k = 0,1,\cdots$. Then formula* (1.9) *holds for the operator* (1.1). *The kernel $G(\phi,\psi;\lambda)$ is defined by* (1.8); *moreover, $G(x) = r^{-n-a}E_{\omega\to\phi}(ia+in/2)^{-1}\Phi(\omega)$, $r = |x|$, $\phi = x/|x|$, and the integral* (1.10) *has to be understood in the sense of the analytic extension described above (cf.* (1.12) *and* (1.13)).

We will now consider the operator (1.1) for those values of a that are excluded in proposition 1.4. Let a be a nonnegative integer. Represent Φ as a sum $\Phi_0+\Phi_1$, where Φ_0 is subject to the conditions

$$\int_{S^{n-1}} \Phi_0(\theta)\theta^\gamma d\theta = 0$$

for all multi-indices γ such that $|\gamma| = a$, and Φ_1 has the form

$$\Phi_1(\theta) = \sum_{j=0}^{[a/2]} h_{a-2j}(\theta);$$

as before, h_j is a harmonic polynomial of degree j. For $\lambda \neq i(k+n/2)$, $\lambda \neq -i(k+a+n/2)$ the operator (1.1) splits into two terms:

$$\mathfrak{U}(\lambda) = E(\lambda+ia)^{-1}\Phi_0(\theta)E(\lambda)+E(\lambda+ia)^{-1}\Phi_1(\theta)E(\lambda). \tag{1.14}$$

P r o p o s i t i o n 1.5. *For the first term on the right in* (1.14) *a formula of the form* (1.11) *holds with $G(x) = r^{-n-a}E_{\omega\to\phi}(ia+in/2)^{-1}\Phi(\omega)$, and the second term is a differential operator of order a on the sphere S^{n-1} in which the parameter λ enters as a polynomial of degree a.*

P r o o f. By proposition 1.4.3, $E(ia+in/2)^{-1}\Phi_1 = 0$. Hence $E(ia+in/2)^{-1}\Phi = E(ia+in/2)^{-1}\Phi_0$, and representation (1.11) for $E(\lambda+ia)^{-1}\Phi_0 E(\lambda)$ follows from proposition 1.4.

Extend Φ_1 to $\mathbb{R}^n$ by means of the equation

$$\Phi_1(\xi) = \sum_{j=0}^{[a/2]} |\xi|^{2j} h_{a-2j}(\xi).$$

Thus, $\xi \mapsto \Phi_1(\xi)$ is a homogeneous polynomial of degree a. Consider the differential operator $P(\partial/\partial x) = F^{-1}\Phi_1(\xi)F$. By proposition 2.3.1,

$$P(\partial/\partial x)u =$$

$$= \frac{1}{\sqrt{2\pi}} \int_{-\infty}^{+\infty} r^{i(\lambda+ia+in/2)} E(\lambda+ia)^{-1}\Phi_1(\theta)E(\lambda)\tilde{u}(\lambda+in/2,\cdot)d\lambda.$$

Note that $P(\partial/\partial x) = r^{-a}\mathcal{P}(r\partial/\partial r, \partial/\partial\phi)$, where $\partial/\partial\phi$ is a differentiation operator in the local coordinates of a point $\phi = x/|x|$ on the sphere and $\mathcal{P}$ is a polynomial. Substituting this expression for P in (1.15) and applying the Mellin transform we find

$$\mathcal{P}(i(\lambda+in/2), \partial/\partial\phi)\tilde{u}(\lambda+in/2,\phi) =$$

$$= E_{\theta\to\phi}(\lambda+ia)^{-1}\Phi_1(\theta)E_{\psi\to\theta}\tilde{u}(\lambda+in/2,\psi),$$

which was required. ■

Finally, let $a = -l-n$, where $l = 0, 1, \cdots$. We require Φ to satisfy the conditions

$$\int \Phi(\theta)\theta^\gamma d\theta = 0$$

for all multi-indices γ such that $|\gamma| = l$. Put $G(x) = r^{-n-a}\hat{E}_{\theta\to\phi}(ia+in/2)^{-1}\Phi(\theta)$, with $\hat{E}(ia+in/2)$ the operator introduced by equation (1.4.3). Then representation (1.11) holds for the operator $\mathfrak{U}(\lambda)$.

Everything that has been said in this section concerning the operator (1.1) can, with obvious changes, be transferred to the operator (1.3) also. In particular, for $a \neq -l-n$ the function G is defined by $G(\phi,x) = r^{-n-a}E_{\theta\to\phi}(ia+in/2)^{-1}\Phi(\phi,\theta)$, and the kernel $G(\phi,\psi;\lambda)$ is defined by

$$G(\phi,\psi;\lambda) = \int_0^\infty t^{-i(\lambda+ia/in/2)-1}G(\phi,t\phi-\psi)dt. \tag{1.16}$$

P r o p o s i t i o n 1.6. *Let* $\eta, \zeta \in C^{\infty}(S^{n-1})$ *with* $\operatorname{supp}\eta \cap \operatorname{supp}\zeta = \varnothing$. *Then for the operator* (1.3) *we have the estimate*

$$\|\zeta \mathfrak{U}(\lambda)\eta; H^{s}(\lambda, S^{n-1}) \to H^{s+p}(\lambda, S^{n-1})\| \leqslant c_{\mathfrak{F}}, \quad \lambda \in \mathfrak{F}, \tag{1.17}$$

where $\mathfrak{F}$ *is an arbitrary closed set in a strip* $|\operatorname{Im}\lambda| < h$ *not containing poles of* $\mathfrak{U}(\lambda)$, p *is an arbitrary real number, and* $c_{\mathfrak{F}}$ *is a constant depending on* $\mathfrak{F}, s$ *and* p.

P r o o f. We will first assume that $a \neq -l-n$, $a \neq l$, where $l = 0, 1, \cdots$. Then $\mathfrak{U}(\lambda)$ allows a representation (1.11), in which the kernel $G(\phi, \psi; \lambda)$ is defined by (1.16). By integration by parts in formulas (1.12), (1.13) we can show that on $\mathfrak{F}$ the kernel $\zeta(\phi) G(\phi, \psi; \lambda)\eta(\psi)$ (as well as its derivatives with respect to the coordinates ϕ, ψ) decreases faster than any power of $|\lambda|$, uniformly with respect to ϕ, ψ. This implies (1.17). Let now a_0 be one of the numbers $-l-n$ or l. If ϵ is sufficiently small, $|(\zeta \mathfrak{U}(\lambda)\eta u, v)| \leqslant c_{\mathfrak{F}} \|u; H^{s}(\lambda, S^{n-1})\| \cdot \|v; H^{-s-p}(\lambda, S^{n-1})\|$, with one and the same constant $c_{\mathfrak{F}}$ for $\mathfrak{U}(\lambda) = \mathrm{E}(\lambda + ia)^{-1}\Phi(\phi, \omega)\mathrm{E}(\lambda)$ for all a subject to $|a - a_0| = \epsilon$. The maximum principle gives the latter inequality for all a in the disk $|a - a_0| \leqslant \epsilon$ (if $\mathfrak{U}(\lambda)$ remains analytic in a). ■

4. An expression for the operator (1.3) in terms of the exterior form ω. Let, as in §1.4,

$$\omega(y) = y_1 dy_2 \wedge \cdots \wedge dy_n + \cdots + (-1)^{n-1} y_n dy_1 \wedge \cdots \wedge dy_{n-1}$$

and let $(x, z) \mapsto \Phi(x, z)$ be a positively homogeneous function of degree a in x as well as in z; here a is a complex number and $x, z \in \mathbb{R}^n \setminus 0$. We will assume that the restriction of Φ is of class $C^{\infty}(S^{n-1} \times S^{n-1})$. Let also $y \mapsto u(y, \lambda)$ be a positively homogeneous function of degree $i\lambda - n/2$. Put

$$\mathfrak{U}_{y \to x}(\lambda) u(y, \lambda) = \mathrm{E}_{z \to x}(\lambda + ia)^{-1} \Phi(x, z) \mathrm{E}_{y \to z}(\lambda) u(y, \lambda), \tag{1.18}$$

where $\mathrm{E}^{\pm 1}$ are understood to mean the operators defined by (1.4.9). It is clear that the operator (1.18) does not depend on the choice of the surface S entering (1.4.9), and maps homogeneous functions of degree $i\lambda - n/2$ to homogeneous functions of the same degree. This way of defining the operator $\mathfrak{U}(\lambda)$ (on homogeneous functions) turns out to be convenient under transition to a manifold.

§2. Operations on canonical meromorphic pseudodifferential operators

In this paragraph we study composition and the operations of taking the adjoint, differentiation, and shift by a parameter, for canonical meromorphic pseudodifferential operators.

We also consider meromorphic operator-functions that have in every strip $|\operatorname{Im}\lambda| < h < \infty$ at most finitely many poles. Any closed set located in a strip $|\operatorname{Im}\lambda| < h$ and not containing poles of the operators considered will be called admissible.

Let $\sigma \in C^\infty(S^{n-1})$ and $\Phi \in C^\infty(S^{n-1} \times S^{n-1})$. Extend σ to $\mathbb{R}^n \setminus 0$ as a homogeneous function of degree v, and extend the function $\theta \mapsto \Phi(\phi,\theta)$ as a homogeneous function of degree a; the numbers a and v are taken arbitrary complex. Introduce the notations

$$\partial^\gamma \sigma(\phi) = \frac{\partial^\gamma \sigma}{\partial \phi_1^{\gamma_1} \cdots \partial \phi_n^{\gamma_n}}(\phi), \quad D_\theta^\gamma \Phi(\phi,\theta) = (-i)^{|\gamma|} \partial_\theta^\gamma \Phi(\phi,\theta),$$

where $(\phi_1, \cdots, \phi_n)$, $(\theta_1, \cdots, \theta_n)$ are the Cartesian coordinates of the vectors ϕ, θ.

P r o p o s i t i o n 2.1. *For arbitrary nonnegative N the formula*

$$\mathfrak{U}(\lambda)\sigma = \tag{2.1}$$

$$= \sum_{|\gamma|=0}^{N} \frac{1}{\gamma!} \partial^\gamma \sigma(\phi) \mathrm{E}(\lambda + iv + i(a - |\gamma|))^{-1} D_\theta^\gamma \Phi(\phi,\theta) \mathrm{E}(\lambda + iv) + \mathfrak{R}(\lambda)$$

holds for the operator (1.3), *where, moreover, the operator-function* $\mathfrak{R}$ *is subject to the inequality*

$$\|\mathfrak{R}(\lambda); H^s(\lambda, S^{n-1}) \to H^{s+N+1-\operatorname{Re} a}(\lambda, S^{n-1})\| \leqslant c_{\mathfrak{F}} \tag{2.2}$$

on every admissible set $\mathfrak{F}$.

(Here and in the sequel the letter $\mathfrak{R}$ refers to distinct operators - remainders in 'asymptotic' formulas.)

P r o o f. Let $\{\zeta_j\}$ be a partition of unity on S^{n-1} subordinate to a sufficiently fine cover of the sphere. Then

$$\mathfrak{U}(\lambda)\sigma = \sum_{j,k} \zeta_j \mathfrak{U}(\lambda) \zeta_k \sigma. \tag{2.3}$$

If $\operatorname{supp}\zeta_j \cap \operatorname{supp}\zeta_k = \varnothing$, by proposition 1.6 the estimate

$$\|\zeta_j \mathfrak{U}(\lambda)\zeta_k; H^s(\lambda, S^{n-1}) \to H^{s+p}(\lambda, S^{n-1})\| \leqslant c_{\mathfrak{F}}(p,s) \tag{2.4}$$

holds on every admissible set $\mathfrak{F}$, where p is an arbitrary number.

Consider those terms in (2.3) for which the supports of ζ_j and ζ_k intersect. For arbitrary $t > 0$,

$$\sigma(\psi) = t^\nu \sigma(\psi t^{-1}) = t^\nu \left[\sum_{|\gamma|=0}^{N} \frac{(-1)^{|\gamma|}}{\gamma!} \partial^\gamma \sigma(\phi)(\phi - \psi t^{-1})^\gamma + \right. \tag{2.5}$$

$$\left. + \sum_{|\gamma|=N+1} \kappa_\gamma(\phi, \psi t^{-1})(\phi - \psi t^{-1})^\gamma \right] =$$

$$= \sum_{|\gamma|=0}^{N} \frac{(-1)^{|\gamma|}}{\gamma!} t^{\nu - |\gamma|} \partial^\gamma \sigma(\phi)(t\phi - \psi)^\gamma +$$

$$+ t^{\nu - N - 1} \sum_{|\gamma|=N+1} \kappa_\gamma(\phi, \psi t^{-1})(t\phi - \psi)^\gamma,$$

where

$$\kappa_\gamma(\phi, \psi t^{-1}) = \frac{(-1)^{N+1}(N+1)}{\gamma!} \int_0^1 (1-\theta)^N \partial^\gamma \sigma(\phi + \theta(\psi t^{-1} - \phi)) d\theta. \tag{2.6}$$

We multiply both sides of (2.5) by $\zeta_j(\phi)\zeta_k(\psi)G(\phi, t\phi - \psi)$ and thus obtain, by applying the Mellin transform with respect to t,

$$\zeta_j(\phi)\zeta_k(\psi)\sigma(\psi) \int_0^\infty t^{-i(\lambda + ia + in/2) - 1} G(\phi, t\phi - \psi) dt = \tag{2.7}$$

$$= \sum_{|\gamma|=0}^{N} \frac{(-1)^{|\gamma|}}{\gamma!} \zeta_j(\phi)\zeta_k(\psi)\partial^\gamma \sigma(\phi) \int_0^\infty t^{-i(\lambda + i\nu + i(a - |\gamma| + n/2)) - 1} \times$$

$$\times G(\phi, t\phi - \psi)(t\phi - \psi)^\gamma dt + \sum_{|\gamma|=N+1} \zeta_j(\phi)\zeta_k(\psi) \times$$

$$\times \int_0^\infty t^{-i(\lambda + i\nu + i(a - N - 1 + n/2)) - 1} \kappa_\gamma(\phi, \psi t^{-1}) G(\phi, t\phi - \psi)(t\phi - \psi)^\gamma dt.$$

Here, each integral is understood to be its regularization using analytic extension, described above proposition 1.4.

Formulas (1.10) and (1.11) imply that the expression

$$(-1)^{|\gamma|}\int_0^\infty t^{-i(\lambda+i\nu+i(a-|\gamma|+n/2))-1}G(t\phi-\psi)(t\phi-\psi)^\gamma dt$$

is the kernel of the operator

$$\mathrm{E}_{\theta\to\phi}(\lambda+i\nu+i(a-|\gamma|))^{-1}D_\theta^\gamma\Phi(\phi,\theta)\mathrm{E}_{\psi\to\theta}(\lambda+i\nu). \tag{2.8}$$

Therefore (2.7) implies the relation

$$\zeta_j\mathfrak{U}(\lambda)\zeta_k\sigma =$$

$$=\sum_{|\gamma|=0}^{N}\zeta_j\frac{1}{\gamma!}\partial^\gamma\sigma(\phi)\mathrm{E}(\lambda+i\nu+i(a-|\gamma|))^{-1}D_\omega^\gamma\Phi(\phi,\omega)\mathrm{E}(\lambda+i\nu)\zeta_k+$$

$$+\zeta_j\mathfrak{R}(\lambda)\zeta_k,$$

where $\mathfrak{R}$ denotes the operator with kernel

$$R(\phi,\psi;\lambda)=\sum_{|\gamma|=N+1}\int_0^\infty t^{-i(\lambda+i\nu+i(a-N-1+n/2))-1}\times \tag{2.9}$$

$$\times\,\kappa_\gamma(\phi,\psi t^{-1})G(\phi,t\phi-\psi)(t\phi-\psi)^\gamma dt.$$

Let η_q, $q=1,2,3$, be nonnegative functions of class $C^\infty[0,\infty)$, $\eta_1+\eta_2+\eta_3=1$, with $\operatorname{supp}\eta_1\subset[0,1/2)$, $\operatorname{supp}\eta_2\subset(1/4,4)$ and $\operatorname{supp}\eta_3\subset[2,\infty)$. We denote by $\zeta_j(\phi)R_q(\phi,\psi;\lambda)\zeta_k(\psi)$ the expression obtained from the righthand side of (2.9) by multiplying the integrand by $\zeta_j(\phi)\eta_q(t)\zeta_k(\psi)$. Using the method for regularization of the integrals occurring in the expression $\zeta_jR_q\zeta_k$ for $q=1,2,3$ (cf. formulas (1.12) and (1.13)), we are led to the estimates

$$\|\zeta_j\mathfrak{R}_q(\lambda)\zeta_k;H^s(\lambda,S^{n-1})\to H^{s+p}(\lambda,S^{n-1})\|\leqslant c_{\mathfrak{F}}(p,s), \tag{2.10}$$

where $\mathfrak{F}$ is an arbitrary admissible set, $\lambda\in\mathfrak{F}$, and $\zeta_j\mathfrak{R}_q(\lambda)\zeta_k$ denotes the operator with kernel $\zeta_j(\phi)R_q(\phi,\psi;\lambda)\zeta_k(\psi)$.

In order to estimate $\zeta_j\mathfrak{R}_2\zeta_k$ we note that by using Taylor's formula of order M ($M>N$) instead of (2.5) we can write the operator $\mathfrak{R}$ as a finite sum of operators of the form (2.8) (for $N<|\gamma|<M$) and a term with kernel (2.9), in which N is replaced by M. It now suffices to apply proposition 1.3 and use inequality (2.10) and the smoothness of the function

$$\sum_{|\gamma|=M+1} \kappa_\gamma(\phi,\psi t^{-1})G(t\phi-\psi)(t\phi-\psi)^\gamma \eta_2(t)\zeta_j(\phi)\zeta_k(\psi). \quad \blacksquare$$

P r o p o s i t i o n 2.2. *Let μ be an arbitrary complex number, and $\phi = x/|x|$. Then for the operator $\mathfrak{U}$ defined by* (1.3) *the formula*

$$\mathfrak{U}(\lambda+\mu)-\mathfrak{U}(\lambda) = \tag{2.11}$$

$$= \sum_{|\gamma|=1}^{N} \frac{1}{\gamma!}(\partial_x^\gamma |x|^{i\mu})|_{|x|=1} \mathrm{E}_{\theta\to\phi}(\lambda+i(a-|\gamma|))^{-1} \times$$

$$\times D_\theta^\gamma \Phi(\phi,\theta)\mathrm{E}_{\psi\to\theta}(\lambda)+\mathfrak{R}(\lambda,\mu)$$

holds, where the operator $\mathfrak{R}$ is subject to the inequality

$$\|\mathfrak{R}(\lambda,\mu);H^s(\lambda,S^{n-1})\to H^{s+N+1-\mathrm{Re}\,a}(\lambda,S^{n-1})\| \leqslant c_{\mathfrak{F}}|\mu| \tag{2.12}$$

on every admissible set $\mathfrak{F}$. If $|\mu|<\delta$, with δ an arbitrary small number, then a set $\mathfrak{F}$ which is admissible for the operators $\mathrm{E}(\lambda+i(a-|\gamma|))^{-1}D_\theta^\gamma\Phi(\phi,\theta)\mathrm{E}(\lambda)$, $0\leqslant|\gamma|\leqslant N$, is admissible also for all operators $\lambda\mapsto\mathfrak{R}(\lambda,\mu)$, and the constant $c_{\mathfrak{F}}$ in (2.12) *may be assumed to be independent of μ.*

P r o o f. It suffices to put $\nu = i\mu$, $\sigma = |x|^{i\mu}$ in proposition 2.1 and then to replace $\lambda-\mu$ by λ. ■

It follows from (2.11) that, in particular, if $\mathfrak{U}$ is an operator of order a, then the difference $\mathfrak{U}(\lambda+\mu)-\mathfrak{U}(\lambda)$ is an operator of order $a-1$.

P r o p o s i t i o n 2.3. *The derivative $\partial_\lambda\mathfrak{U}\equiv\partial\mathfrak{U}/\partial\lambda$ of the operator* (1.3) *has, for arbitrary natural N, the representation*

$$\partial_\lambda\mathfrak{U}(\lambda) = \sum_{|\gamma|=1}^{N}\frac{1}{\gamma!}\rho_\gamma(x)|_{|x|=1} \times \tag{2.13}$$

$$\times\, \mathrm{E}_{\theta\to\phi}(\lambda+i(a-|\gamma|))^{-1}D_\theta^\gamma\Phi(\phi,\theta)\mathrm{E}(\lambda)+\mathfrak{R}(\lambda),$$

where $\rho_\gamma(x) = \lim_{\mu\to 0}\mu^{-1}\partial_x^\gamma|x|^{i\mu}$, while the estimate (2.2) *is valid for the operator $\mathfrak{R}(\lambda)$ (which depends on N).*

The p r o o f follows from formulas (2.11) and (2.12). ■

We now consider composition of canonical meromorphic pseudodifferential operators. Let

$$\mathfrak{U}(\lambda) = \mathrm{E}(\lambda+ia)^{-1}\Phi(\phi,\theta)\mathrm{E}(\lambda),$$

$$\mathfrak{B}(\lambda) = \mathrm{E}(\lambda+ib)^{-1}\Psi(\phi,\theta)\mathrm{E}(\lambda).$$

We will assume that the function Ψ is extended to $\mathbb{R}^n \setminus 0$ in each argument ϕ, θ as a homogeneous function of degree b, and Φ — as a homogeneous function of degree a.

Proposition 2.4. *For arbitrary nonnegative integers N the formula*

$$\mathfrak{U}(\lambda)\mathfrak{B}(\lambda) = \sum_{|\gamma|=0}^{N} \frac{1}{\gamma!}\mathrm{E}(\lambda+i(a+b-|\gamma|))^{-1} \times \tag{2.14}$$

$$\times D_\theta^\gamma \Phi(\phi,\theta)\partial_\phi^\gamma \Psi(\phi,\theta)\mathrm{E}(\lambda)+\mathfrak{R}(\lambda)$$

holds; on every admissible set $\mathfrak{F}$ we have the estimate

$$\|\mathfrak{R}(\lambda); H^s(\lambda,S^{n-1})\to H^{s-\mathrm{Re}(a+b)+N+1}(\lambda,S^{n-1})\| \leqslant c_{\mathfrak{F}}.$$

Proof. Expand Ψ is a series in spherical harmonics, $\Psi(\phi,\theta) = \Sigma a_{mk}(\phi)Y_{mk}(\phi)$. The coefficients a_{mk} are taken to be homogeneous functions of degree b. It is clear that

$$\mathfrak{B}(\lambda) = \sum a_{mk}(\phi)\mathrm{E}(\lambda+ib)^{-1}Y_{mk}(\theta)\mathrm{E}(\lambda); \tag{2.15}$$

moreover, this series converges in the norm of operators from $H^s(\lambda,S^{n-1})$ to $H^{s-\mathrm{Re}\,b}(\lambda,S^{n-1})$. By formula (2.1),

$$\mathfrak{U}(\lambda)a_{mk} = \sum_{|\gamma|=0}^{N} \frac{1}{\gamma!}\partial_\phi^\gamma a_{mk}(\phi)\mathrm{E}(\lambda+i(a+b-|\gamma|))^{-1} \times \tag{2.16}$$

$$\times D_\theta^\gamma \Phi(\phi,\theta)\mathrm{E}(\lambda+ib)+\mathfrak{R}(\lambda).$$

It remains to combine (2.15) and (2.16). ■

Finalising this paragraph, we consider operators adjoint to canonical meromorphic pseudodifferential operators with respect to the scalar product in $L_2(S^{n-1})$.

Proposition 2.5. *Let $\mathfrak{U}(\lambda) = \mathrm{E}(\lambda+ia)^{-1}\Phi(\phi,\theta)\mathrm{E}(\lambda)$ and let $\mathfrak{U}(\lambda)^*$ be the operator adjoint to $\mathfrak{U}(\lambda)$ with respect to duality in $L_2(S^{n-1})$. Then*

$$\mathfrak{U}(\lambda)^* = \sum_{|\gamma|=0}^{N} \frac{1}{\gamma!} \mathrm{E}(\bar{\lambda}+i(\bar{a}-|\gamma|))^{-1} \partial_\phi^\gamma D_\theta^\gamma \overline{\Phi(\phi,\theta)} \mathrm{E}(\bar{\lambda}) + \mathfrak{R}(\bar{\lambda}), \tag{2.17}$$

where N is an arbitrary nonnegative integer and the operator $\mathfrak{R}$ is subject to inequality (2.2). The function Φ is taken to be extended to $\mathbb{R}^n \setminus 0$ in both arguments ϕ,θ as a homogeneous function of degree a.

P r o o f. Expand Φ in a series

$$\Phi(\phi,\theta) = \sum a_{mk}(\phi) Y_{mk}(\theta),$$

where the a_{mk} can be extended to $\mathbb{R}^n \setminus 0$ as homogeneous functions of degree a. Then

$$\mathfrak{U}(\lambda) = \sum a_{mk} \mathrm{E}(\lambda+ia)^{-1} Y_{mk}(\theta) \mathrm{E}(\lambda).$$

Clearly,

$$(a_{mk} \mathrm{E}(\lambda+ia)^{-1} Y_{mk} \mathrm{E}(\lambda))^* = \mathrm{E}(\lambda)^* \overline{Y_{mk}} (\mathrm{E}(\lambda+ia)^{-1})^* \overline{a_{mk}}.$$

By proposition 1.4.4,

$$(\mathrm{E}(\lambda))^* = \mathrm{E}(\bar{\lambda})^{-1},\ (\mathrm{E}(\lambda+ia)^{-1})^* = \mathrm{E}(\bar{\lambda}-i\bar{a}).$$

This and proposition 2.1 imply that

$$\mathrm{E}(\lambda)^* \overline{Y_{mk}} (\mathrm{E}(\lambda+ia)^{-1})^* \overline{a_{mk}} = \mathrm{E}(\bar{\lambda})^{-1} \overline{Y_{mk}} \mathrm{E}(\bar{\lambda}-i\bar{a}) \overline{a_{mk}} =$$

$$= \sum_{|\gamma|=0}^{N} \frac{1}{\gamma!} \mathrm{E}(\bar{\lambda}+i(\bar{a}-|\gamma|))^{-1} \partial_\phi^\gamma \overline{a_{mk}}(\phi) D_\theta^\gamma \overline{Y_{mk}}(\theta) \mathrm{E}(\bar{\lambda}) + \mathfrak{R}(\bar{\lambda}).$$

Summation over m and k leads to (2.17). ■

§3. General meromorphic pseudodifferential operators

The results of the previous paragraph imply that a composite of canonical meromorphic pseudodifferential operators is not, in general, a canonical meromorphic pseudodifferential operator. The operations of taking the adjoint and differentiation with respect to a parameter also lead outside the class of canonical operators. In this paragraph we define a larger class of meromorphic pseudodifferential operators which is invariant already under composition, differentiation, and the operation $\mathfrak{U}(\lambda)\mapsto\mathfrak{U}(\bar{\lambda})^*$.

1. Definition of general meromorphic pseudodifferential operators. Let $a_0, a_1, \cdots$ be a sequence of complex numbers, let $\alpha_j = \operatorname{Re} a_j$, and let $\alpha_0 > \alpha_1 \geqslant \alpha_2 \geqslant \cdots$; moreover, let $\alpha_j \to -\infty$. We denote by $\{\Phi_j\}_{j=0}^{\infty}$ a sequence of functions of class $C^\infty(S^{n-1}\times S^{n-1})$. We will assume that each function $(\phi,\theta)\mapsto\Phi_j(\phi,\theta)$ is extendable in each argument ϕ and θ to $\mathbf{R}^n\setminus 0$ as a homogeneous function of degree a_j.

D e f i n i t i o n 3.1. An operator-function $\lambda\mapsto\mathfrak{U}(\lambda)$ that is meromorphic in the complex plane is a *meromorphic pseudodifferential operator of order* a_0 if every strip of the form $|\operatorname{Im}\lambda| < h$ contains only finitely many poles of $\mathfrak{U}$ and if for all nonnegative integers N the inequality

$$\left\|\mathfrak{U}(\lambda)-\sum_{j=0}^{N}\mathfrak{U}_j(\lambda); H^s(\lambda,S^{n-1})\to H^{s-\alpha_{N+1}}(\lambda,S^{n-1})\right\|\leqslant c_{\mathfrak{F}} \tag{3.1}$$

holds an every set $\mathfrak{F}$ that is admissible for the operators $\mathfrak{U}$ and $\mathfrak{U}_j$, $j = 1,\cdots,N$. Here $\mathfrak{U}_j(\lambda) = \mathrm{E}_{\theta\to\phi}(\lambda+ia_j)^{-1}\Phi_j(\phi,\theta)\mathrm{E}_{\psi\to\theta}(\lambda)$ are canonical meromorphic pseudodifferential operators and $c_{\mathfrak{F}}$ is a constant depending on $\mathfrak{F}, N$ and s.

The formal series

$$\sigma(\phi,\theta) = \sum_{j=0}^{\infty}\Phi_j(\phi,\theta) \tag{3.2}$$

will be called the *(complete) symbol* of $\mathfrak{U}$, and the function Φ_0 will be called the *principal symbol.*

A series $\Sigma_{j=0}^{\infty}\mathfrak{U}_j$ consisting of canonical meromorphic pseudodifferential operators is called an *asymptotic series* for the meromorphic pseudodifferential operator $\mathfrak{U}$ if inequalities of the type (3.1) hold for all N and $\mathfrak{F}$. In the sequel the notation $\mathfrak{U}\sim\Sigma\mathfrak{U}_j$ means that $\Sigma\mathfrak{U}_j$ is an asymptotic series for $\mathfrak{U}$.

Note that by proposition 2.2 the function $\lambda\mapsto\mathfrak{U}(\lambda+\mu)$, with $\mathfrak{U}$ defined by (1.3), is a meromorphic pseudodifferential operator of order a with symbol

$$\sum_{|\gamma|=0}^{\infty}\frac{1}{\gamma!}|x|^{-i\mu}\partial_x^{\gamma}|x|^{i\mu}D_\theta^{\gamma}\Phi(x,\theta)$$

(the function Φ is homogeneous of degree a both in x and in θ).

By proposition 2.3 the derivative $\partial_\lambda \mathfrak{U}$ is also a meromorphic pseudodifferential operator, and its symbol is

$$\sum_{|\gamma|=0}^{\infty} \frac{1}{\gamma!} \rho_\gamma(x) D_\theta^\gamma \Phi(x,\theta).$$

Proposition 3.2. *In definition* 3.1 *the expansion*

$$\mathfrak{U}(\lambda) \sim \sum_{j=0}^{\infty} \mathrm{E}(\lambda + ia_j)^{-1} \Phi_j(\phi,\theta) \mathrm{E}(\lambda) \tag{3.3}$$

may be replaced by an expansion

$$\mathfrak{U}(\lambda) \sim \sum_{j=0}^{\infty} \mathrm{E}(\lambda)^{-1} \Psi_j(\phi,\omega) \mathrm{E}(\lambda - ib_j), \tag{3.4}$$

where $b_0 = a_0$, $\beta_0 > \beta_1 \geqslant \beta_2 \geqslant \cdots$, $\beta_j = \operatorname{Re} b_j \to -\infty$ *and* $\{\Psi_j\}$ *is a sequence of* $C^\infty(S^{n-1} \times S^{n-1})$ *functions,* $\Psi_0 = \Phi_0$.

Proof. Suppose, e.g., expansion (3.3) holds. By proposition 2.2, for $\rho_\gamma(\phi,\mu) = |x|^{|\gamma|-\mu} \partial_x^\gamma |x|^\mu$ we have

$$\mathrm{E}(\lambda + ia_j)^{-1} \Phi_j(\phi,\omega) \mathrm{E}(\lambda) \sim \mathrm{E}(\lambda)^{-1} \Phi_j(\phi,\omega) \mathrm{E}(\lambda - ia_j) +$$

$$+ \sum_{|\gamma|=1}^{\infty} \frac{\rho_\gamma(\phi,-a_j)}{\gamma!} \mathrm{E}(\lambda - i|\gamma|)^{-1} D_\omega^\gamma \Phi_j(\phi,\omega) \mathrm{E}(\lambda - ia_j).$$

By successively applying this proposition to operators of the form $\mathrm{E}(\lambda - ik)^{-1} \chi(\phi,\omega) \mathrm{E}(\lambda - ic_k)$ we obtain the expansion

$$\mathrm{E}(\lambda + ia)^{-1} \Phi_j(\phi,\omega) \mathrm{E}(\lambda) \sim \sum_{q=0}^{\infty} \mathrm{E}(\lambda)^{-1} \chi_q(\phi,\omega) \mathrm{E}(\lambda - i(a_j - q)). \tag{3.5}$$

Hence, in order to pass from (3.3) to (3.4) it suffices to replace the terms in the series (3.3) by their asymptotic expansions of the form (3.5). Transition from (3.4) to (3.5) is performed similarly. ■

2. Existence of meromorphic pseudodifferential operators.

T h e o r e m 3.3. *For every given series* (3.2) *there is a meromorphic pseudodifferential operator having it as symbol.*

P r o o f. By proposition 3.2 it suffices to prove that there is a meromorphic operator-function $\lambda \mapsto \mathfrak{U}(\lambda)$ having only finitely many poles in any strip $|\operatorname{Im}\lambda| < h$ and allowing the asymptotic expansion

$$\mathfrak{U}(\lambda) \sim \sum_{j=0}^{\infty} \mathfrak{B}_j(\lambda),$$

where $\mathfrak{B}_j(\lambda) = E(\lambda)^{-1}\Phi_j(\phi,\omega)E(\lambda - ib_j)$, $\{\Psi_j\}$ is a given sequence from $C^\infty(S^{n-1}\times S^{n-1})$ and $\{b_j\}$ is a sequence of complex numbers such that $\beta_0 > \beta_1 \geqslant \beta_2 \geqslant \cdots, \beta_j = \operatorname{Re} b_j \to -\infty$.

The proof is divided into several steps

A) We introduce auxiliary operators. Denote by Π_l (l a real number) the projection operator that annihilates the spherical harmonics of degree at most l and that is the identity on the remaining harmonics.

Recall that the function $\lambda \mapsto E(\lambda)u$ is singularity free at the point $\lambda = i(k+n/2)$ if

$$\int u(\theta) Y_{m\mathcal{K}}(\theta)d\theta = 0, \quad 0 \leqslant m \leqslant k, \quad m \equiv k(\operatorname{mod} 2)$$

(cf. (1.2.2); the fact that conditions (1.2.3) and (1.3.13) are equivalent must be taken into account here). Hence the function $\lambda \mapsto E(\lambda - ib_j)\Pi_l u$ is singularity free in the halfplane $\operatorname{Im}\lambda < h$ if $l \geqslant h - \operatorname{Re} b_j - n/2$. Put $l_j = j - \operatorname{Re} b_j - n/2$. Clearly, $l_j \leqslant l_{j+1}$ and $l_j \to +\infty$. The operator-function $\lambda \mapsto \mathfrak{B}_j(\lambda)\Pi_{l_j}$ is analytic everywhere in the halfplane $\operatorname{Im}\lambda < j$, possibly except at the points $\lambda = -i(k+n/2)$, $k = 0, 1, \cdots$.

So, in the halfplane $\operatorname{Im}\lambda < q$ the operators $\mathfrak{B}_j(\lambda)\Pi_{l_j}$, $j = 0, \cdots, q$, can have poles at the points $\lambda = -i(k+n/2)$ only. Hence, every strip of the form $|\operatorname{Im}\lambda| < h < \infty$ contains at most finitely many such points, which coincide with the poles of at least one of the operators $\mathfrak{B}_j\Pi_{l_j}$, $j = 0, 1, \cdots$.

Let $\mathfrak{F}$ be an admissible set for the sequence $\{\mathfrak{B}_j\Pi_{l_j}\}_{j=0}^{\infty}$. Denote by $\lambda_k^{(j)}$ the poles of the operator $E(\lambda - ib_j)(I - \Pi_{l_j})$ and put

$$q_j(\lambda) \equiv q_j(\lambda;\mathfrak{F},\delta) = \prod_{\lambda_k^{(j)} \in \mathfrak{F}} \{1-\exp(-\delta(\lambda-\lambda_k^{(j)})^2)\}, \quad \delta>0.$$

(Note that the function $\lambda \mapsto \exp(-\delta\lambda^2)$ decreases exponentially as $\lambda \to \infty$ within angles $|\arg\lambda| < \pi/4$, $|\arg\lambda - \pi| < \pi/4$, and that it differs by an arbitrarily small amount from the identity in any disk $|\lambda| < R$, for $\delta = \delta(R)$ sufficiently small.)

On $\mathfrak{F}$ only the points $\lambda = -i(k+n/2)$, $k = 0,1,\cdots$, can be poles of $\mathfrak{B}_j(\lambda)q_j(\lambda;\mathfrak{F},\delta)$.

B) We now prove that we can choose an increasing sequence $\{l_j'\}$ such that $l_j' \geq l_j$ and

$$\left\| \sum_{j=N}^{\infty} \mathfrak{B}_j(\lambda)\Pi_{l_j'};H^s(\lambda,S^{n-1}) \to H^{s-\beta_N}(\lambda,S^{n-1}) \right\| \leq c(\mathfrak{F},N), \tag{3.6}$$

where N is an arbitrary natural number.

Let $\mathfrak{F}_j'$ be the set obtained from $\mathfrak{F}$ by deleting from the latter small neighborhoods of the poles of $\mathfrak{B}_j$ that are not poles of $\mathfrak{B}_j\Pi_{l_j}$. Let also $c(\mathfrak{F}_j')$ be a constant such that

$$\|\mathfrak{B}_j(\lambda)v;H^{s-\beta_j-1}(\lambda,S^{n-1})\| \leq c(\mathfrak{F}_j')\|v;H^{s-1}(\lambda,S^{n-1})\|, \quad \lambda \in \mathfrak{F}_j'.$$

For $l_j' \geq l_j$ and $\lambda \in \mathfrak{F}_j'$ we have

$$\|\mathfrak{B}_j(\lambda)\Pi_{l_j'}v;H^{s-\beta_j-1}(\lambda,S^{n-1})\|^2 \leq \tag{3.7}$$

$$\leq c(\mathfrak{F}_j')^2\|\Pi_{l_j'}v;H^{s-1}(\lambda,S^{n-1})\|^2 =$$

$$= c(\mathfrak{F}_j')^2 \sum_{m=l_j'}^{\infty} \sum_{k=1}^{k_m} (1+m^2+|\lambda|^2)^{s-1}|v_{mk}|^2 \leq$$

$$\leq c(\mathfrak{F}_j')^2(1+l_j'^2)^{-1}\|v;H^s(\lambda,S^{n-1})\|^2.$$

If $\mu \in \mathfrak{F}$ is a pole of $\mathfrak{B}_j(\lambda)$, then using the maximum principle, applied to the function $(\mathfrak{B}_j(\lambda)\Pi_{l_j'}v,w)$ (duality in $L_2(S^{n-1})$), and estimate (3.7), the inequality

$$\|\mathfrak{B}_j(\lambda)\Pi_{l_j'}v;H^{s-\beta_j-1}(\lambda,S^{n-1})\|^2 \leq \tag{3.8}$$

$$\leq c(\mathfrak{F}_j')^2(1+l_j'^2)^{-1}\|v;H^s(\lambda,S^{n-1})\|^2$$

can be proved for all λ in a neighborhood of μ.

Choose l_j' such that the condition

$$(1+l_j'^2)^{-1}c(\mathfrak{F}_j')^2 < 2^{-2j},\ j = 0,1,\cdots,$$

holds.

Note that for all $j = N+1,\ N+2,\ \cdots$, with the possible exception of some initial terms, the inequalities $\beta_j+1<\beta_N$ hold. Hence by (3.8),

$$\|\mathfrak{B}_j(\lambda)\Pi_{l_j}'v;H^{s-\beta_N}(\lambda,S^{n-1})\| \leqslant$$

$$\|\mathfrak{B}_j(\lambda)\Pi_{l_j}'v;H^{s-\beta_j-1}(\lambda,S^{n-1})\| \leqslant 2^{-j}\|v;H^s(\lambda,S^{n-1})\|.$$

This implies inequality (3.6).

C) In this step we determine a sequence of positive numbers $\delta_j = \delta_j(\mathfrak{F})$ such that the estimate

$$\left\|\sum_{j=N}^{\infty}\mathfrak{B}_j(\lambda)(I-\Pi_{l_j}')q_j(\lambda;\mathfrak{F},\delta_j)v;H^s(\lambda,S^{n-1})\to H^{s-\beta_N}(\lambda,S^{n-1})\right\| \leqslant \tag{3.9}$$

$$\leqslant c(\mathfrak{F},N)$$

holds for all natural numbers N.

For $\lambda\in\mathfrak{F}$ we have

$$\|\mathfrak{B}_j(\lambda)(I-\Pi_{l_j}')q_j(\lambda;\mathfrak{F},\delta_j)v;H^{s-\beta_j-1}(\lambda,S^{n-1})\|^2 \leqslant \tag{3.10}$$

$$\leqslant |q_j(\lambda,\mathfrak{F},\delta_j)|^2(1+|\lambda|^2)^{-1}c(\mathfrak{F}_j')^2\|v;H^s(\lambda,S^{n-1})\|^2.$$

Choose a number $\delta_j = \delta_j(\mathfrak{F})$ small such that for $\lambda\in\mathfrak{F}$ the inequality

$$|q_j(\lambda;\mathfrak{F},\delta_j)|^2(1+|\lambda|^2)^{-1}c(\mathfrak{F}_j')^2 < 2^{-2j}$$

holds.

Under the condition $\beta_j+1<\beta_N$ we obtain

$$\|\mathfrak{B}_j(\lambda)(I-\Pi_{l_j}')q_jv;H^{s-\beta_N}(\lambda,S^{n-1})\| \leqslant$$

$$\leqslant \|\mathfrak{B}_j(\lambda)(I-\Pi_{l_j}')q_jv;H^{s-\beta_j-1}(\lambda;S^{n-1})\| \leqslant$$

$$\leqslant 2^{-j}\|v;H^s(\lambda,S^{n-1})\|.$$

This implies the estimate (3.9).

D) We define the operator $\mathfrak{U}$. Let $\{\mathfrak{F}_k\}_{k=0}^{\infty}$ be an increasing sequence of sets admissible for the operators $\mathfrak{B}_j(\lambda)\Pi_{l_j}$, and having the following properties:

1. The union $\cup_k \mathfrak{F}_k$ coincides with the complex plane from which the poles of the operators $\mathfrak{B}_j(\lambda)\Pi_{l_j}$ are deleted;

2. Whatever a set $\mathfrak{F}$ admissible for the sequence $\{\mathfrak{B}_j(\lambda)\Pi_{l_j}\}$, there is a number $N = N(\mathfrak{F})$ such that all $j \geqslant N$ the inclusion $\mathfrak{F} \subset \mathfrak{F}_j$ holds. (E.g., we may take

$$\mathfrak{F}_k = \left\{\lambda: |\operatorname{Im}\lambda| \leqslant k\} \setminus \bigcup_{|\lambda_q| \leqslant k} \{\lambda: |\lambda - \lambda_q| < k^{-1}\right\},$$

where $\{\lambda_q\}$ is a set of points each of which is pole of at least one operator $\mathfrak{B}_j(\lambda)\Pi_{l_j}$, $j = 0, 1, \cdots$. Recall that every strip $|\operatorname{Im}\lambda| \leqslant k$ contains only finitely many points λ_q, by the choice of l_j.)

Denote by $\{l_j'(\mathfrak{F}_k)\}$, $\{\delta_j(\mathfrak{F}_k)\}$ sequences as constructed in B) and C) for the set $\mathfrak{F}_p$. Put $L_j = l_j'(\mathfrak{F}_j)$, $\Delta_j = \delta_j(\mathfrak{F}_j)$. The operators $\mathfrak{B}_j(\lambda)(I - \Pi_{l_j})q_j(\lambda; \mathfrak{F}_j, \Delta_j)$ and $\mathfrak{B}_j(\lambda)\Pi_{L_j}$, $j \geqslant N$, do not have poles in $\mathfrak{F}$ and satisfy the inequalities

$$\left\| \sum_{j=M}^{\infty} \mathfrak{B}_j(\lambda)\Pi_{L_j}; H^s(\lambda, S^{n-1}) \to H^{s-\beta_M}(\lambda, S^{n-1}) \right\| \leqslant c(\mathfrak{F}, M), \tag{3.11}$$

$$\left\| \sum_{j=M}^{\infty} \mathfrak{B}_j(\lambda)(I - \Pi_{L_j})q_j(\lambda; \mathfrak{F}, \Delta_j); H^s(\lambda, S^{n-1}) \to H^{s-\beta_M}(\lambda, S^{n-1}) \right\| \leqslant$$

$$\leqslant c(\mathfrak{F}, M)$$

for all natural numbers M.

We introduce the operators

$$\mathfrak{U}(\lambda) = \sum_{j=0}^{\infty} \mathfrak{B}_j(\lambda)\Pi_{L_j} + \sum_{j=0}^{\infty} \mathfrak{B}_j(\lambda)(I - \Pi_{L_j})q_j(\lambda; \mathfrak{F}_j, \Delta_j). \tag{3.13}$$

E) We will show that the symbol of (3.13) is $\Sigma\Psi_j$. Let M be an arbitrary natural number and let $\mathfrak{F}$ be a set admissible for the operators $\mathfrak{U}$ and $\mathfrak{B}_j$, $j = 0, \cdots, M-1$. We have

$$\mathfrak{U}(\lambda) - \sum_{j=0}^{M-1} \mathfrak{B}_j(\lambda) = \sum_{j=M}^{\infty} \mathfrak{B}_j(\lambda)\Pi_{L_j} + \tag{3.14}$$

$$+ \sum_{j=M}^{\infty} \mathfrak{B}_j(\lambda)(I-\Pi_{L_j})q_j(\lambda;\mathfrak{F}_j,\Delta_j) + \sum_{j=0}^{M-1} \mathfrak{B}_j(\lambda)(I-\Pi_{L_j})(q_j(\lambda;\mathfrak{F}_j,\Delta_j)-1).$$

For $\lambda \in \mathfrak{F}$ we have

$$\left\| \sum_{j=0}^{M-1} \mathfrak{B}_j(\lambda)(I-\Pi_{L_j})(q_j(\lambda;\mathfrak{F}_j,\Delta_j)-1)u;H^{s-\beta_M}(\lambda,S^{n-1})\right\|^2 \leqslant \tag{3.15}$$

$$\leqslant \sum_{j=0}^{M-1} c_{j,\mathfrak{F}}^2 \|(I-\Pi_{L_j})(q_j(\lambda;\mathfrak{F}_j,\Delta_j)-1)u;H^{s+\beta_j-\beta_M}(\lambda,S^{n-1})\|^2 \leqslant$$

$$\leqslant \sum_{j=0}^{M-1} c_{j,\mathfrak{F}}^2 \sum_{m=0}^{L_j} \sum_{k=1}^{k_m} (1+m^2+|\lambda|^2)^{s+\beta_j-\beta_M} |q_j(\lambda;\mathfrak{F}_j,\Delta_j)-1|^2 |u_{mk}|^2 \leqslant$$

$$\leqslant c_{\mathfrak{F}}(M)^2 \|u;H^s(\lambda,S^{n-1})\|^2,$$

where

$$c_{\mathfrak{F}}(M)^2 = \sum_{j=0}^{M-1} c_{j,\mathfrak{F}}^2 \sup_{\substack{\lambda\in\mathfrak{F}\\ 0\leqslant m\leqslant L_j}} (1+m^2+|\lambda|^2)^{\beta_j-\beta_M} |q_j(\lambda;\mathfrak{F}_j,\Delta_j)-1|^2$$

Combining (3.11), (3.12), (3.15), and formula (3.14) leads to the inequality

$$\left\| \mathfrak{U}(\lambda) - \sum_{j=0}^{M-1} \mathfrak{B}_j(\lambda);H^s(\lambda,S^{n-1}) \to H^{s-\beta_M}(\lambda,S^{n-1})\right\| \leqslant$$

$$\leqslant c(\mathfrak{F},M) = \text{const.} \quad \blacksquare$$

3. The algebra of meromorphic pseudodifferential operators. The following two theorems follow immediately from propositions 2.4, 2.5, and definition 3.1.

Theorem 3.4. *Let $\mathfrak{U}$ and $\mathfrak{B}$ be meromorphic pseudodifferential operators of orders a_0 and b_0 with symbols*

$$\sigma_{\mathfrak{U}}(\phi,\omega) = \sum_{j=0}^{\infty} \Phi_j(\phi,\omega), \quad \sigma_{\mathfrak{B}}(\phi,\omega) = \sum_{k=0}^{\infty} \Psi_k(\phi,\omega),$$

where Φ_j and Ψ_k are functions homogeneous in ϕ of degree a_j and in ω of degree a_j (resp. b_k). Then $\mathfrak{U}(\lambda)\mathfrak{B}(\lambda)$ is a meromorphic pseudodifferential operator of order a_0+b_0, whose symbol is defined by

$$\sigma_{\mathfrak{U}\mathfrak{B}}(\phi,\omega) = \sum_{j,k} \sum_{|\gamma|=0}^{\infty} \frac{1}{\gamma!} D_\omega^\gamma \Phi_j(\phi,\omega) \partial_\phi^\gamma \Psi_k(\phi,\omega).$$

Theorem 3.5. *Let $\mathfrak{U}$ be a meromorphic pseudodifferential operator of order a_0 with symbol*

$$\sigma_{\mathfrak{U}}(\phi,\omega) = \sum_{j=0}^{\infty} \Phi_j(\phi,\omega),$$

where the Φ_j are homogeneous functions of degree a_j in ϕ and in ω. Let $\mathfrak{U}(\lambda)^*$ be the operator adjoint to $\mathfrak{U}$ with respect to the scalar product in $L_2(S^{n-1})$. Then the operator $\mathfrak{U}^*(\lambda)$ defined by $\mathfrak{U}^*(\bar{\lambda}) = \mathfrak{U}(\lambda)^*$ is a meromorphic pseudodifferential operator of order $\bar{a}_0$, and the asymptotic expansion

$$\mathfrak{U}^*(\lambda) \sim \sum_{j=0}^{\infty} \sum_{|\gamma|=0}^{\infty} \frac{1}{\gamma!} \mathrm{E}(\lambda + i(\bar{a}_j - |\gamma|))^{-1} \partial_\phi^\gamma D_\omega^\gamma \overline{\Phi_j(\phi,\omega)} \mathrm{E}(\lambda)$$

holds.

Definition 3.6. A meromorphic pseudodifferential operator $\mathfrak{U}$ is called *elliptic* if its principal symbol Φ_0 does not vanish anywhere on $S^{n-1} \times S^{n-1}$.

Definition 3.7. We say that the *order of a meromorphic pseudodifferential operator $\mathfrak{L}$ is equal to* $-\infty$ if the estimate

$$\|\mathfrak{L}(\lambda); H^s(\lambda, S^{n-1}) \to H^{s+p}(\lambda, S^{n-1})\| \leqslant c(\mathfrak{F}, p) = \text{const}$$

holds on every admissible set $\mathfrak{F}$, for any positive p.

Note that any strip $|\operatorname{Im}\lambda| \leqslant h$ contains only finitely many poles of $\mathfrak{L}$ and that the complete symbol of $\mathfrak{L}$ vanishes identically.

Theorem 3.8. *Let $\mathfrak{U}$ be an elliptic meromorphic pseudodifferential operator of order a. Then there is a meromorphic pseudodifferential operator $\mathfrak{R}$ of order $-a$ such that both differences $\mathfrak{R}\mathfrak{U} - \mathrm{I}$ and $\mathfrak{U}\mathfrak{R} - \mathrm{I}$, where I is the identity operator, are meromorphic pseudodifferential operators of order $-\infty$.*

If $\mathfrak{R}'$ is an arbitrary meromorphic pseudodifferential operator 'inverting' $\mathfrak{U}$ in this sense from one side, then it inverts $\mathfrak{U}$ from the other side also, and $\mathfrak{R}'(\lambda) = \mathfrak{R}(\lambda) + \mathfrak{L}(\lambda)$, where $\mathfrak{L}$ is a meromorphic pseudodifferential operator of order $-\infty$.

Proof. Put $\mathfrak{B}(\lambda) = \mathrm{E}(\lambda - ia)^{-1} \Phi_0(\phi,\omega)^{-1} \mathrm{E}(\lambda)$. Then $\mathfrak{B}(\lambda)\mathfrak{U}(\lambda) = \mathrm{I} - \mathfrak{J}(\lambda)$, and $\mathfrak{J}$ is a meromorphic pseudodifferential operator whose order has negative real part. Denote by $\sigma_{\mathfrak{J}}$ the complete symbol of $\mathfrak{J}$ and introduce the formal series $\Sigma_{j=0}^{\infty}(\sigma_{\mathfrak{J}})^j$, where the operation of taking a power is understood in the sense of multiplication of symbols under composition of operators. It is clear that this series can be regarded as the complete symbol of a meromorphic pseudodifferential operator $\mathfrak{C}(\lambda)$. Now put $\mathfrak{R}_{\text{left}} = \mathfrak{C}(\lambda)\mathfrak{B}(\lambda)$. We have

$$\mathfrak{R}_{\text{left}}(\lambda)\mathfrak{U}(\lambda) = \mathfrak{C}(\lambda)\mathfrak{B}(\lambda)\mathfrak{U}(\lambda) = \mathfrak{C}(\lambda)(\mathrm{I}-\mathfrak{J}(\lambda)) = \mathrm{I}+\mathfrak{L}_{\text{left}}(\lambda),$$

where $\mathfrak{L}_{\text{left}}$ is a meromorphic pseudodifferential operator of order $-\infty$.

We can similarly construct an operator $\mathfrak{R}_{\text{right}}$, 'inverting' $\mathfrak{U}$ from the right.

Let $\mathfrak{R}'_{\text{left}}$ and $\mathfrak{R}'_{\text{right}}$ be arbitrary meromorphic pseudodifferential operators inverting (in the same sense) $\mathfrak{U}$ from the left and from the right, respectively. Then

$$\mathfrak{R}'_{\text{left}}(\lambda)\mathfrak{U}(\lambda)\mathfrak{R}'_{\text{right}}(\lambda) = \mathfrak{R}'_{\text{left}}(\lambda)(\mathrm{I}+\mathfrak{L}'_{\text{right}}(\lambda)) = (\mathrm{I}+\mathfrak{L}'_{\text{left}}(\lambda))\mathfrak{R}'_{\text{right}}(\lambda),$$

where $\mathfrak{L}'_{\text{left}}$ and $\mathfrak{L}'_{\text{right}}$ are meromorphic pseudodifferential operators of order $-\infty$. Hence the order of the difference $\mathfrak{R}'_{\text{left}}-\mathfrak{R}'_{\text{right}}$ is also $-\infty$. This implies both assertions in the theorem (we may put, e.g., $\mathfrak{R}(\lambda) = \mathfrak{R}_{\text{left}}(\lambda)$). ∎

§4. Traces of meromorphic pseudodifferential operators

1. The trace of a canonical meromorphic pseudodifferential operator. We first recall some well-known notions and facts (see, e.g., [14], [16]). A linear operator A on a Hilbert space is called *nuclear* if for arbitrary orthonormal systems $\{f_n\},\{g_n\}$ the series $\Sigma(\mathrm{A}f_n,g_n)$ converges. For a nuclear operator the sum of the series $\Sigma(\mathrm{A}f_n,f_n)$ does not depend on the choice of an orthonormal basis $\{f_n\}$; it is called the *trace* of A, and is denoted by Sp A.*) If A and B are bounded operators such that AB and BA are nuclear, then Sp BA = Sp AB. The equality $\mathrm{Sp}\,\mathrm{A} = \Sigma\lambda_n$, where $\{\lambda_n\}$ are the eigenvalues of A counted with multiplicities, holds. Finally, if an integral operator A is nuclear,

$$(\mathrm{A}u)(x) = \int_{\Omega} G(x,y)u(y)dy,$$

and G is a continuous function on $\Omega \times \Omega$, then

*) In English literature often Tr A is used instead of Sp A.

$$\mathrm{Sp}\,A = \int_{\Omega} G(x,x)dx. \tag{4.1}$$

(In the sequel this result is applied with $\Omega = S^{n-1}$.) An imbedding operator $H^l(S^{n-1}) \subset H^{l-N}(S^{n-1})$, where H^s is a Sobolev-Slobodetskii space, is nuclear for $N > n-1$.

P r o p o s i t i o n 4.1. *Let* $G(\phi,x) = r^{-a-n} f(\phi,\psi)$, *where* $f \in C^\infty(S^{n-1} \times S^{n-1})$, $\psi = x/|x|$, $r = |x|$, $\phi \in S^{n-1}$. *Let also*

$$G(\phi,\psi;\lambda) = \int_0^\infty t^{-i(\lambda+ia+in/2)-1} G(\phi, t\phi-\psi)dt \tag{4.2}$$

(the integral is understood as the analytic extension with respect to λ*; cf.* (1.12), (1.13)*). If* $\mathrm{Re}\,a < 1-n$, *then the function* $(\phi,\psi) \mapsto G(\phi,\psi;\lambda)$ *is continuous on* $S^{n-1} \times S^{n-1}$.

P r o o f. Let χ be a nonnegative smooth function, equal to 1 in a small neighborhood of $t = 1$ and equal to 0 outside an interval $[1-\epsilon, 1+\epsilon]$. We have

$$G(\phi,\psi;\lambda) = \int_0^\infty (1-\chi(t)) t^{-i(\lambda+ia+in/2)-1} G(\phi, t\phi-\psi)dt +$$

$$+ \int_{1-\epsilon}^{1+\epsilon} \chi(t) t^{-i(\lambda+ia+in/2)-1} G(\phi, t\phi-\psi)dt.$$

The first integral is a smooth function on $S^{n-1} \times S^{n-1}$. Hence it suffices to prove that for small ϵ the second term is small.

Rewrite the second integral as

$$\int_{1-\epsilon}^{1} t^{-i(\lambda+ia+in/2)-1} (1-t)^{-n-a} \chi(t) G(\phi, (\phi-\psi)(1-t)^{-1} - \phi)dt +$$

$$+ \int_{1}^{1+\epsilon} t^{-i(\lambda+ia+in/2)-1} (t-1)^{-n-a} \chi(t) G(\phi, \phi + (\phi-\psi)(t-1)^{-1})dt.$$

If $2|t-1| < |\phi-\psi|$ or if $|\phi-\psi| < |t-1|/2$, then $|(\phi-\psi)(1-t)^{-1} - \phi| \geqslant 1/2$ and $|G(\phi, \pm(\phi-\psi)(1-t)^{-1} \mp \phi)| \leqslant \mathrm{const}$. If now $|t-1|/2 < |\phi-\psi| < 2|t-1|$, $|t-1| < \epsilon$, the vectors ϕ and $\phi-\psi$ are almost orthogonal, hence again $|(\phi-\psi)(1-t)^{-1}-\phi| \geqslant 1/2$ and $|G| \leqslant \text{const}$. It remains to note that the integrals

$$\int_{1-\epsilon}^{1} t^{-i(\lambda+ia+in/2)-1}(1-t)^{-n-a}dt,$$

$$\int_{1}^{1+\epsilon} t^{-i(\lambda+ia+in/2)-1}(t-1)^{-n-a}dt$$

converge. ■

Let $\mathfrak{U}(\lambda) = \mathrm{E}_{\theta\to\phi}(\lambda+ia)^{-1}\Phi(\phi,\theta)\mathrm{E}_{\psi\to\theta}(\lambda)$ be a canonical meromorphic pseudodifferential operator with symbol Φ, $\operatorname{Re} a < n-1$. The map $\mathfrak{U}(\lambda)$: $L_2(S^{n-1}) \to H^{-\operatorname{Re} a}(\lambda, S^{n-1})$ is continuous, hence $\mathfrak{U}(\lambda)$ viewed as an operator on $L_2(S^{n-1})$ is nuclear. By §1.3,

$$(\mathfrak{U}(\lambda)u)(\phi) = \int G(\phi,\psi;\lambda)u(\psi)d\psi,$$

where $G(\phi,\psi;\lambda)$ is defined by (4.2), $G(\phi,x) = r^{-a-n}f(\phi,\psi)$, $f(\phi,\psi) = \mathrm{E}_{\theta\to\psi}(ia+in/2)^{-1}\Phi(\phi,\theta)$. By proposition 4.1 the function $(\phi,\psi) \to G(\phi,\psi;\lambda)$ is continuous on $S^{n-1}\times S^{n-1}$, hence formula (4.1) applies. Thus,

$$\operatorname{Sp}\mathfrak{U}(\lambda) = \int G(\phi,\phi;\lambda)d\phi.$$

It is clear that

$$G(\phi,\phi;\lambda) = \int_0^\infty t^{-i(\lambda+ia+in/2)-1}G(\phi,(t-1)\phi)dt =$$

$$= \int_0^1 t^{-i(\lambda+ia+in/2)-1}(1-t)^{-n-a}dt\,G(\phi,-\phi)+$$

$$+ \int_1^\infty t^{-i(\lambda+ia+in/2)-1}(t-1)^{-n-a}dt\,G(\phi,\phi).$$

Therefore,

$$\operatorname{Sp}\mathfrak{U}(\lambda) = \frac{\Gamma(a-i\lambda+n/2)\Gamma(1-n-a)}{\Gamma(1-i\lambda-n/2)}\int G(\phi,-\phi)d\phi+ \tag{4.3}$$

$$+\frac{\Gamma(i\lambda+n/2)\Gamma(1-n-a)}{\Gamma(i\lambda+1-a-n/2)}\int G(\phi,\phi)d\phi.$$

Using analytic extension this formula can be generalized to all complex a and λ, except for the poles.

We will now express the trace of an operator $\mathfrak{U}(\lambda)$ in terms of its symbol.

Proposition 4.2. *Let* $\mathfrak{U}(\lambda) = \mathrm{E}_{\theta\to\phi}(\lambda+ia)^{-1}\Phi(\phi,\theta)\mathrm{E}_{\psi\to\theta}(\lambda)$ *be a canonical meromorphic pseudodifferential operator. For the trace* $\mathrm{Sp}\,\mathfrak{U}(\lambda)$, *defined using analytic extension, the formula*

$$\mathrm{Sp}\,\mathfrak{U}(\lambda) = \frac{1}{(2\pi)^{n/2}} e^{i\pi(a+n)/2} \frac{\pi}{\sin\pi(a+n)} \left[\frac{\Gamma(a-i\lambda+n/2)}{\Gamma(1-i\lambda-n/2)} \times \right. \tag{4.4}$$

$$\times \int d\phi \int (-\phi\theta+i0)^{-a-n}\Phi(\phi,\theta)d\theta +$$

$$\left. + \frac{\Gamma(i\lambda+n/2)}{\Gamma(i\lambda+1-a-n/2)} \int d\phi \int (\phi\theta+i0)^{-a-n}\Phi(\phi,\theta)d\theta \right]$$

holds for $a \neq \pm k-n$, $k = 0,1,\cdots$; *while for* $a = -k-n$, $k = 0,1,\cdots$, *the formula*

$$\mathrm{Sp}\,\mathfrak{U}(\lambda) = \frac{i^k \pi}{(2\pi)^{n/2}(i\lambda+n/2)\cdots(i\lambda+n/2+k)} \times \tag{4.5}$$

$$\times \left[i\int d\phi \int (\phi\theta)^k \,\mathrm{sgn}\,(\phi\theta)\Phi(\phi,\theta)d\theta - \mathrm{cotan}\,\pi(i\lambda+n/2) \times \right.$$

$$\left. \times \int d\phi \int (\phi\theta)^k \Phi(\phi,\theta)\,d\theta \right]$$

holds.

Proof. If $a \neq \pm k - n$, then, substituting in (4.3) the expression

$$G(\phi,\psi) = \mathrm{E}_{\theta\to\psi}(ia+in/2)^{-1}\Phi(\phi,\theta) =$$

$$= \frac{1}{(2\pi)^{n/2}} e^{i\pi(a+n)/2}\Gamma(a+n)\int(\psi\theta+i0)^{-a-n}\Phi(\phi,\theta)d\theta,$$

we obtain (4.4). (We must use the fact $\Gamma(z)\Gamma(1-z) = \pi/\sin\pi z$.) It is relatively easy to see that in (4.4) the factor in front of $1/\sin\pi(a+n)$ vanishes at the points $a = -k-n$, $k = 0,1,\cdots$. Hence by limit transition in (4.4) as $a \to -k-n$ we obtain

$$\mathrm{Sp}\,\mathfrak{U}(\lambda) = \left[\frac{\Gamma(i\lambda+n/2)\Gamma'(i\lambda+n/2+k+1)}{\Gamma^2(i\lambda+n/2+k+1)} + \right. \tag{4.6}$$

$$+(-1)^k \frac{\Gamma'(-k-i\lambda-n/2)}{\Gamma(1-i\lambda-n/2)}\Bigg] \int d\phi \int (\phi\theta)^k \Phi(\phi,\theta)d\theta +$$

$$-\left[\frac{\Gamma(i\lambda+n/2)}{\Gamma(i\lambda+n/2+k+1)}+(-1)^{k+1}\frac{\Gamma(-k-i\lambda-n/2)}{\Gamma(1-i\lambda-n/2)}\right]\times$$

$$\times \int d\phi \int (\phi\theta)^k (\ln(\phi\theta+i0)-\ln(-\phi\theta+i0))\Phi(\phi,\theta)d\theta.$$

Further (cf., e.g., [5]),

$$\frac{\Gamma(i\lambda+n/2)}{\Gamma(i\lambda+n/2+k+1)} = (-1)^k \frac{\Gamma(-k-i\lambda-n/2)}{\Gamma(1-i\lambda-n/2)} =$$

$$= \frac{1}{(i\lambda+n/2)\cdots(i\lambda+n/2+k)},$$

$$\frac{\Gamma'(i\lambda+n/2+k+1)}{\Gamma(i\lambda+n/2+k+1)} - \frac{\Gamma'(-k-i\lambda-n/2)}{\Gamma(-k-i\lambda-n/2)} = -\pi \text{cotan}\, \pi(i\lambda+n/2).$$

Using these equations (4.5) can be derived from (4.6). ■

We note that also the residue in (4.4) at the points $a = 1-n, 2-n, \cdots$ vanishes, under certain restrictions regarding the symbol (of the type $\int d\phi \int \delta^{(k)}(\phi\theta) I(\phi,\theta)d\theta = 0$). If such restrictions are satisfied, an expression for $\mathrm{Sp}\, \mathfrak{U}(\lambda)$ at $a = 1-n, 2-n, \cdots$ can be obtained from (4.4) also.

2. Traces of general meromorphic pseudodifferential operators. Let now $\mathfrak{U}$ be a general meromorphic pseudodifferential operator of order a_0 in the sense of definition 3.1. Then for arbitrary N the formula

$$\mathfrak{U}(\lambda) = \sum_{j=0}^{N-1} \mathfrak{U}_j(\lambda) + \mathfrak{R}_N(\lambda)$$

holds, and for $\mathfrak{R}_N$ the estimate

$$\|\mathfrak{R}_N(\lambda); L_2(S^{n-1}) \to H^{-\alpha_N}(\lambda, S^{n-1})\| \leq c(N, \mathfrak{F}) \tag{4.7}$$

holds an every admissible set $\mathfrak{F}$. Here $\alpha_j = \mathrm{Re}\, a_j \downarrow -\infty$, a_j being the order of the canonical meromorphic pseudodifferential operator $\mathfrak{U}_j$. It is clear that for sufficiently large N the operator $\mathfrak{R}_N(\lambda)$ is nuclear on $L_2(S^{n-1})$.

We define the trace of $\mathfrak{U}$ by

$$\operatorname{Sp}\mathfrak{U}(\lambda) = \sum_{j=0}^{N-1} \operatorname{Sp}\mathfrak{U}_j(\lambda) + \operatorname{Sp}\mathfrak{R}_N(\lambda).$$

§4.1 implies that $\operatorname{Sp}\mathfrak{U}$ is a meromorphic function on the λ-plane.

Let $\mathfrak{U}$ and $\mathfrak{B}$ be meromorphic pseudodifferential operators of orders a_0 and b_0. Then the operators $\mathfrak{U}\mathfrak{B}$ and $\mathfrak{B}\mathfrak{U}$ are also meromorphic pseudodifferential operators (theorem 3.4).

Theorem 4.3. *The equality*

$$\operatorname{Sp}(\mathfrak{U}(\lambda)\mathfrak{B}(\lambda)) = \operatorname{Sp}(\mathfrak{B}(\lambda)\mathfrak{U}(\lambda)) \tag{4.8}$$

holds.

Proof. Write $\mathfrak{U}$ and $\mathfrak{B}$ as

$$\mathfrak{U}(\lambda) = \sum_{j=0}^{N-1} \mathfrak{U}_j(\lambda) + \mathfrak{R}_N(\lambda),$$

$$\mathfrak{B}(\lambda) = \sum_{k=0}^{N-1} \mathfrak{B}_k(\lambda) + \mathfrak{C}_N(\lambda),$$

where $\mathfrak{U}_j$, $\mathfrak{B}_k$ are canonical meromorphic pseudodifferential operators of orders a_j, b_k; for $\mathfrak{R}_N$ the estimate (4.7) holds, while for $\mathfrak{C}_N$ the analogous estimate holds with α_N replaced by $\beta_N = \operatorname{Re} b_N$. Choose N large such that $\beta_N - \alpha_0 > n-1$, $\alpha_N - \beta_0 > n-1$. Then the composites of $\mathfrak{B}_j$ and $\mathfrak{R}_N$, $\mathfrak{U}_j$ and $\mathfrak{C}_N$, and $\mathfrak{R}_N$ and $\mathfrak{C}_N$ are nuclear, so their traces do not depend on the orders of the factors. Hence it suffices to prove (4.8) under the assumption that $\mathfrak{U}$ and $\mathfrak{B}$ are canonical operators.

So, let $\mathfrak{U}$ and $\mathfrak{B}$ be canonical meromorphic pseudodifferential operators of orders a and b with symbols Φ and Ψ, which may be assumed to be homogeneous of orders a and b, resp., in each argument ϕ and θ.

By the rule for composition (proposition 3.2.4)

$$\mathfrak{U}(\lambda)\mathfrak{B}(\lambda) = \tag{4.9}$$

$$= \sum_{0 \leq |\gamma| \leq N} \mathrm{E}(\lambda + i(a+b-|\gamma|))^{-1} \frac{1}{\gamma!} D_\theta^\gamma \Phi(\phi,\theta) \partial_\phi^\gamma \Psi(\phi,\theta) \mathrm{E}(\lambda) + \mathfrak{M}(\lambda),$$

$$\mathfrak{B}(\lambda)\mathfrak{U}(\lambda) =$$

$$= \sum_{0 \le |\gamma| \le N} E(\lambda + i(a+b-|\gamma|))^{-1} \frac{1}{\gamma!} D_\theta^\gamma \Psi(\phi,\theta) \partial_\phi^\gamma \Phi(\phi,\theta) E(\lambda) + \mathfrak{N}(\lambda),$$

while estimates of the type (4.7) hold for $\mathfrak{M}$ and $\mathfrak{N}$.

We are to convince ourselves that for $k = 0, 1, \cdots,$

$$\mathrm{Sp}\left[\sum_{|\gamma|=k} E(\lambda + i(a+b-k))^{-1} \frac{1}{\gamma!} \partial_\theta^\gamma \Phi(\phi,\theta) \partial_\phi^\gamma \Psi(\phi,\theta) E(\lambda)\right] = \tag{4.10}$$

$$= \mathrm{Sp}\left[\sum_{|\gamma|=k} E(\lambda + i(a+b-k))^{-1} \frac{1}{\gamma!} \partial_\theta^\gamma \Psi(\phi,\theta) \partial_\phi^\gamma \Phi(\phi,\theta) E(\lambda)\right].$$

Since both sides of this equation are meromorphic functions of, say, the parameter a we may assume that $k-a-b-n+1$ is not a natural number and apply formula (4.4). Hence the situation reduces to checking the relation

$$\sum_{|\gamma|=k} \int_{S^{n-1}} d\phi \int_{S^{n-1}} (\pm\phi\theta + i0)^{k-a-b-n} \frac{1}{\gamma!} \partial_\theta^\gamma \Phi(\phi,\theta) \partial_\phi^\gamma \Psi(\phi,\theta) d\theta = \tag{4.11}$$

$$= \sum_{|\gamma|=k} \int_{S^{n-1}} d\phi \int_{S^{n-1}} (\pm\phi\theta + i0)^{k-a-b-n} \frac{1}{\gamma!} \partial_\theta^\gamma \Psi(\phi,\theta) \partial_\phi^\gamma \Phi(\phi,\theta) d\theta.$$

Note that the integrands in (4.11) are homogeneous of degree $-n$ in ϕ and in θ, and that (4.11) is equivalent to

$$\int_{1<|x|,|y|<2} (\pm xy + i0)^{k-a-b-n} \sum_{i,\cdots,i_k=1}^{n} \frac{\partial^k \Phi(x,y)}{\partial x_{i_1} \cdots \partial x_{i_k}} \times \tag{4.12}$$

$$\times \frac{\partial^k \Psi(x,y)}{\partial y_{i_1} \cdots \partial y_{i_k}} dxdy = \int_{1<|x|,|y|<2} (\pm xy + i0)^{k-a-b-n} \times$$

$$\times \sum_{i_1,\cdots,i_k=1}^{n} \frac{\partial^k \Psi(x,y)}{\partial x_{i_1} \cdots \partial x_{i_k}} \cdot \frac{\partial^k \Phi(x,y)}{\partial y_{i_1} \cdots \partial y_{i_k}} dxdy.$$

We show, e.g., the 'plus' relation (4.12). Put

$$J_{m,k-m} = \int_{1<|x|,|y|<2} (xy + i0)^{k-a-b-n} \times$$

$$\times \sum_{\substack{i_1,\cdots,i_m \\ j_1,\cdots,j_{k-m}}} \frac{\partial^k \Phi}{\partial x_{i_1} \cdots \partial x_{i_m} \partial y_{j_1} \cdots \partial y_{j_{k-m}}} \frac{\partial^k \Psi}{\partial y_{i_1} \cdots \partial y_{i_m} \partial x_{j_1} \cdots \partial x_{j_{k-m}}} dxdy$$

(the summation indices run over 1, $\cdots$, n). The result of partial integration gives for $1 \leqslant m \leqslant k$,

$$J_{m,k-m} = J_{m-1,k-m+1} + (k-a-b-n) \times \tag{4.13}$$

$$\times \int_{1<|x|,|y|<2} \sum_{\substack{i_1,\cdots,i_{m-1} \\ j_1,\cdots,j_{k-m}}} \frac{\partial^{k-1}\Phi}{\partial x_{i_1} \cdots \partial x_{i_{m-1}} \partial y_{i_1} \cdots \partial y_{j_{k-m}}} \times$$

$$\times \sum_j \left\{ x_j \frac{\partial}{\partial x_j} - y_j \frac{\partial}{\partial y_j} \right\} \frac{\partial^{k-1}\Psi}{\partial y_{i_1} \cdots \partial y_{i_{m-1}} \partial x_{j_1} \cdots \partial x_{j_{k-m}}} dxdy.$$

The function $(xy) \mapsto \Psi(x,y)$ is homogeneous of degree b both in x and in y. Hence (4.13) can be written as

$$J_{m,k-m} = J_{m-1,k-m+1} + (k-a-b-n)(2m-k-1)h_m, \tag{4.14}$$

where

$$h_m = \int_{1<|x|,|y|<2} (xy+i0)^{k-a-b-n-1} \times$$

$$\times \sum_{\substack{i_1,\cdots,i_{m-1} \\ j_1,\cdots,j_{k-m}}} \frac{\partial^{k-1}\Phi}{\partial x_{i_1} \cdots \partial x_{i_{m-1}} \partial y_{j_1} \cdots \partial y_{j_{k-m}}} \frac{\partial^{k-1}\Psi}{\partial y_{i_1} \cdots \partial y_{i_{m-1}} \partial x_{j_1} \cdots \partial x_{j_{k-m}}} dxdy.$$

From this we derive that

$$J_{k,0} - J_{0,k} = (k-a-b-n) \sum_{m=1}^{k} (2m-k-1)h_m = \tag{4.15}$$

$$= (k-a-b-n) \sum_{m=1}^{[(k+1)/2]} (2m-k-1)(h_m - h_{k-m+1}).$$

The equality

$$J_{p,0} = J_{0,p} \tag{4.16}$$

for $p = 1$ follows from (4.14). Assume that the relations (4.16) hold for $p = 1, \cdots, k-1$. We show that (4.16) also holds for $p = k$. The induction hypothesis implies that $h_m = h_{k-m+1}$ for $m = 1, \cdots, [(k+1)/2]$. Indeed, put

$$\Phi_m = \frac{\partial^{2m-2}\Phi}{\partial x_{i_1} \cdots \partial x_{i_{m-1}} \partial y_{j_1} \cdots \partial y_{j_{m-1}}},$$

$$\Psi_m = \frac{\partial^{2m-2}\Psi}{\partial y_{i_1} \cdots \partial y_{i_{m-1}} \partial x_{j_1} \cdots \partial x_{j_{m-1}}},$$

where $i_1, \cdots, i_{m-1}, j_1, \cdots, j_{m-1}$ is an arbitrary set of indices. The functions Φ_m and Ψ_m are homogeneous of degrees $a_m = a - m + 1$ and $b_m = b - m + 1$, resp. (in each of the arguments x and y). The formula $h_m = h_{k-m+1}$ follows from the equations

$$\int_{1<|x|,|y|<2} (xy + i0)^{k-2m+1-a_m-b_m-n} \times$$
$$\times \sum_{l_1, \cdots, l_{k-2m+1}} \left[\frac{\partial^{k-2m+1}\Phi}{\partial x_{l_1} \cdots \partial x_{l_{k-2m+1}}} \frac{\partial^{k-2m+1}\Psi}{\partial y_{l_1} \cdots \partial y_{l_{k-2m+1}}} + \right.$$
$$\left. - \frac{\partial^{k-2m+1}\Phi}{\partial y_{l_1} \cdots \partial y_{l_{k-2m+1}}} \frac{\partial^{k-2m+1}\Psi}{\partial x_{l_1} \cdots \partial x_{l_{k-2m+1}}} \right] dxdy = 0,$$

which are satisfied by assumption. By now applying (4.15) we are led to (4.12), which was required. So, (4.12) holds, and with it (4.10).

It remains to convince ourselves that the traces of the operators $\mathfrak{M}$ and $\mathfrak{N}$ from (4.9) coincide. The number N in these formulas is subject to $N > \mathrm{Re}(a+b)+n-2$, ensuring that $\mathfrak{M}$ and $\mathfrak{N}$ are nuclear.

Let $\{f_j\}$ be some orthogonal basis in $L_2(S^{n-1})$. Since the operator

$$\mathfrak{N}(\lambda) = \mathfrak{U}(\lambda)\mathfrak{B}(\lambda) +$$
$$- \sum_{0 \le |\gamma| \le N} \mathrm{E}(\lambda + i(a+b-|\gamma|))^{-1} \frac{1}{\gamma!} D_\theta^\gamma \Phi(\phi,\theta) \partial_\phi^\gamma \Psi(\phi,\theta) \mathrm{E}(\lambda)$$

is nuclear, the series $\Sigma_j(\mathfrak{N}(\lambda)f_j, f_j)$ converges. Every term in it is a meromorphic function of a in the halfplane $N > \mathrm{Re}(a+b)+n-2$, hence the sum of the series, i.e. $\mathrm{Sp}\,\mathfrak{N}(\lambda)$, has the same property. Everything said remains true for the operator $\mathfrak{M}(\lambda)$ also. The equality

$$\mathrm{Sp}\,\mathfrak{N}(\lambda) = \mathrm{Sp}\,\mathfrak{M}(\lambda) \tag{4.17}$$

holds for $\mathrm{Re}(a+b) < 1-n$. In fact, this equality ensures that the composite of $\mathfrak{U}(\lambda)$ and $\mathfrak{B}(\lambda)$ is nuclear, hence that $\mathrm{Sp}\,(\mathfrak{U}(\lambda)\mathfrak{B}(\lambda))$ and $\mathrm{Sp}(\mathfrak{B}(\lambda)\mathfrak{U}(\lambda))$ coincide. This and (4.9), (4.10) imply (4.17). Uniqueness of analytic extension implies that (4.17) holds for $\mathrm{Re}(a+b) < N-n+2$ also. ■

R e m a r k 4.4. Theorem 4.3 holds also for matrix operators $\mathfrak{U}$ and $\mathfrak{B}$. Indeed, if $\mathfrak{U}$ is a canonical matrix operator, then (4.4) or (4.5) holds for its trace (replacing $\Phi(\phi,\theta)$ by $\mathrm{Sp}\,\Phi(\phi,\theta)$). Equality (4.12) is also preserved, and the sign of Sp must be specified in it.

§5. Meromorphic pseudodifferential operators on strongly oscillating functions

Here we clarify the asymptotic behavior of an expression $e^{-i\mu g(\phi)}\mathfrak{U}_{\psi\to\phi}(\lambda)e^{i\mu g(\psi)}u(\psi)$, where $\mathfrak{U}(\lambda)$ is a meromorphic pseudodifferential operator, as $|\lambda|^2+\mu^2\to\infty$, $|\mathrm{Im}\lambda|\leqslant h$, $\mu\in\mathbf{R}$. In particular, theorem 5.1 justifies the terminology 'pseudodifferential' for operators $\mathfrak{U}(\lambda)$ (cf. the definition of pseudodifferential operator in the article [72]). This theorem will be repeatedly used in the sequel.

Let $g,u\in C^\infty(S^{n-1})$, with g a real-valued function. Extend u and g to $\mathbf{R}^n\setminus 0$ as homogeneous functions of degree 0. Let also $\mathfrak{U}(\lambda)=\mathrm{E}_{\omega\to\phi}(\lambda+ia)^{-1}\Phi(\phi,\omega)\mathrm{E}_{\psi\to\omega}(\lambda)$ be a canonical meromorphic pseudodifferential operator of order a, let $\sigma=\mathrm{Re}\lambda$ and $\mu\in\mathbf{R}$.

T h e o r e m 5.1. *For any $P\geqslant 0$ there are nonnegative integers N and Q such that the inequality*

$$\Big| e^{-i\mu g(\phi)}\mathfrak{U}_{\psi\to\phi}(\lambda)e^{i\mu g(\psi)}u-\sum_{|\alpha|\leqslant N}\frac{1}{\alpha!}\Phi^{(\alpha)}(\phi,\mu\nabla g(\phi)+\sigma\phi)\times \tag{5.1}$$

$$\times D_y^\alpha\{u(\psi)\rho^{i(\lambda+in/2)}e^{i\mu(g(\psi)-g(\phi))-i(y-\phi,\mu\nabla g(\phi)+\sigma\phi)}\}\Big|_{\substack{\rho=1\\ \phi=\psi}}\Big|\leqslant$$

$$\leqslant c(\mu^2+\sigma^2)^{-P/2}\|u;C^Q(S^{n-1})\|$$

holds, where $\rho=|y|$, $\psi=y/|y|$, $\Phi^{(\alpha)}(\phi,\omega)=D_\omega^\alpha\Phi(\phi,\omega)$.

The p r o o f is divided into several steps. In order to simplify notations we write $\Phi(\omega)$ instead of $\Phi(\phi,\omega)$.

A) In this step we derive the basic asymptotic formula, but not yet estimate the remainder. Let $\zeta\in C_0^\infty(\mathbf{R})$ with $\zeta(t)=1$ for $|t|<1/4$ and $\zeta(t)=0$ for $|t|>1/2$. Put

$$v(r,\rho,\phi,\psi,\mu)=$$

$$= \frac{u(\psi)}{\sqrt{2\pi}} \int\limits_{\operatorname{Im}\lambda=\tau} \rho^{i(\lambda+in/2)}\zeta(\nu\operatorname{dist}(\phi,\psi))\zeta(\nu(1-r/\rho))\tilde{w}(\lambda+in/2)d\lambda,$$

where $\nu = (\mu^2+\sigma^2)^{1/4}$, $\sigma = \operatorname{Re}\lambda$, $\mu \in \mathbb{R}$, $\operatorname{dist}(\phi,\psi)$ is the distance on S^{n-1} between two points ϕ and ψ, and $w \in C_0^\infty(\mathbb{R}_+)$. Permuting the operator $A_{y\to x} = F^{-1}_{\xi\to x}\Phi(\xi)F_{y\to\xi}$ and the integral we obtain

$$A(e^{i\mu g(\psi)}v(r,\rho,\phi,\psi,\mu)) = \frac{1}{\sqrt{2\pi}} \times \tag{5.2}$$

$$\times \int\limits_{\operatorname{Im}\lambda=\tau} A[e^{i\mu g(\psi)}\rho^{i(\lambda+in/2)}u(\psi)\tilde{w}(\lambda+in/2)\times$$

$$\times \zeta(\nu\operatorname{dist}(\phi,\psi))\zeta(\nu(1-r/\rho))]d\lambda.$$

Rewrite the integrand as

$$\iint e^{i(x-y)\xi}\Phi(\xi)e^{i(\chi(y)-\chi(x)-(y-x,\xi_x))}e^{i(\chi(x)+(y-x,\xi_x))}q(\lambda,\mu,x,y)dyd\xi, \tag{5.3}$$

where $\chi(y) = \mu g(\psi)+\sigma\ln\rho$, $q(\lambda,\mu,x,y) = u(\psi)\tilde{w}(\lambda+in/2)\times \times\zeta(\nu\operatorname{dist}(\phi,\psi))\zeta(\nu(1-r/\rho))\rho^{-\tau-n/2}$, $\xi_x = \nabla\chi(x) = \mu\nabla g(x)+\sigma x/|x|^2 = = r^{-1}(\mu\nabla g(\phi)+\sigma\phi)$. Put $h(x,y) = \chi(y)-\chi(x)-(y-x,\xi_x)$ and note that (5.3) is equal to

$$\iint e^{i(x-y)\eta}e^{ih(x,y)+i\chi(x)}\Phi(\xi_x+\eta)q(\lambda,\mu,x,y)dyd\eta.$$

We will assume that $\operatorname{Re}a$ is a sufficiently large positive number. By expanding $\Phi(\xi_x+\eta)$ using Taylor's formula we obtain

$$A[e^{i\mu g(\psi)}\rho^{i(\lambda+in/2)}u(\psi)\tilde{w}(\lambda+in/2)\zeta(\nu\operatorname{dist}(\phi,\psi)\zeta(\nu(1-r/\rho))] = \tag{5.4}$$

$$= \iint e^{i(x-y)\eta}e^{ih(x,y)+i\chi(x)} \sum_{|\alpha|\leqslant N} \frac{1}{\gamma!}\Phi^{(\alpha)}(\xi_x)\eta^\alpha q(\lambda,\mu,x,y)dyd\eta+$$

$$+ \sum_{|\alpha|=N+1} \int_0^1 (1-\theta)^N R_\alpha(x,\lambda,\mu,\theta)d\theta,$$

where

$$R_\alpha(x,\lambda,\mu,\theta) = (N+1)\iint e^{i(x-y)\eta+ih(x,y)+i\chi(x)}\frac{1}{\alpha!}\eta^\alpha\Phi^{(\alpha)}(\xi_x+\theta\eta)qdyd\eta.$$

Consider the terms with $|\alpha| \leqslant N$. Their sum can be written as

$$\sum_{|\alpha|\leq N}\frac{1}{\alpha!}\Phi^{(\alpha)}(\xi_x)e^{i\chi(x)}\int\int e^{i(x-y)\eta}\eta^{\alpha}e^{ih(x,y)}q(\lambda,\mu,x,y)dyd\eta = \tag{5.5}$$

$$= \sum_{|\alpha|\leq N}\frac{1}{\alpha!}\Phi^{(\alpha)}(\chi_x)e^{i\chi(x)}D_y^{\alpha}(e^{ih(x,y)}q(\lambda,\mu,x,y))|_{y=x}.$$

Recall that for sufficiently close ϕ and ψ (r and ρ) the quantity $\zeta(\nu\mathrm{dist}(\phi,\psi))\zeta(\nu(1-r/p))$ equals 1. Moreover,

$$e^{i\chi(x)} = e^{i\mu g(\phi)}r^{i\sigma} = e^{i\mu g(\phi)}r^{i(\lambda+in/2)}r^{\tau+n/2}.$$

Hence

$$D_y^{\alpha}(e^{ih(x,y)}q(\lambda,\mu,x,y))|_{y=x} =$$

$$= D_y^{\alpha}(e^{ih(x,y)}u(\psi)(r/\rho)^{\tau+n/2})|_{y=x}\tilde{w}(\lambda+in/2)$$

(we assume that the function $\psi\mapsto u(\psi)$ has been extended to $\mathbb{R}^n\setminus 0$ as a homogeneous function of degree zero). Thus, the righthand side of (5.5) takes the form

$$e^{i\mu g(\phi)}r^{-a}\sum_{|\alpha|\leq N}\frac{1}{\alpha!}r^{i(\lambda+in/2)}\Phi^{(\alpha)}(\mu\nabla g(\phi)+\sigma\phi)\times \tag{5.6}$$

$$\times r^{|\alpha|}D_y^{\alpha}(e^{ih(x,y)}u(\psi)(r/\rho)^{\tau+n/2})|_{y=x}\tilde{w}(\lambda+in/2).$$

Taking (5.4) into consideration, we can rewrite (5.2) as

$$r^a e^{-i\mu g(\phi)}A_{y\to x}(e^{i\mu g(\psi)}v(r,\rho,\phi,\psi,\mu)) = \tag{5.7}$$

$$= \frac{1}{\sqrt{2\pi}}\int_{\mathrm{Im}\lambda=\tau} r^{i(\lambda+in/2)}\sum_{|\alpha|\leq N}\frac{1}{\alpha!}\Phi^{(\alpha)}(\mu\nabla g(\phi)+\sigma\phi)r^{|\alpha|}\times$$

$$\times D_y^{\alpha}(e^{ih(x,y)}u(\psi)(r/\rho)^{\tau+n/2})|_{y=x}\tilde{w}(\lambda+in/2)+$$

$$+\frac{1}{\sqrt{2\pi}}\int_{\mathrm{Im}\lambda=\tau}\int_0^1(1-\theta)^N r^a e^{-i\mu g(\phi)}\sum_{|\alpha|=N+1}R_{\alpha}(x,\lambda,\mu,\theta)d\lambda d\theta.$$

Note that (5.6) becomes, by multiplication with $r^a e^{-i\mu g(\phi)}$, a homogeneous function of degree zero (in the variable x). Hence, after applying the Mellin transform $M_{r\to\lambda+in/2}$ to (5.7), the terms at the righthand side with $|\alpha|\leq N$ coincide with

$$\sum_{|\alpha|\leq N}\frac{1}{\alpha!}\Phi^{(\alpha)}(\mu\nabla g(\phi)+\sigma\phi)r^{|\alpha|}\times \tag{5.8}$$

$$\times D_y^\alpha(e^{ih(x,y)}u(\psi)(r/\rho)^{\tau+n/2})|_{y=x}\tilde{w}(\lambda+in/2).$$

In view of the fact that this expression is independent of r (it is a homogeneous function of degree zero) we may assume that $r=1$, hence $|x|=|y|=1$, $\phi=\psi$ (after differentiation). Finally we take into account that

$$e^{ih(x,y)} = e^{i\mu g(\psi)-i\mu g(\phi)-r^{-1}(y-x,\mu\nabla g(\phi)+\sigma\phi)}(r/\rho)^{-i\sigma}.$$

As the result the sum (5.8) takes the form

$$\sum_{|\alpha|\leqslant N}\frac{1}{\alpha!}\Phi^{(\alpha)}(\mu\nabla g(\phi)+\sigma\phi)\times \tag{5.9}$$

$$\times D_y^\alpha[u(\psi)\rho^{i(\lambda+in/2)}e^{i\mu(g(\psi)-g(\phi))}e^{-i(y-\phi,\mu\nabla g(\phi)+\sigma\phi)}]|_{\substack{|y|=1\\ \psi=\phi}}\times$$

$$\times\tilde{w}(\lambda+in/2).$$

Now we have to clarify what happens with the remaining quantities in (5.7) under the action of the Mellin transform. We first consider the lefthand side. Write A as convolution with the function $G=\mathrm{F}^{-1}\Phi$,

$$(Af)(x) = \int G(x-y)f(y)dy.$$

Then

$$A(e^{i\mu g(\psi)}v(r,\rho,\phi,\psi,\mu)) = \frac{1}{\sqrt{2\pi}}\int_{S^{n-1}}u(\psi)e^{i\mu g(\psi)}d\psi\times$$

$$\times\int_{\mathrm{Im}\lambda=\tau}\zeta(\nu\,\mathrm{dist}(\phi,\psi))d\lambda\int_0^\infty\rho^{i(\lambda+in/2)+n-1}\zeta(\nu(1-r/(\rho))G(r\phi-\rho\psi)d\rho.$$

(In this and the previous equality we may assume that Rea is a negative number whose absolute value is sufficiently large, so that all transformations are justified. The final formulas are generalized to the values of a required by analytic extension.) Putting $t=r/\rho$ we obtain

$$\int_0^\infty\rho^{i\lambda+n/2-1}\zeta(\nu(1-r/\rho))G(r\phi-\rho\psi)d\rho =$$

$$= r^{i(\lambda+ia+in/2)}\int_0^\infty G(t\phi-\psi)t^{-i(\lambda+in/2+ia)-1}\zeta(\nu(1-t))dt.$$

This implies that the lefthand side of (5.7) is equal to

$$e^{-i\mu g(\phi)}\frac{1}{\sqrt{2\pi}}\int_{\operatorname{Im}\lambda=\tau} r^{i(\lambda+in/2)}\tilde{w}(\lambda+in/2)d\lambda \times$$

$$\times \int_{S^{n-1}} u(\psi)e^{i\mu g(\psi)}\zeta(\nu\operatorname{dist}(\phi,\psi))d\psi \times$$

$$\times \int_0^\infty G(t\phi-\psi)t^{-i(\lambda+in/2+ia)-1}\zeta(\nu(1-t))dt.$$

After Mellin transformation this expression takes the form

$$e^{-i\mu g(\phi)}\tilde{w}(\lambda+in/2)\times \tag{5.10}$$

$$\times \int_{S^{n-1}} \left[\int_0^\infty G(t\phi-\psi)\zeta(\nu(1-t))t^{-i(\lambda+in/2+ia)-1}dt\right]\times$$

$$\times\, u(\psi)e^{i\mu g(\psi)}\zeta(\nu\operatorname{dist}(\phi,\psi))d\psi.$$

We turn to the remainder in (5.7). By putting

$$\Psi_\alpha(\xi_x,\eta) = (N+1)\int_0^1(1-\theta)^N\Phi^{(\alpha)}(\xi_x+\theta\eta)d\theta$$

and by taking into account that $e^{i\chi(x)} = e^{i\mu g(\phi)}r^{i(\lambda+in/2)}r^{\tau+n/2}$, we can rewrite it as

$$\frac{1}{\sqrt{2\pi}}r^a\int_{\operatorname{Im}\lambda=\tau} r^{i(\lambda+in/2)}\tilde{w}(\lambda+in/2)d\lambda\times \tag{5.11}$$

$$\times\iint e^{i(x-y)\eta+ih(x,y)}u(\psi)\zeta(\nu\operatorname{dist}(\phi,\psi))\zeta(\nu(1-r/\rho))(r/\rho)^{\tau+n/2}\times$$

$$\times\sum_{|\alpha|=N+1}\frac{1}{\alpha!}\eta^\alpha\Psi_\alpha(\xi_x,\eta)dyd\eta.$$

Since $\xi_x = r^{-1}(\mu\nabla g(\phi)+\sigma\phi)$, we have

$$\Psi_\alpha(\xi_x,\eta) = (N+1)r^{|\alpha|-a}\int_0^1(1-\theta)^N\Phi^{(\alpha)}(\mu\nabla g(\phi)+\sigma\phi+\theta\eta r)d\theta =$$

$$= r^{|\alpha|-a}\Psi_\alpha(\mu\nabla g(\phi)+\sigma\phi,\eta r).$$

By denoting ηr by η again and r/ρ by t, we see that (5.11) equals

$$\frac{1}{\sqrt{2\pi}}e^{-i\mu g(\phi)}\int_{\operatorname{Im}\lambda=\tau} r^{i(\lambda+in/2)}\tilde{w}(\lambda+n/2)d\lambda\times$$

$$\times \left\{ \int_{\mathbf{R}^n} \sum_{|\alpha|=N+1} \frac{1}{\alpha!} \eta^\alpha \Psi_\alpha(\mu \nabla g(\phi) + \sigma\phi, \eta) d\eta \times \right.$$

$$\times \int_0^\infty t^{-i(\lambda - n/2)-1} \zeta(\nu(1-t)) dt \times$$

$$\left. \times \int_{S^{n-1}} e^{i\mu g(\psi)} \zeta(\nu \mathrm{dist}(\phi,\psi)) e^{i(\phi - t^{-1}\psi)(\eta + \mu\nabla g(\phi) + \sigma\phi)} d\psi \right\}.$$

The expression between brackets is independent of r. Hence we obtain after Mellin transformation,

$$\tilde{w}(\lambda + in/2) e^{-i\mu g(\phi)} \int_{\mathbf{R}^n} \sum_{|\alpha|=N+1} \frac{1}{\alpha!} \eta^\alpha \Psi_\alpha(\mu\nabla g(\phi) + \sigma\phi, \eta) d\eta \times \tag{5.12}$$

$$\times \int_0^1 t^{-i(\lambda - in/2)-1} \zeta(\nu(1-t)) dt \times$$

$$\times \int_{S^{n-1}} e^{i\mu g(\psi)} u(\psi) \zeta(\nu \mathrm{dist}(\phi,\psi)) e^{i(\phi - t^{-1}\psi)(\eta + \mu\nabla g(\phi) + \sigma\phi)} d\psi.$$

So, as the result of Mellin transformation, the lefthand side of (5.7) coincides with (5.10), and the righthand side transforms to the sum of (5.9) and (5.12). Consequently, after division by $\tilde{w}(\lambda + in/2)$ we are led to

$$e^{-i\mu g(\phi)} \mathfrak{U}_{\psi\to\phi}(\lambda) e^{i\mu g(\psi)} u(\psi) = \sum_{|\alpha| \leqslant N} \frac{1}{\alpha!} \Phi^{(\alpha)}(\mu\nabla g(\phi) + \sigma\phi) \times \tag{5.13}$$

$$\times D_y^\alpha \{ u(\psi) \rho^{i(\lambda + in/2)} e^{i\mu g(\psi) - g(\phi)) - i(y - \phi, \mu\nabla g(\phi) + \sigma\phi)} \} \big|_{\substack{\rho = 1 \\ \phi = \psi}} +$$

$$+ \int_0^1 (1-\theta)^N \sum_{|\alpha|=N+1} \frac{1}{\alpha!} \mathfrak{R}_\alpha(\theta) d\theta + e^{-i\mu g(\phi)} \int_{S^{n-1}} \left[\int_0^\infty G(t\phi - \psi) \times \right.$$

$$\left. \times t^{-i(\lambda + in/2 + ia)-1} [1 - \zeta(\nu \mathrm{dist}(\phi,\psi)) \zeta(\nu(1-t))] dt \right] e^{i\mu g(\psi)} u(\psi) d\psi,$$

where

$$\mathfrak{R}_\alpha(\theta) = \tag{5.14}$$

$$= e^{-i\mu g(\phi)} (N+1) \sum_{|\alpha|=N+1} \int \frac{1}{\alpha!} \eta^\alpha \Phi^{(\alpha)}(\mu\nabla g(\phi) + \sigma\phi + \theta\eta) d\eta \times$$

$$\times \int_0^\infty t^{-i(\lambda - in/2)-1} \zeta(\nu(1-t)) dt \int e^{i(\phi - t^{-1}\psi)(\eta + \mu\nabla g(\phi) + \sigma\phi)} \times$$

$$\times e^{i\mu g(\psi)}\zeta(\nu \mathrm{dist}(\phi,\psi))u(\psi)d\psi.$$

B) We will prove that the last term on the righthand side in (5.13) allows the estimate $O((\mu^2+\sigma^2)^{-q})$ as $\mu^2+\sigma^2 \to \infty$, for any q. Recall that the inner integral (with respect to the variable t) is understood as the analytic extension of a convergent integral, i.e. it equals the sum of the expressions (1.12) and (1.13) in which $\chi_j(t)$ must be replaced by $\chi_j(t)[1-\zeta(\nu \mathrm{dist}(\phi,\psi))\zeta(\nu(1-t))]^j$, $j = 1,2$. We consider one of these expressions, e.g.

$$\frac{(-1)^p}{(a+in/2-i\lambda)\cdots(a+n/2+p-1-i\lambda)}\times \tag{5.15}$$

$$\times \int_{S^{n-1}}\left[\int_0^\infty t^{-i(\lambda+in/2+ia)+p-1}\frac{d^{p-1}}{dt^{p-1}}\times\right.$$

$$\left.\times \{\chi_1(t)[1-\zeta(\nu \mathrm{dist}(\phi,\psi))\zeta(\nu(1-t))]G(t\phi-\psi)\}dt\right]e^{i\mu(g(\psi)-g(\phi))}u(\psi)d\psi,$$

where p can be assumed sufficiently large. We introduce the new variables $s = -\ln t$, $z = g(\psi)$ (without loss of generality we assume that the z variables are local coordinates on the support of the function u). Using the equality

$$e^{-i\sigma\ln t+i\mu g(\psi)} = (-1)^q(\sigma^2+\mu^2)^{-q}\Delta^q_{s,z}e^{i\sigma s+i\mu z},$$

we rewrite (5.15) as

$$\int_{S^{n-1}}\int f(\phi,z,s,\lambda,\mu)(\sigma^2+\mu^2)^{-q}\Delta^q_{s,z}e^{i\sigma s+i\mu z}dsdz;$$

here $f(\phi,z,s,\lambda,\mu) = 0$ for $s < -N$ and sufficiently large N (in view of the presence of the factor $\chi_1(e^{-s})$), while f is rapidly decreasing as $s \to +\infty$ (since p is a large number). Transferring the operator Δ^q to f and recalling that $\nu = (\mu^2+\sigma^2)^{1/4}$ we obtain the estimate required for (5.15). The second expression (containing χ_2) is treated similarly.

C) We now turn to the integral

$$\int_0^1(1-\theta)^N\sum_{|\alpha|=N+1}\frac{1}{\alpha!}\mathfrak{R}_\alpha(\theta)d\theta, \tag{5.16}$$

where $\mathfrak{R}_\alpha$ is defined by (5.14). Represent in (5.14) the function $\Phi^{(\alpha)}$ as a sum $\chi_1\Phi^{(\alpha)}+\chi_2\Phi^{(\alpha)}$, where $\chi_j \in C^\infty(\mathbf{R}^n)$, $\chi_j \geqslant 0$ $(j=1,2)$, $\chi_1+\chi_2 = 1$, and $\chi_1(x) = 1$ for $|x| < 1/2$ and $\chi_1(x) = 0$ for $|x| > 3/4$. Then $\mathfrak{R}_\alpha = \mathfrak{R}^{(1)}_\alpha+\mathfrak{R}^{(2)}_\alpha$, where $\mathfrak{R}_\alpha^{(j)}$ is obtained from $\mathfrak{R}_\alpha$ by replacing $\Phi^{(\alpha)}$ by $\chi_j\Phi^{(\alpha)}$. In this step we estimate the integral (5.16) with $\mathfrak{R}_\alpha^{(2)}$ instead of $\mathfrak{R}_\alpha$.

Put $z = (t^{-1}\psi_1, \cdots, t^{-1}\psi_n) \in \mathbf{R}^n$, $|\psi| = 1$, $t \in \mathbf{R}_+$. We have

$$(N+1)^{-1}\mathfrak{R}_\alpha^{(2)}(\theta) = \tag{5.17}$$

$$= \int \sum_{|\alpha|=N+1} \frac{1}{\alpha!}\eta^\alpha(\chi_2\Phi^{(\alpha)})(\mu\nabla g(\phi)+\sigma\phi+\theta\eta)d\eta \times$$

$$\times \int e^{i(\phi-z)\eta+ih(\phi,z)}|z|^{-\tau-n/2}\zeta(\nu(1-|z|^{-1}))\zeta(\nu\mathrm{dist}(\phi,\psi))u(z)dz,$$

where $h(\phi,z) = \mu(g(\psi)-g(\phi))-(z-\phi)(\mu\nabla g(\phi)+\sigma\phi)+\sigma\ln|z|$, while u is, as before, a homogeneous function of degree zero. Introduce new coordinates $\xi = (\phi-z)\nu$, $\kappa = \eta/\nu$ and rewrite (5.17) as

$$(N+1)^{-1}\mathfrak{R}_\alpha^{(2)}(\theta) = \nu^{N+1}\int e^{i\xi\kappa}(\chi_2\Phi^{(\alpha)})(\mu\nabla g(\phi)+\sigma\phi+\theta\nu\kappa)\times$$

$$\times D_\xi^\alpha\{e^{ih(\phi,z)}\zeta(\nu(1-|z|^{-1}))\zeta(\nu\mathrm{dist}(\phi,\psi))u(z)|z|^{-\tau-n/2}\}d\xi d\kappa,$$

with $z = \phi-\xi\nu^{-1}$. Represent this expression as a sum $(N+1)^{-1}\mathfrak{R}_{\alpha,1}^{(2)}+(N+1)^{-1}\mathfrak{R}_{\alpha,2}^{(2)}$, by regarding $|\kappa|<(2\nu)^{-1}|\mu\nabla g(\phi)+\sigma\phi|$ in the first and $|\kappa|>(2\nu)^{-1}|\mu\nabla g(\phi)+\sigma\phi|$ in the second term. We first consider the first term. Obviously, $|\mu\nabla(\phi)+\sigma\phi+\theta\nu k|>c\nu^2$ for a certain constant $c>0$ and all $\theta \in [0,1]$. Hence

$$|(\chi_2\Phi^{(\alpha)})(\mu\nabla g(\phi)+\sigma\phi+\theta\nu k)| \leqslant c\,|\mu\nabla g(\phi)+\sigma\phi+\theta\nu k|^{\mathrm{Re}\,a-|\alpha|} \leqslant$$

$$\leqslant c\nu^{2(\mathrm{Re}\,a-N-1)} \leqslant c\nu^{2(\mathrm{Re}\,a-N-1)+n+1}(1+|\kappa|)^{-n-1}.$$

This implies the inequality

$$|\mathfrak{R}_{\alpha,1}^{(2)}(\theta)| \leqslant c\nu^{2\mathrm{Re}\,a-N+n}\times$$

$$\times \int |D_\xi^\alpha\{e^{ih(\phi,z)}\zeta(\nu(1-|z|^{-1}))\zeta(\nu\mathrm{dist}(\phi,\psi))|z|^{-\tau-n/2}u(z)\}|d\xi.$$

The estimate $h(\phi,z) = \nu^2O(|\phi-z|^2)$, the definition of ζ, and the relation $D_\xi = -\nu^{-1}D_z$ imply that the derivative

$$D_\xi^\gamma\{e^{ih(\phi,z)}\zeta(\nu(1-|z|^{-1}))\zeta(\nu\mathrm{dist}(\phi,\psi))|z|^{-\tau-n/2}\} \tag{5.19}$$

has support not containing the point $z=0$ and located in a ball $|\xi|\leqslant K$, and remains bounded as $\nu\to\infty$. Together with (5.18) this leads to the inequality

$$|\mathfrak{R}^{(2)}_{\alpha,1}(\theta)| \leqslant c\nu^{2\mathrm{Re}\,a-N+n}\|u;C^{N+1}(S^{n-1})\|. \tag{5.20}$$

Further,

$$(N+1)^{-1}\mathfrak{R}^{(2)}_{\alpha,2}(\theta) =$$

$$= \nu^{N+1}\int e^{i\xi\kappa}|\kappa|^{-2M}(\chi_2\Phi^{(\alpha)})(\mu\nabla g(\phi)+\sigma\phi+\theta\nu\kappa)\times$$

$$\times\Delta_\xi^M D_\xi^\alpha\{e^{ih(\phi,z)}\zeta(\nu(1-|z|^{-1}))\zeta(\nu\mathrm{dist}(\phi,\psi))|z|^{-\tau-n/2}u(z)\}d\xi d\kappa,$$

where M is an arbitrary natural number and integration is over the domain $\{(\kappa,\xi)\colon \xi\in\mathbb{R}^n,\ \ |\kappa|>|\mu\nabla g(\phi)+\sigma\phi|/(2\nu)\}$. Taking into account that $\mathrm{Re}\,a-N-1\leqslant 0$, we obtain

$$|\mathfrak{R}^{(2)}_{\alpha,2}(\theta)| \leqslant c\nu^{N+1}\times$$

$$\times\int_{|\kappa|>c\nu}|\kappa|^{-2M}d\kappa\int|\Delta_\xi^M D_\xi^\alpha\{e^{ih(\phi,z)}\zeta(\nu(1-|z|^{-1}))\times$$

$$\times\zeta(\nu\mathrm{dist}(\phi,\psi))|z|^{-\tau-n/2}u(z)\}|d\xi.$$

Using the property of the derivatives (5.19) already mentioned we can thus derive the inequality

$$|\mathfrak{R}^{(2)}_{\alpha,2}(\theta)| \leqslant c\nu^{-2M+n+N+1}\|u;C^{2M+N+1}(S^{n-1})\| \tag{5.21}$$

for all $\theta\in[0,1]$.

So, choosing first N in (5.20) and then M in (5.21) we see that

$$\left|\int_0^1(1-\theta)^N\sum_{|\alpha|=N+1}\frac{1}{\alpha!}\mathfrak{R}^{(2)}_\alpha(\theta)d\theta\right| \leqslant c\nu^{-2P}\|u;C^Q(S^{n-1})\| \tag{5.22}$$

for any given P and certain Q.

D) We estimate the other 'part' of the integral (5.16), obtained by replacing $\mathfrak{R}_\alpha$ with $\mathfrak{R}^{(1)}_\alpha$. Recall that at the begin of the proof the number $\mathrm{Re}\,a$ was assumed to be large. The estimates in the previous step do not depend on this assumption. In order to finish the proof it suffices to perform an analytic extension with respect to a of (5.16) (with $\mathfrak{R}_\alpha$ replaced by $\mathfrak{R}^{(1)}_\alpha$) and to obtain an estimate analogous to (5.22) for this extension.

If $\mathrm{Re}\,a-N-1>-n$ then, as in C), we have

$$\int_0^1(1-\theta)^N\sum_{|\alpha|=N+1}\frac{1}{\alpha!}\mathfrak{R}^{(1)}_\alpha(\theta)d\theta = \tag{5.23}$$

$$= (N+1)\nu^{N+1}\int_0^1 (1-\theta)^N d\theta \int \sum_{|\alpha|=N+1} \frac{1}{\alpha!} |\kappa|^{-2M}(\chi_1\Phi^{(\alpha)}) \times$$

$$\times (\mu\nabla g(\phi)+\sigma\phi+\theta\nu\kappa)v(x,\nu,\kappa)d\kappa,$$

where $v(x,\nu,\kappa) = \int e^{i\kappa\xi}\Delta_\xi^M D_\xi^\alpha\{e^{ih(\phi,z)}|z|^{-\tau-n/2}\zeta(\nu(1-|z|^{-1}))\zeta(\nu\mathrm{dist}(\phi,\psi))u(z)\}d\xi$. Put $e(\nu) = \mu\nabla g(\phi)+\sigma\phi$ and introduce a new variable $s = e(\nu)+\theta\nu\kappa$, after which the righthand side of (5.23) takes the form

$$(N+1)(|e(\nu)|/\nu)^{-2M}\nu^{N+1-n}\int_0^1 (1-\theta)^N\theta^{2M-n}d\theta \times \tag{5.24}$$

$$\times \int \sum_{|\alpha|=N+1} \frac{1}{\alpha!}(\chi_1\Phi^{(\alpha)})(s)w(x,\nu,\theta,s)ds,$$

where $w(x,\nu,\theta,s) = |1-s/|e(\nu)||^{-2M}v(x,\nu,(s-e(\nu))\theta^{-1}\nu^{-1})$. The inner integral is equal to

$$(\Gamma+1)\int \sum_{|\alpha|=N+1} \frac{1}{\alpha!}(\chi_1\Phi^{(\alpha)})(s) \sum_{|\gamma|=\Gamma+1} s^\gamma ds \times \tag{5.25}$$

$$\times \int_0^1 (1-t)^\gamma \frac{1}{\gamma!} w^{(\gamma)}(x,\nu,\theta,ts)dt +$$

$$+ \sum_{|\alpha|=N+1} \sum_{|\gamma|=0} \frac{1}{\alpha!\gamma!} w^{(\gamma)}(x,\nu,\theta,0)\int (\chi_1\Phi^{(\alpha)})(s)s^\gamma ds,$$

where $w^{(\gamma)}$ is the derivative of $w(x,\nu,\theta,s)$ with respect to s.

The first integral converges for $\mathrm{Re}\,\alpha - N+1+\Gamma > -n$, while the integrals of the product $(\chi_1\Phi^{(\alpha)})(s)s^\gamma$ can be extended in the usual way as meromorphic functions to the whole a-plane (see, e.g., [13]). Thus, substituting (5.25) in (5.24) we obtain the looked-for analytic extension of the integral (5.16) (with $\mathcal{R}_\alpha^{(1)}$ instead of $\mathcal{R}_\alpha$).

The properties of the derivatives (5.19) (indicated after (5.19)) ensure the estimate

$$|v_\kappa^{(\gamma)}(x,\nu,\kappa)| \leq c\|u;C^{2M+N+1}(S^{n-1})\|.$$

Hence

$$|w_\kappa^{(\gamma)}(x,\nu,\theta,s)| \leqslant c\|u;C^{2M+N+1}(S^{n-1})\|\nu^{-|\gamma|}\theta^{-|\gamma|} \tag{5.26}$$

for $|s| \leqslant 1$. Note that $c_1\nu \leqslant |e(\nu)|/\nu \leqslant c_2\nu$ for $\nu \geqslant 1$ (with positive constants c_1 and c_2). Choose Γ such that the inequality $\mathrm{Re}\,a-N-1-\Gamma > -n$ holds for a given a. Combining (5.23) - (5.26) we obtain for sufficiently large M relation (5.22) with $\mathfrak{R}_\alpha^{(2)}$ replaced by $\mathfrak{R}_\alpha^{(1)}$. This implies that (5.22) remains valid if $\mathfrak{R}_\alpha$ is written instead of $\mathfrak{R}_\alpha^{(2)}$. It remains to use the estimate proved in part B) and formula (5.13). ■

§6. Estimates for meromorphic pseudodifferential operators

In this paragraph we will prove analogs for meromorphic pseudodifferential operators of inequalities which are well-known for classic pseudodifferential operators. An analogue of Gårding's inequality will be established, the essential norm of an operator will be computed, etc.

T h e o r e m 6.1 (generalized Gårding's inequality). *Let* $\mathfrak{U}(\lambda) = \mathrm{E}_{\theta\to\phi}(\lambda+ia)^{-1}\Phi(\phi,\theta)\mathrm{E}_{\psi\to\theta}(\lambda)$ *be a canonical meromorphic pseudodifferential operator, let* a *be a real number, let* $\mathrm{Re}\,\Phi(\phi,\theta) \geqslant c = \mathrm{const}$ *for* $\phi,\theta \in S^{n-1}$. *Then for every positive number* ϵ *and every set* $\mathfrak{F}$ *that is admissible for* $\mathfrak{U}$ *there is a constant* $c' = c'(\epsilon,\mathfrak{F})$ *such that the inequality*

$$\mathrm{Re}(\mathfrak{U}(\lambda)u,u)+c'\|u;H^{(a-1)/2}(\lambda,S^{n-1})\|^2 \geqslant \tag{6.1}$$

$$\geqslant (c-\epsilon)\|\mathrm{E}(\sigma-ia/2)u;H^0(S^{n-1})\|^2$$

holds for all $\lambda \in \mathfrak{F}$ *and* $u \in C^\infty(S^{n-1})$; *here* $\sigma = \mathrm{Re}\lambda$ *if* $|\mathrm{Re}\lambda| \geqslant 1$ *and* $\sigma = 1$ *if* $|\mathrm{Re}\lambda| < 1$.

(Taking $\|\mathrm{E}(\sigma-ia/2)u;H^0(S^{n-1})\|$ as (equivalent) norm in $H^{a/2}(\lambda,S^{n-1})$, relation (6.1) can be given a form better resembling the usual Gårding inequality.)

P r o o f. Denote by $\mathfrak{U}_I(\lambda)$ the canonical meromorphic pseudodifferential operator with symbol $i\,\mathrm{Im}\,\Phi(\phi,\theta)$. Then the adjoint operator $\mathfrak{U}_I(\lambda)^*$ equals, by proposition 2.5, $-i\mathrm{E}(\bar\lambda+ia)^{-1}\mathrm{Im}\,\Phi(\phi,\theta)\mathrm{E}(\bar\lambda)$ (up to terms of order $a-1$). Applying proposition 2.2 and again neglecting terms whose orders do not exceed $a-1$, we may change $\bar\lambda$ to λ in this formula. So, up to such terms, $\mathfrak{U}_I(\lambda)^* = -i\mathrm{E}(\lambda+ia)^{-1}\times$ $\times\,\mathrm{Im}\,\Phi(\phi,\theta)\mathrm{E}(\lambda)$. This and the equality $2\mathrm{Re}(\mathfrak{U}_I(\lambda)u,u) = (\mathfrak{U}(\lambda)u,u)+(\mathfrak{U}_I(\lambda)^*u,u)$

implies the estimate

$$|\operatorname{Re}(\mathfrak{U}_I(\lambda)u,u)| \leqslant c(\mathfrak{F})\|u;H^{(a-1)/2}(\lambda,S^{n-1})\|^2$$

on every admissible set $\mathfrak{F}$. Hence the symbol Φ may be regarded as being real in the proof of (6.1). Moreover, proposition 2.2 implies that it suffices to check (6.1) for real λ satisfying $|\lambda| \geqslant 1$.

We denote by Ψ the positive function defined by $\Psi^2(\phi,\theta) = \Phi(\phi,\theta)-c+\epsilon$. Let $\mathfrak{B}(\lambda)$ be the canonical meromorphic pseudodifferential operator of order $a/2$ with symbol Ψ. The operator $\mathfrak{B}(\lambda)^*$ coincides (up to terms of order $a/2-1$) with the operator $E(\lambda+ia/2)^{-1}\Psi(\phi,\theta)E(\lambda)$. By theorem 3.4 the asymptotic series for $\mathfrak{B}(\lambda)^*\mathfrak{B}(\lambda)$ has as initial term $E(\lambda+ia)^{-1}\Psi^2(\phi,\theta)E(\lambda)$, while the second term is an operator of order $a-1$. Hence the order of

$$\mathfrak{U}(\lambda)-(c-\epsilon)E(\lambda+ia)^{-1}E(\lambda)-\mathfrak{B}(\lambda)^*\mathfrak{B}(\lambda)$$

is also $a-1$. Since the symbol of $\mathfrak{U}(\lambda)$ is real, proposition 2.5 applied to $\mathfrak{U}(\lambda)$ leads to the inequality

$$\operatorname{Re}(\mathfrak{U}(\lambda)u,u)-(c-\epsilon)(E(\lambda+ia/2)^{-1}E(\lambda-ia/2)u,u)-(\mathfrak{B}(\lambda)^*\mathfrak{B}(\lambda)u,u) \geqslant$$
$$\geqslant c'\|u;H^{(a-1)/2}(\lambda,S^{n-1})\|^2$$

(in order to replace $E(\lambda+ia)^{-1}E(\lambda)$ by $E(\lambda+ia/2)^{-1}E(\lambda-ia/2)$ we use proposition 2.2). It remains to note that $(E(\lambda+ia/2)^{-1})^* = E(\lambda-ia/2)$ (proposition 1.4.4) and that $(\mathfrak{B}(\lambda)^*\mathfrak{B}(\lambda)u,u) = \|\mathfrak{B}(\lambda)u;H^0(S^{n-1})\|^2 \geqslant 0$. ■

T h e o r e m 6.2. *Let* $\Phi \in C^\infty(S^{n-1}\times S^{n-1})$, $K = \|\Phi;C(S^{n-1}\times S^{n-1})\|$, *and* $\mathfrak{U}(\lambda) = E(\lambda+ia)^{-1}\Phi(\phi,\theta)E(\lambda)$, *where a is a real number. Then for every positive number ϵ and every set $\mathfrak{F}$ that is admissible for $\mathfrak{U}$ there is a constant c such that*

$$\|\mathfrak{U}(\lambda)u;H^0(S^{n-1})\| \leqslant \tag{6.2}$$
$$\leqslant (K+\epsilon)\|E(\sigma-ia)u;H^0(S^{n-1})\|+c\,\|u;H^{a-1/2}(\lambda,S^{n-1})\|;$$

here, as before, $\sigma = \operatorname{Re}\lambda$ if $|\operatorname{Re}\lambda| \geqslant 1$, and $\sigma = 1$ if $|\operatorname{Re}\lambda| < 1$.

P r o o f. Proposition 2.2 implies that it suffices to check (6.2) for real λ satisfying $|\lambda| \geqslant 1$ only. Put $\Psi(\phi,\theta) = \overline{\Phi(\phi,\theta)}\Phi(\phi,\theta)$ and $\mathfrak{B}(\lambda) = E(\lambda+ia)^{-1}\Psi(\phi,\theta)E(\lambda-ia)$. By theorem 3.4 and propositions 2.2 and 2.5, the order of $\mathfrak{B}(\lambda)-\mathfrak{U}(\lambda)^*\mathfrak{U}(\lambda)$ is

$2a-1$. Hence

$$\mathrm{Re}(\mathfrak{B}(\lambda)u,u)-\|\mathfrak{U}(\lambda)u;H^0(S^{n-1})\|^2 \geqslant c\|u;H^{a-1/2}(\lambda,S^{n-1})\|^2. \qquad (6.3)$$

Since $K^2-\Psi(\phi,\theta)\geqslant 0$ for $\phi,\theta\in S^{n-1}$, by applying theorem 3.1 to $\mathrm{E}(\lambda+ia)^{-1}(K^2-\Psi(\phi,\theta))\mathrm{E}(\lambda-ia)$ we obtain the inequality

$$K^2\mathrm{Re}(\mathrm{E}(\lambda+ia)^{-1}\mathrm{E}(\lambda-ia)u,u)-\mathrm{Re}(\mathfrak{B}(\lambda)u,u)+ \qquad (6.4)$$

$$+c'\|u;H^{a-1/2}(\lambda,S^{n-1})\|\geqslant -\epsilon\|\mathrm{E}(\sigma-ia)u;H^0(S^{n-1})\|^2.$$

Since

$$(\mathrm{E}(\lambda+ia)^{-1}\mathrm{E}(\lambda-ia)u,u) = \|\mathrm{E}(\lambda-ia)u;H^0(S^{n-1})\|^2,$$

we finish the proof by combining formulas (6.3) and (6.4). ■

R e m a r k 6.3. If $a\neq\pm(k+n/2)$, where $k = 0,1,\cdots$, then we may put $\sigma = \mathrm{Re}\lambda$ in (6.1) and (6.2) also if $|\mathrm{Re}\lambda|<1$.

P r o p o s i t i o n 6.4. *Let $\mathfrak{U}(\lambda)$ be a canonical meromorphic pseudodifferential operator of order a, with symbol Φ, and let $c_0 = |\Phi(\phi_0,\theta_0)|$, with (ϕ_0,θ_0) an arbitrary point of $S^{n-1}\times S^{n-1}$. Then for any sequence $\{\epsilon_j\}$ of positive arbitrarily small numbers there are a sequence $\{\lambda_j\}$ of complex numbers and a sequence $\{u_j\}$ of $C^\infty(S^{n-1})$ functions such that* $\mathrm{Im}\lambda_j = \tau$ *(τ a given number), $|\lambda_j|\to\infty$, and*

$$\big|\,\|\mathfrak{U}(\lambda_j)u;H^0(S^{n-1})\|-c_0\|\mathrm{E}(\sigma_j-ia)u_j;H^0(S^{n-1})\|\,\big|\leqslant \qquad (6.5)$$

$$\leqslant\epsilon_j\|u_j;H^a(\lambda_j,S^{n-1})\|,$$

where $\sigma_j = \mathrm{Re}\lambda_j$.

P r o o f. First of all we note that it suffices to prove (6.5) for $\tau = 0$. Indeed, for $\lambda = \sigma+i\tau$ we have $\mathfrak{U}(\lambda) = \mathfrak{U}(\sigma)+\mathrm{T}(\lambda)$, where $\mathrm{T}(\lambda):H^s(\lambda,S^{n-1})\to H^{s-a+1}(\lambda,S^{n-1})$ is a continuous map with bounded norm on the line $\mathrm{Im}\lambda = \tau$ (proposition 2.2). Further, $\|\mathrm{T}(\lambda_j)u;H^0(S^{n-1})\|\leqslant c\|u;H^{a-1}(\lambda_j,S^{n-1})\|\leqslant c|\lambda_j|^{-1}\|u;H^a(\lambda_j,S^{n-1})\|$. This and the truth of the proposition for $\tau = 0$ imply that the proposition holds for all τ.

Let first $\mathfrak{U}$ be a canonical meromorphic pseudodifferential operator of order zero. Choose a real-valued function g homogeneous of degree zero such that the vector $\nabla g(\phi_0)$ lies in the (ϕ_0,θ_0)-plane and is nonzero. Obviously, the vectors ϕ_0 and $\nabla g(\phi_0)$ are orthogonal. Expand θ_0 with respect to the vectors ϕ_0 and $\nabla g(\phi_0)$, $\theta_0 = \mu_0 \nabla g(\phi_0) + \sigma_0 \phi_0$, and put $\mu = t\mu_0$, $\sigma = t\sigma_0$, $\lambda = \sigma + i\tau$. By applying theorem 5.1 we find that for a function $u \in C^\infty(S^{n-1})$ with support in a small neighborhood of ϕ_0 the inequality

$$|e^{-i\mu g(\phi)} \mathfrak{U}_{\psi\to\phi}(\lambda) e^{i\mu g(\psi)} u(\psi) - \Phi(\phi_0,\theta_0) u(\phi)| < \epsilon$$

holds for sufficiently large positive t and given $\epsilon > 0$. This means that for any sequence $\{\epsilon_j\}$ of arbitrarily small numbers there are a sequence $\{u_j\} \subset C^\infty(S^{n-1})$ and a sequence $\{t_j\}$ of arbitrarily large positive numbers such that $\|u_j; H^0(S^{n-1})\| = 1$ and

$$|e^{-i\mu_j g(\phi)} \mathfrak{U}_{\psi\to\phi}(\lambda_j) e^{i\mu_j g(\psi)} u_j(\psi) - \Phi(\phi_0,\theta_0) u_j| < \epsilon_j,$$

where $\mu_j = t_j \mu_0$, $\sigma_j = t_j \sigma_0$, $\lambda_j = \sigma_j + i\tau$. This implies that for the sequence $\{v_j = e^{i\mu_j g} u_j\}$ the relation

$$\|\mathfrak{U}(\lambda_j) v_j - \Phi(\phi_0,\theta_0) v_j; H^0(S^{n-1})\| \leqslant \epsilon_j \|v_j; H^0(S^{n-1})\| \tag{6.6}$$

holds, which immediately leads to (6.5) with $a = 0$.

Let now a be an arbitrary number. Using proposition 2.2 the operator $\mathfrak{U}(\lambda) = \mathrm{E}(\lambda + ia)\Phi(\phi,\theta)\mathrm{E}(\lambda)$ is written as

$$\mathfrak{U}(\lambda) = \mathrm{E}(\lambda)^{-1} \Phi(\phi,\theta) \mathrm{E}(\lambda - ia) + \mathrm{T}(\lambda),$$

where $\operatorname{Im}\lambda = 0$, while the operator T allows the estimate $\|\mathrm{T}(\lambda); H^s(\lambda, S^{n-1}) \to H^{s-a+1}(\lambda, S^{n-1})\| \leqslant$ const. This allows us to check (6.5) for $\mathrm{E}(\lambda)^{-1}\Phi(\phi,\theta)\mathrm{E}(\lambda - ia)$ (cf. the begin of the proof). Clearly, the latter operator equals $\mathfrak{U}_0(\lambda)\mathrm{E}(\lambda)^{-1}\mathrm{E}(\lambda - ia)$, where $\mathfrak{U}_0(\lambda) = \mathrm{E}(\lambda)^{-1}\Phi(\phi,\theta)\mathrm{E}(\lambda)$ is a canonical meromorphic pseudodifferential operator of order zero. Put $w_j = \mathrm{E}(\lambda_j - ia)^{-1}\mathrm{E}(\lambda_j) v_j$, where λ_j and v_j are the same as in (6.6). Since $\mathrm{E}(\lambda_j)$ is unitary for $\operatorname{Im}\lambda_j = 0$ we have

$$\begin{aligned} &| \|\mathfrak{U}(\lambda_j) w_j; H^0(S^{n-1})\| - c_0 \|\mathrm{E}(\lambda_j - ia) w_j; H^0(S^{n-1})\| | \leqslant \\ &\leqslant \|(\mathfrak{U}_0(\lambda_j) - \Phi(\phi_0,\theta_0)) \mathrm{E}(\lambda_j)^{-1} \mathrm{E}(\lambda_j - ia) w_j; H^0(S^{n-1})\| = \\ &= \|(\mathfrak{U}_0(\lambda_j) - \Phi(\phi_0,\theta_0)) v_j; H^0(S^{n-1})\|. \end{aligned}$$

Using formula (6.6) for $\mathfrak{U}_0$ (it holds for this operator) we are led to the estimate

$$\left| \|\mathfrak{U}(\lambda_j)w_j;H^0(S^{n-1})\| - c_0\|\mathrm{E}(\lambda_j - ia)w_j;H^0(S^{n-1})\| \right| \leqslant$$
$$\leqslant \epsilon_j\|v_j;H^0(S^{n-1})\|.$$

Note that

$$\|v_j;H^0(S^{n-1})\| = \|\mathrm{E}(\lambda_j)^{-1}\mathrm{E}(\lambda_j - ia)w_j;H^0(S^{n-1})\| =$$
$$= \|\mathrm{E}(\lambda_j - ia)w_j;H^0(S^{n-1})\| \leqslant c\|w_j;H^a(\lambda_j,S^{n-1})\|$$

(proposition 1.5.5). Hence $\{w_j\}$ can play the role of the sequence whose existence is asserted in the theorem. ■

Theorem 6.5. *Let* $\mathfrak{U}(\lambda) = \mathrm{E}(\lambda+ia)^{-1}\Phi(\phi,\theta)\mathrm{E}(\lambda)$, *let a be a real number, and let* $K = \|\Phi;C(S^{n-1},S^{n-1})\|$. *Assume also that the line* $\{\lambda:\mathrm{Im}\lambda=\tau\}$ *does not contain poles of* $\mathfrak{U}$. *Then for every positive* ϵ *there is a uniformly bounded operator* $\mathfrak{J}(\lambda):H^s(\lambda,S^{n-1})\to H^{s-a+1}(\lambda,S^{n-1})$ *such that on this line*

$$\|(\mathfrak{U}(\lambda)-\mathfrak{J}(\lambda))u;H^0(S^{n-1})\| \leqslant (K+\epsilon)\|\mathrm{E}(\sigma-ia)u;H^0(S^{n-1})\|, \qquad (6.7)$$

where σ *is defined as in theorems 6.1, 6.2 (cf. remark 6.3); the constant K is sharp, i.e. it cannot be replaced by a smaller number while preserving inequality* (6.7).

Proof. Proposition 2.2 implies that it suffices to prove (6.7) for $\mathrm{Im}\lambda = 0$.

Denote by χ a function from $C_0^\infty(\mathbb{R})$ such that $\chi(t) = 1$ for $|t|<N$ and $\chi(t) = 0$ for $|t|\geqslant 2N$. Introduce the operator $\hat{\chi}(\lambda)$ by

$$\hat{\chi}(\lambda)u = \sum_{m=0}^{\infty}\sum_{k=1}^{k_m}\chi(m^2+\lambda^2)u_{mk}Y_{mk}$$

and put $\mathfrak{J}(\lambda) = \mathfrak{U}(\lambda)\hat{\chi}(\lambda)$. By theorem 2.2 we have

$$\|(\mathfrak{U}(\lambda)-\mathfrak{J}(\lambda))u;H^0(S^{n-1})\| = \|\mathfrak{U}(\lambda)(1-\hat{\chi}(\lambda))u;H^0(S^{n-1})\| \leqslant \qquad (6.8)$$
$$\leqslant (K+\epsilon/2)\|\mathrm{E}(\sigma-ia)(1-\hat{\chi}(\lambda))u;H^0(S^{n-1})\| +$$
$$+c\|(1-\hat{\chi}(\lambda))u;H^{a-1/2}(\lambda,S^{n-1})\|.$$

Taking into account that $\mathrm{E}(\lambda)$ has the representation

$$\mathrm{E}(\lambda) = \sum_{m,k} \mu_m(\lambda) u_{mk} Y_{mk},$$

where the μ_m are complex-valued functions, we obtain

$$\|\mathrm{E}(\sigma - ia)(1-\hat{\chi}(\lambda))u; H^0(S^{n-1})\| =$$
$$= \|(1-\hat{\chi}(\lambda))\mathrm{E}(\sigma - ia)u; H^0(S^{n-1})\| \leqslant \|\mathrm{E}(\sigma - ia)u; H^0(S^{n-1}))\|.$$

The definitions of the norm and the function χ imply the estimate

$$\|(1-\hat{\chi}(\lambda))u; H^{a-1/2}(\lambda, S^{n-1})\| \leqslant (1+N^2)^{-1/4}\|u; H^a(\lambda, S^{n-1})\|. \quad (6.9)$$

By proposition 1.5.5,

$$\|u; H^a(\lambda, S^{n-1})\| \leqslant c\|\mathrm{E}(\sigma - ia)u; H^0(S^{n-1})\|.$$

This and (6.8), (6.9) imply for sufficiently large N inequality (6.7).

It remains to convince ourselves that K is sharp. Suppose this is not so, and that for some operator T inequality (6.7) holds with K replaced by $Q < K$. Then we have for the sequence $\{u_j\}$ from proposition 6.4,

$$(c_0 - \epsilon_j)\|\mathrm{E}(\lambda_j - ia)u_j; H^0(S^{n-1})\| \leqslant$$
$$\leqslant \|\mathfrak{U}(\lambda_j)u; H^0(S^{n-1})\| \leqslant$$
$$\leqslant \|(\mathfrak{U}(\lambda_j)\mathfrak{I}(\lambda_j))u_j; H^0(S^{n-1})\| + \|\mathfrak{I}(\lambda_j)u; H^0(S^{n-1})\| \leqslant$$
$$\leqslant (Q+\epsilon)\|\mathrm{E}(\lambda_j - ia)u_j; H^0(S^{n-1})\| + c\|u_j; H^{a-1}(\lambda, S^{n-1})\|.$$

From the definition of the norms in the spaces $H^s(\lambda, S^{n-1})$ and proposition 1.5.5 we obtain that for sufficiently large j,

$$c\|u_j; H^{a-1}(\lambda_j, S^{n-1})\| \leqslant c|\lambda_j|^{-1}\|u; H^a(\lambda_j, S^{n-1})\| \leqslant$$
$$\leqslant \epsilon\|\mathrm{E}(\lambda_j - ia)u_j; H^0(S^{n-1})\|.$$

Thus,

$$(c_0 - \epsilon)\|\mathrm{E}(\lambda_j - ia)u_j; H^0(S^{n-1})\| \leqslant (Q+2\epsilon)\|\mathrm{E}(\lambda_j - ia)u_j; H^0(S^{n-1})\|.$$

Taking $c_0 > Q$, the number j large, and ϵ small, we obtain a contradiction. ∎

§7. Periodic meromorphic pseudodifferential operators

A canonical meromorphic pseudodifferential operator arises as the result of applying the Mellin transform to an operator $F^{-1}_{\xi\to x}\Phi(\phi,\xi)F_{y\to\xi}$, where $\phi = x/|x|$ (cf., e.g., (2.3.6)). Dependence of the symbol on ϕ means that it has a discontinuity 'of the first kind' at the coordinate origin (the limit as $x\to 0$ depends on the direction of approach). Now we consider an operator $F^{-1}_{\xi\to x}\Phi(r,\phi,\xi)F_{y\to\xi}$, $r = |x|$. We will assume that for any value of $r>0$ the equality $\Phi(re,\phi,\xi) = \Phi(r,\phi,\xi)$ holds, where e is the base of the natural logarithm. Thus, Φ can have a discontinuity 'of the second kind' at $x=0$. Applying a certain integral transform to $F^{-1}_{\xi\to x}\Phi(r,\phi,\xi)F_{y\to\xi}$ will lead to a periodic meromorphic pseudodifferential operator. This transform is obtained by a change of variable from the transform used by I.M. Gel'fand in the study of differential operators with periodic coefficients [15] (cf. also [42]).

We will first describe this integral transform. Using it we are led to a periodic meromorphic pseudodifferential operator, starting from an operator $F^{-1}\Phi(r,\phi,\xi)F$. We then establish for periodic meromorphic pseudodifferential operators analogs of the results given in §3 for meromorphic pseudodifferential operators (concerning composition, transition to the adjoint operator, etc.).

1. The transform G. For a $v\in C_0^\infty(\mathbb{R}_+)$ we put

$$(Gv)(\mu,r) = \frac{1}{(2\pi)^{1/2}}\sum_{j=-\infty}^{+\infty}(re^j)^{-i\mu}v(re^j), \tag{7.1}$$

where μ is an arbitrary complex number, and $0<r<\infty$. It is obvious that $V(\mu,r) = (Gv)(\mu,r)$ is a 'periodic' function, i.e. $V(\mu,re) = V(\mu,r)$. The inversion formula for (7.1) can be verified immediately:

$$v(r) = (G^{-1}V)(r) = \frac{1}{(2\pi)^{1/2}}\int_{I_\tau} r^{i\mu}V(\mu,r)d\mu; \tag{7.2}$$

where by I_τ we denote the set $\{\mu\in\mathbb{C}\colon \operatorname{Im}\mu=\tau,\ 0\leqslant\operatorname{Re}\mu\leqslant 2\pi\}$.

A Plancherel formula holds for G:

$$\int_{I_\tau} d\mu\int_1^e |Gv(\mu,r)|^2\frac{dr}{r} = \int_0^\infty |v(r)|^2 r^{2\tau-1}dr. \tag{7.3}$$

Let's prove this formula. Without loss of generality we may assume that $\tau=0$.

Putting $\psi(t) = v(e^t)$ we can rewrite (7.3) as

$$\int_0^{2\pi} d\mu \int_0^1 |\Psi(\mu,t)|^2 dt = \int_{-\infty}^{+\infty} |\Psi(t)|^2 dt,$$

where

$$\Psi(\mu,t) = \frac{1}{(2\pi)^{1/2}} \sum_{j=-\infty}^{+\infty} e^{-i\mu(t+j)}\psi(t+j).$$

Clearly, $\Psi(\mu,t+1) = \Psi(\mu,t)$. Denote by $\Psi_n(\mu)$ the Fourier coefficients of $\Psi(\mu,\cdot)$:

$$\Psi_n(\mu) = \int_0^1 \Psi(\mu,t)e^{-2\pi int}dt =$$

$$= \frac{1}{(2\pi)^{1/2}} \int_0^1 \sum_{j=-\infty}^{+\infty} \psi(t+j)e^{-i\mu(t+j)-2\pi int}dt =$$

$$= \frac{1}{(2\pi)^{1/2}} \int_{-\infty}^{+\infty} \psi(t)e^{-i(\mu+2\pi n)t}dt.$$

We have

$$\int_0^{2\pi} d\mu \int_0^1 |\Psi(\mu,t)|^2 dt = \sum_{n=-\infty}^{+\infty} \int_0^{2\pi} |\Psi_n(\mu)|^2 d\mu =$$

$$= \frac{1}{2\pi} \int_{-\infty}^{+\infty} \left| \int_{-\infty}^{+\infty} \psi(t)e^{-i\mu t}dt \right|^2 d\mu = \int_{-\infty}^{+\infty} |\psi(t)|^2 dt,$$

which was required.

2. Canonical operators. In chapter 2 we have proved that the equation

$$F^{-1}_{\xi\to x}\Phi(\xi)F_{y\to\xi}v(y) = \frac{1}{\sqrt{2\pi}} \int_{\operatorname{Im}\lambda=\beta-s} r^{i(\lambda+ia+in/2)}\mathfrak{U}(\lambda)\tilde{v}(\lambda+in/2,\cdot)d\lambda,$$

holds on an everywhere dense set (in $H^s_\beta(\mathbb{R}^n)$). Here $\mathfrak{U}(\lambda) = E_{\theta\to\phi}(\lambda+ia)^{-1}\Phi(\theta)E_{\psi\to\theta}(\lambda)$, $r = |x|$, $\psi = y/|y|$, $\phi = x/|x|$. Taking into consideration operators of more general nature we introduce the operator

$$(Av)(x) = \frac{1}{\sqrt{2\pi}} \int_{\operatorname{Im}\lambda=\tau} r^{i(\lambda+ia+in/2)}\mathfrak{U}_{\psi\to\phi}(r,\lambda)\tilde{v}(\lambda+in/2,\psi)d\lambda; \qquad (7.4)$$

here $\mathfrak{U}(r,\lambda) = \mathrm{E}(\lambda+ia)^{-1}\Phi(r,\phi,\theta)\mathrm{E}(\lambda)$, $\Phi(re,\cdot) = \Phi(r,\cdot)$.

P r o p o s i t i o n 7.1. *For* $v \in C_0^\infty(\mathbf{R}^n \setminus 0)$ *the equation*

$$\mathrm{A}v = (\mathrm{G}_{r\to(r,\mu+ia+in/2)})^{-1}\mathrm{A}(r,\mu)\mathrm{G}_{r\to(r,\mu+in/2)}v \tag{7.5}$$

holds, where $\mathrm{Im}\,\mu = \tau$,

$$\mathrm{A}_{\psi\to\phi}(r,\mu)V(\mu+in/2,r,\psi) = \tag{7.6}$$

$$= \sum_{j=-\infty}^{+\infty} r^{2\pi ij}\mathfrak{U}_{\psi\to\phi}(r,\mu+2\pi j)V_j(\mu+in/2,\psi),$$

$$V_j(v,\psi) = \int_1^e r^{-2\pi ij}V(v,r,\psi)r^{-1}dr, \quad V(v,r,\psi) = G_{r\to(r,v)}v(r,\psi).$$

P r o o f. Using the 'periodicity' of Φ in r and formulas (7.1), (7.4) we find

$$\mathrm{G}_{r\to(r,\mu+ia+in/2)}\mathrm{A}v = \tag{7.7}$$

$$= \frac{1}{2\pi}\sum_{j=-\infty}^{+\infty}(re^j)^{-i\mu}\int_{\mathrm{Im}\lambda=\tau}(re^j)^{i\lambda}\mathfrak{U}(r,\lambda)v(\lambda+in/2,\cdot)d\lambda =$$

$$= \frac{1}{2\pi}\int_{\mathrm{Im}\lambda=\tau} r^{i(\lambda-\mu)}\sum_{j=-\infty}^{+\infty}e^{i(\lambda-\mu)j}\mathfrak{U}(r,\lambda)\tilde{v}(\lambda+in/2,\cdot)d\lambda.$$

It is well-known (cf. e.g., [13]) that

$$\sum_{j=-\infty}^{+\infty}e^{i(\lambda-\mu)j} = 2\pi\sum_{j=-\infty}^{+\infty}\delta(\lambda-\mu-2\pi j).$$

Substituting this in (7.7) we obtain

$$\mathrm{G}_{r\to(r,\mu+ia+in/2)}\mathrm{A}v = \tag{7.8}$$

$$= \sum_{j=-\infty}^{+\infty}r^{2\pi ij}\mathfrak{U}(r,\mu+2\pi j)\tilde{v}(\mu+2\pi j+in/2,\cdot).$$

The definition of Mellin transform and formula (7.2) imply that

$$\tilde{v}(\mu,\cdot) = \frac{1}{2\pi}\int_0^\infty r^{-i\mu-1}dr\int_{I_\tau} r^{i\lambda}V(\lambda,r,\cdot)d\lambda = \tag{7.9}$$

$$= \frac{1}{2\pi} \int_{I_\tau} d\lambda \int_0^\infty r^{-i(\mu-\lambda)-1} V(\lambda, r, \cdot) d\lambda.$$

Expanding V in a series,

$$V(\lambda, r, \cdot) = \sum_{k=-\infty}^{+\infty} V_k(\lambda, \cdot) r^{2\pi i k},$$

we obtain

$$\int_0^\infty r^{-i(\mu-\lambda)-1} V(\lambda, r, \cdot) dr = 2\pi \sum_{k=-\infty}^{+\infty} V_k(\lambda, \cdot) \delta(\lambda - \mu + 2k\pi).$$

Hence (7.9) implies $\tilde{v}(\mu, \cdot) = V_k(\mu - 2k\pi, \cdot)$, where k is such that $\operatorname{Re}\mu - 2k\pi \in [0, 2\pi]$. Hence $\tilde{v}(\mu + 2\pi j + in/2, \cdot) = V_j(\mu + in/2, \cdot)$. Now formula (7.8) takes the form

$$\mathbf{G}_{r \to (r, \mu + ia + in/2)} \mathbf{A} v = \sum_{j=-\infty}^{+\infty} r^{2\pi i j} \mathfrak{U}(r, \mu + 2\pi j) V_j(\mu + in/2, \cdot).$$

From this we can derive (7.5) immediately. ■

Denote by $H^s(\Pi)$, where $\Pi = [1, e] \times S^{n-1}$, the Sobolev-Slobodetskii space obtained by completion of the set of smooth functions V satisfying the condition $V(1, \theta) = V(e, \theta)$, $\theta \in S^{n-1}$.

P r o p o s i t i o n 7.2. *Let $\mathfrak{F}$ be an arbitrary closed set lying within the rectangle* $\{\mu: 0 \leq \operatorname{Re}\mu \leq 2\pi, \ |\operatorname{Im}\mu| \leq h\}$, $h < \infty$, *and not containing the points* $2\pi j - i(a + k + n/2)$, $2\pi j + i(k + n/2)$, *where* $k = 0, 1, \cdots$, $j = 0, \pm 1, \cdots$. *Then the estimate*

$$\|\mathbf{A}(\cdot, \mu) V; H^{s - \operatorname{Re} a}(\Pi)\| \leq c(\mathfrak{F}) \|V; H^s(\Pi)\| \tag{7.10}$$

holds for the operator $\mathbf{A}$ *defined by* (7.6) *on the elements* $V \in H^s(\Pi)$.

P r o o f. If $\mathfrak{U}$ is independent of r (i.e. Φ is independent of r), then (7.10) follows from proposition 1.3. This estimate can be generalized to the case of an arbitrary function $\Phi \in C^\infty([1, e] \times S^{n-1} \times S^{n-1})$ satisfying $\Phi(r, \cdot) = \Phi(re, \cdot)$ using the series expansion $\Phi(r, \phi, \theta) = \Sigma_j r^{2\pi i j} \Phi_j(\phi, \theta)$ (cf. the proof of proposition 1.3). ■

The operator $\mathbf{A}$ defined by (7.6) on smooth functions $\Pi \ni (r, \psi) \mapsto V(r, \psi)$

satisfying $V(1,\psi) = V(e,\psi)$ will be called a *canonical periodic meromorphic pseudodifferential operator of order a*; the function Φ is called the *symbol* of this operator.

Formula (7.6) implies the relation

$$r^{2\pi i}\mathbf{A}(r,\mu+2\pi)r^{-2\pi i} = \mathbf{A}(r,\mu). \tag{7.11}$$

3. General periodic meromorphic pseudodifferential operators. Next we will consider meromorphic operator-functions which either have finitely many poles in any strip $|\mathrm{Im}\lambda| < h < \infty$, or satisfy (7.11) and have finitely many poles in any rectangle $\{\mu \in \mathbf{C}: 0 \leqslant \mathrm{Re}\,\mu \leqslant 2\pi,\ |\mathrm{Im}\,\mu| < h\}$. Every closed set lying in a strip (rectangle) of this kind and not containing poles of an operator considered will be called admissible.

A meromorphic operator-function $\mu \mapsto \mathbf{A}(\cdot,\mu)$ satisfying (7.11), with $r \in [1,e]$, is called a *periodic meromorphic pseudodifferential operator of order* a_0 if for any $N = 0, 1, \cdots$, the inequality

$$\left\|\mathbf{A}(\cdot,\mu) - \sum_{j=0}^{N} \mathbf{A}_j(\cdot,\mu); H^s(\Pi) \to H^{s-\alpha_{N+1}}(\Pi)\right\| \leqslant c_{\mathfrak{F}}(s,N)$$

holds on every set $\mathfrak{F}$ that is admissible for the operators $\mathbf{A}(r,\mu)$, $\mathbf{A}_j(r,\mu)$, $j = 0, \cdots, N$. Here $\mathbf{A}_j$ is the canonical periodic meromorphic pseudodifferential operators of order a_j with symbol $\Phi^{(j)}(r,\phi,\theta)$, $\alpha_j = \mathrm{Re}\,a_j$, $\alpha_0 \geqslant \alpha_1 \geqslant \alpha_2 \geqslant \cdots$, $\alpha_j \to -\infty$ as $j \to \infty$. We will assume that the symbol $\Phi^{(j)}$ of a canonical operator can be extended to $\mathbb{R}^n \setminus 0$ in each argument ϕ, θ as a homogeneous function of degree a_j. The formal series $\Sigma_j \Phi^{(j)}(r,\phi,\theta)$ is called the *symbol* of $\mathbf{A}(r,\mu)$.

T h e o r e m 7.3. *For a given series σ there is a periodic meromorphic pseudodifferential operator* **A** *with σ as symbol.*

P r o o f. Using theorem 3.3 we determine a meromorphic pseudodifferential operator $\mathfrak{U}(r,\lambda)$ with symbol $\sigma(r,\phi,\theta)$ (the variable r is regarded as parameter). The periodic meromorphic pseudodifferential operator looked for is defined, using these operators $\mathfrak{U}$, by (7.6). ■

4. The algebra of periodic meromorphic pseudodifferential operators.

Theorem 7.4. *Let* **A**,**B** *be periodic meromorphic pseudodifferential operators of orders* a_0, b_0, *with symbols*

$$\sigma_{\mathbf{A}}(r,\phi,\theta) = \sum_{j=0}^{\infty} \Phi^{(j)}(r,\phi,\theta),$$

$$\sigma_{\mathbf{B}}(r,\phi,\theta) = \sum_{k=0}^{\infty} \Psi^{(k)}(r,\phi,\theta),$$

where $\Phi^{(j)}$ *and* $\Psi^{(k)}$ *are functions homogeneous in* ϕ *and* θ *of degree* a_j *and* b_k, *respectively. Then* **AB** *is a periodic meromorphic pseudodifferential operator, of order* a_0+b_0 *and with symbol defined by*

$$\sigma_{\mathbf{AB}}(r,\phi,\theta) = \sum_{j,k=0}^{\infty} \sum_{|\gamma|\geqslant 0} \frac{1}{\gamma!} D_\theta^\gamma \Phi^{(j)}(r,\phi,\theta) r^{|\gamma|-b_k} \partial_x^\gamma \Psi^{(k)}(r,x,\theta). \qquad (7.12)$$

Here $\partial^\gamma = \partial^{|\gamma|}/\partial x_1^{\gamma_1}\cdots\partial^{\gamma_n} x_n$, $D_\theta^\gamma = (-i)^{|\gamma|}\partial_\theta^\gamma$, $\partial_x^\gamma \Psi^{(k)}(r,x,\theta) = \Sigma_j \partial_x^\gamma (r^{2\pi ij}\Psi_j^{(k)}(x,\theta))$, $r = |x|$, $\phi = x/|x|$; *moreover,* $\Psi^{(k)}$ *is homogeneous of degree* b_k *both in* x *and in* θ. *Thus, the terms of the series* (7.12) *are periodic functions in* r, *i.e.* $\sigma_{\mathbf{AB}}(r,\phi,\theta) = \sigma_{\mathbf{AB}}(re,\phi,\theta)$.

Proof. It suffices to assume **A** and **B** to be canonical operators of orders a and b with symbols $\Phi(r,\phi,\theta)$ and $\Psi(r,\phi,\theta)$ respectively. We have

$$\mathbf{A}(r,\mu)\mathbf{B}(r,\mu)V = \sum_{j=-\infty}^{+\infty} r^{2\pi ij}\mathfrak{U}(r,\mu+2j\pi)\times \qquad (7.13)$$

$$\times \sum_{k=-\infty}^{+\infty} \mathfrak{B}_{j-k}(\mu+2k\pi)V_k,$$

where $\mathfrak{B}_q(\mu) = \mathrm{E}(\mu+ib)\Psi_q(\phi,\theta)\mathrm{E}(\mu)$. By proposition 2.2 the asymptotic series expansion

$$\mathfrak{U}(r,\mu+2j\pi) \sim \mathfrak{U}(r,\mu+2k\pi)+ \qquad (7.14)$$

$$+\sum_{|\gamma|\geqslant 1}\frac{1}{\gamma!}\rho_\gamma(\phi,2i(j-k)\pi)\mathrm{E}_{\theta\to\phi}(\mu+2k\pi+i(a-|\gamma|))^{-1}\times$$

$$\times D_\theta^\gamma \Phi(r,\phi,\theta)\mathrm{E}_{\psi\to\theta}(\mu+2k\pi)$$

is valid, where $\rho_\gamma(\phi,\mu) = |x|^{|\gamma|-\mu}\partial_x^\gamma |x|^\mu$. Applying proposition 2.4 we obtain

$$E_{\theta\to\phi}(\mu+2k\pi+i(a-|\gamma|))^{-1}D_\theta^\gamma\Phi(r,\phi,\theta)E_{\psi\to\theta}(\mu+2k\pi)\times \tag{7.15}$$
$$\times E_{\omega\to\psi}(\mu+2k\pi+ib)^{-1}\Psi_{j-k}(\psi,\omega)E_{\chi\to\omega}(\mu+2k\pi)\sim$$
$$\sim\sum_{|\kappa|\geq 0}E(\mu+2k\pi+i(a+b-|\gamma+\kappa|))^{-1}\times$$
$$\times\frac{1}{\kappa!}D_\theta^{\gamma+\kappa}\Phi(r,\phi,\theta)\partial_x^\kappa\Psi_{j-k}(\phi,\theta)E(\mu+2k\pi).$$

Taking into account cancellation because of the homogeneity of the extension Ψ_{j-k}, we have

$$\sum_{\{\gamma,\kappa:\gamma+\kappa=\text{const}\}}\frac{1}{\gamma!\kappa!}\rho_\gamma(\phi,2i(j-k)\pi)(\partial_x^\kappa\Psi_{j-k})(\phi,\theta)= \tag{7.16}$$
$$=\sum\frac{1}{\gamma!\kappa!}r^{|\gamma|+|\kappa|-b-2i(j-k)\pi}\partial_x^\gamma r^{2i(j-k)\pi}\partial_x^\kappa\Psi_{j-k}(x,\theta)=$$
$$=\frac{1}{(\gamma+\kappa)!}r^{|\gamma+\kappa|-b-2i(j-k)\pi}\partial_x^{\gamma+\kappa}(r^{2i(j-k)\pi}\Psi_{j-k}(x,\theta)).$$

Relations (7.13) - (7.16) imply

$$\mathbf{A}(r,\mu)\mathbf{B}(r,\mu)\sim\sum_{k=-\infty}^{+\infty}\sum_{|\gamma|\geq 0}r^{2k\pi i}E_{\theta\to\phi}(\mu+2k\pi+i(a+b-|\gamma|))^{-1}\times$$
$$\times\frac{1}{\gamma!}D_\theta^\gamma\Phi(r,\phi,\theta)r^{|\gamma|-b}\partial_x^\gamma\Psi(r,x,\theta)E_{\psi\to\theta}(\mu+2k\pi)V_k.\ \blacksquare$$

Denote by $\mathbf{A}(r,\mu)^*$ the operator adjoint to $\mathbf{A}(r,\mu)$ with respect to the scalar product

$$<U,V>=\int_1^e\int_{S^{n-1}}U(r,\phi)\overline{V(r,\phi)}\frac{drd\phi}{r}. \tag{7.17}$$

Put $\mathbf{A}^*(r,\overline{\mu})=\mathbf{A}(r,\mu)^*$.

Theorem 7.5. *Let $\mathbf{A}(r,\mu)$ be the periodic meromorphic pseudodifferential operator of order a_0 with symbol $\sigma_{\mathbf{A}}(r,\phi,\theta)=\Sigma\Phi^{(j)}(r,\phi,\theta)$, where the $\Phi^{(j)}$ are functions homogeneous of degree a_j both in ϕ and in θ. Then $\mathbf{A}^*(r,\mu)$ is a periodic meromorphic pseudodifferential operator of order $\overline{a}_0$, and its symbol is*

$$\sum_{j=0}^{\infty}\sum_{|\gamma|\geq 0}\frac{1}{\gamma!}r^{|\gamma|-\overline{a}_j}\partial_x^\gamma D_\theta^\gamma\overline{\Phi^{(j)}(r,x,\theta)}$$

(the operations ∂_x^γ and D_θ^γ are understood as is done in theorem 7.4).

P r o o f. Assume, without loss of generality, that **A** is the canonical operator of order a with symbol $\Phi(r,\phi,\theta)$. Then

$$<\mathbf{A}(\cdot,\mu)U,V> \;=\; \sum_{j,k=0}^{\infty} (\mathfrak{U}_{k-j}(\mu+2j\pi)U_j, V_k), \tag{7.18}$$

where (f,g) is the scalar product in $L_2(S^{n-1})$, $\mathfrak{U}_k(\mu) = \mathrm{E}_{\theta\to\phi}(\mu+ia)^{-1}\Phi_k(\phi,\theta)\mathrm{E}_{\psi\to\theta}(\mu)$, and Φ_k are the Fourier coefficients of $\Phi(\cdot,\phi,\theta)$. By proposition 2.5 the asymptotic formula

$$\mathfrak{U}_{k-j}^{*}(\mu+2j\pi) \sim \sum_{|\gamma|\geqslant 0}\frac{1}{\gamma!}\mathrm{E}(\mu+2j\pi+i(\bar{a}-|\gamma|))^{-1}\times \tag{7.19}$$

$$\times\, \partial_\phi^\gamma D_\theta^\gamma \overline{\Phi_{k-j}(\phi,\theta)}\mathrm{E}(\mu+2j\pi)$$

holds for $\mathfrak{U}_{k-j}(\mu+2\pi j)^* = \mathfrak{U}_{k-j}^*(\bar{\mu}+2\pi j)$. Applying proposition 2.2 we obtain

$$\mathrm{E}(\mu+2j\pi+i(\bar{a}-|\gamma|))^{-1}\partial_\phi^\gamma D_\theta^\gamma\overline{\Phi_{k-j}(\phi,\theta)}\mathrm{E}(\mu+2j\pi) \sim$$

$$\sim \sum_{|\kappa|\geqslant 0}\frac{1}{\kappa!}\rho_\kappa(\phi,2i(j-k)\pi)\times$$

$$\times\, \mathrm{E}(\mu+2k\pi+i(\bar{a}-|\gamma|-|\kappa|))^{-1}\partial_\phi^\gamma D_\theta^{\gamma+\kappa}\overline{\Phi_{k-j}(\phi,\theta)}\mathrm{E}(\mu+2k\pi).$$

Substitute this formula in (7.19) and use a relation of the form (7.16). Then

$$\mathfrak{U}_{k-j}^{*}(\mu+2j\pi) \sim \sum_{|\gamma|\geqslant 0}\frac{1}{\gamma!}r^{|\gamma|-\bar{a}-2i(j-k)\pi}\times \tag{7.20}$$

$$\times\, \mathrm{E}(\mu+2k\pi+i(\bar{a}-|\gamma|))^{-1}\partial_x^\gamma(r^{2i(j-k)\pi}D_\theta^\gamma\overline{\Phi_{k-j}(x,\theta)})\mathrm{E}(\mu+2k\pi).$$

It follows from (7.18) that

$$\mathbf{A}^*(r,\mu)V = \sum_{j,k} r^{2ij\pi}\mathfrak{U}_{k-j}^*(\mu+2j\pi)V_k =$$

$$= \sum_k r^{2ik\pi}\sum_j r^{2i(j-k)\pi}\mathfrak{U}_{k-j}^*(\mu+2j\pi)V_k.$$

It remains to combine this equation with (7.20). ■

A periodic meromorphic pseudodifferential operator $\mathbf{A}(r,\mu)$ is called *elliptic* if its principal symbol $\Phi^{(0)}$ does not vanish anywhere. Theorems 7.3 and 7.4 allow us to prove the following statement (cf. theorem 3.8).

Theorem 7.6. *Let $\mathbf{A}(r,\mu)$ be an elliptic periodic meromorphic pseudodifferential operator of order a_0. Then there is a periodic meromorphic pseudodifferential operator $\mathbf{R}(r,\mu)$ of order $-a_0$ such that the symbols of the operators $\mathbf{A}(r,\mu)\mathbf{R}(r,\mu)$ and $\mathbf{R}(r,\mu)\mathbf{A}(r,\mu)$ are equal to one.*

§8. Change of variables in meromorphic pseudodifferential operators

Let g be a diffeomorphism of the sphere S^{n-1}, $g(\sigma) = (g_1(\sigma), \cdots, g_n(\sigma))$, $\sigma \in \mathbf{R}^n$, $|\sigma| = 1$, $|g(\sigma)| = 1$. We will assume that the functions g_j are defined on $\mathbf{R}^n \setminus 0$ and are homogeneous of degree one, that $g'(\sigma) = \|\partial g_j / \partial \sigma_k\|_{j,k=1}^n$, and that $|g'(\sigma)|$ is the modulus of the determinant $\det g'(\sigma)$.

Theorem 8.1. *Let $\mathfrak{U}(\lambda) = \mathrm{E}(\lambda + ia)^{-1}\Phi(\phi,\omega)\mathrm{E}(\lambda)$. Then $\mathfrak{U}_{\psi\to\phi}(\lambda)u(\psi) = |g'(\sigma)^{-1}|\hat{\mathfrak{U}}_{\tau\to\sigma}(\lambda)|g'(\tau)|\hat{u}(\tau)$, where $\phi = g(\sigma)$, $\psi = g(\tau)$, $\hat{u}(\tau) = u(g(\tau))$, and $\hat{\mathfrak{U}}_{\tau\to\sigma}$ is the meromorphic pseudodifferential operator with symbol*

$$\sum_{|\gamma|=0}^{\infty} \frac{1}{\gamma!} \partial_\tau^\gamma e^{ih(\omega,\sigma,\tau)}\Big|_{\tau=\sigma} D_\omega^\gamma \Phi(g(\sigma),(g'(\sigma)^{-1})^*\omega),$$

where, moreover, $h(\omega,\sigma,\tau) = (\omega, \tau - g'(\sigma)^{-1}g(\tau)) = O(|\sigma-\tau|^2)$.

Proof. Recall that the kernel of $\mathfrak{U}_{\psi\to\phi}(\lambda)$ is given by

$$G(\phi,\psi;\lambda) = \int_0^\infty t^{-i(\lambda+ia+in/2)-1} G(\phi, t\phi - \psi)dt, \tag{8.1}$$

where $G(\phi,t\xi) = t^{-n-a}G(\phi,\xi)$, $t > 0$ (§1.3). For simplicity of notation the dependence on ϕ of $G(\phi,\xi)$ will not be indicated. Since the functions g_j are homogeneous we have $g(\sigma) = g'(\sigma)\sigma$. Hence the equation $g(\tau) = g(\sigma) + g'(\sigma)(\tau-\sigma) + R(\sigma,\tau)$ can be rewritten as $g(\tau) = g'(\sigma)\tau + R(\sigma,\tau)$, $R(\sigma,\tau) = O(|\sigma-\tau|^2)$. This implies that

$$G(t\phi-\psi) = G(tg(\sigma)-g(\tau)) = G(g'(\sigma)(t\sigma-\tau)-R(\sigma,\tau)) = \tag{8.2}$$

$$= \sum_{|\gamma|=0}^{N} \frac{(-1)^{|\gamma|}}{\gamma!} \partial^\gamma G(g'(\sigma)(t\sigma-\tau))R(\sigma,\tau)^\gamma + T_{N+1}(\sigma,\tau;t).$$

The term T_{N+1} is a sufficiently smooth function of its arguments (for N large), having only singularity of power type at infinity. Whatever given numbers p,s and an admissible set $\mathfrak{F}$, we can choose N large such that the inequality

$$\|T_{N+1,\lambda};H^s(\lambda,S^{n-1})\to H^{s+p}(\lambda,S^{n-1})\| \leqslant c(p,s,\mathfrak{F}) \tag{8.3}$$

holds; here $T_{N+1,\lambda}$ is the operator on S^{n-1} with kernel

$$\tilde{T}_{N+1}(\sigma,\tau;\lambda) = \int_0^\infty t^{-i(\lambda+in/2+ia)-1}T_{N+1}(\sigma,\tau;t)dt.$$

Apply the Mellin transform to both sides of (8.2). Using (8.1) we obtain

$$G(\phi,\psi;\lambda) = \sum_{|\gamma|=0}^{N}\frac{(-1)^{|\gamma|}}{\gamma!}\int_0^\infty t^{-i(\lambda+in/2+ia)-1}\times \tag{8.4}$$

$$\times\,\partial^\gamma G(g'(\sigma)(t\sigma-\tau))dtR(\sigma,\tau)^\gamma+\tilde{T}_{N+1}(\sigma,\tau;\lambda).$$

Since

$$\int G(\phi,\psi;\lambda)u(\psi)d\psi = \int G(g(\sigma),g(\tau);\lambda)u(g(\tau))|g'(\tau)|d\tau,$$

relation (8.4) between kernels corresponds to the following equation for the operators:

$$\mathrm{E}_{\omega\to\phi}(\lambda+ia)^{-1}\Phi(\omega)\mathrm{E}_{\phi\to\omega}(\lambda)u(\psi) = \tag{8.5}$$

$$= \sum_{|\gamma|=0}^{N}\frac{(-1)^{|\gamma|}}{\gamma!}\mathrm{E}_{\omega\to\sigma}(\lambda-i|\gamma|+i(a+|\gamma|))^{-1}\Phi_\gamma(\omega)\times$$

$$\times\,\mathrm{E}_{\tau\to\omega}(\lambda-i|\gamma|)|g'(\tau)|R(\sigma,\tau)^\gamma\hat{u}(\tau)+\mathrm{T}_{N+1,\lambda}\hat{u}$$

where $\hat{u}(\tau) = u(g(\tau))$, and $\Phi_\gamma(\omega)$ is the Fourier transform $\mathrm{F}_{z\to\omega}$ of $z\mapsto\partial^\gamma G(g'(\sigma)z)$. In view of the estimate (8.3) for $\mathrm{T}_{N+1,\lambda}$ it suffices to consider the other terms at the righthand side of (8.5). Put $\zeta = g'(\sigma)z$. Then

$$\Phi_\gamma(\omega) = \frac{1}{(2\pi)^{n/2}}\int e^{-i\omega z}\partial^\gamma G(g'(\sigma)z)dz =$$

$$= \frac{1}{(2\pi)^{n/2}}|g'(\sigma)^{-1}|\int e^{-i(g'(\sigma)^{-1})^*\omega\zeta}\partial^\gamma G(\zeta)d\zeta =$$

$$= \frac{1}{(2\pi)^{n/2}}|g'(\sigma)^{-1}|\,i^{|\gamma|}[(g'(\sigma)^{-1})^*\omega]^\gamma\int e^{-i(g'(\sigma)^{-1})^*\omega\zeta}G(\zeta)d\zeta =$$

$$= i^{|\gamma|}\, |g'(\sigma)^{-1}|\, [((g'(\sigma)^{-1})^{*}\omega]^{\gamma}\Phi(((g'(\sigma))^{-1})^{*}\omega).$$

Hence the righthand side of (8.5) can be rewritten as

$$|g'(\sigma)^{-1}| \sum_{|\gamma|=0}^{N} \mathrm{E}_{\omega\to\sigma}(\lambda+ia)^{-1}\Phi((g'(\sigma)^{-1})^{*}\omega) \times \tag{8.6}$$

$$\times\, \mathrm{E}_{\tau\to\omega}(\lambda - i\,|\gamma|)\frac{(-i)^{|\gamma|}}{\gamma!}[(g'(\sigma)^{-1})^{*}\omega]^{\gamma}R(\sigma,\tau)^{\gamma}\,|g'(\tau)|\,\hat{u}(\tau).$$

Note that for vectors $\xi,\eta \in \mathbf{R}^n$ the formula

$$\frac{1}{s!}(\xi\eta)^s = \sum_{|\gamma|=s}\frac{1}{\gamma!}\xi^{\gamma}\eta^{\gamma} \tag{8.7}$$

holds. Thus,

$$\sum_{|\gamma|=s}\frac{1}{\gamma!}[(g'(\sigma)^{-1})^{*}\omega]^{\gamma}R(\sigma,\tau)^{\gamma} =$$

$$= \frac{1}{s!}((g'(\sigma)^{-1})^{*}\omega R(\sigma,\tau))^s = \frac{1}{s!}(\omega g'(\sigma)^{-1}R(\sigma,\tau))^s.$$

This and formulas (8.5), (8.6) imply

$$\mathrm{E}_{\omega\to\phi}(\lambda+ia)^{-1}\Phi(\omega)\mathrm{E}_{\psi\to\omega}(\lambda)u(\psi) = \tag{8.8}$$

$$= |g'(\sigma)^{-1}| \sum_{s=0}^{N}\frac{(-i)^s}{s!}\mathrm{E}_{\omega\to\sigma}(\lambda+ia)^{-1}\Phi((g'(\sigma)^{-1})^{*}\omega) \times$$

$$\times\, \mathrm{E}_{\tau\to\omega}(\lambda - is)(\omega g'(\sigma)^{-1}R(\sigma,\tau))^s\,|g'(\tau)|\,\hat{u}(\tau) + \mathrm{T}_{N+1,\lambda}\hat{u}.$$

By proposition 3.2.1,

$$\mathrm{E}_{\omega\to\sigma}(\lambda+ia)^{-1}\Phi((g'(\sigma)^{-1})^{*}\omega)\mathrm{E}_{\tau\to\omega}(\lambda-is)(\omega g'(\sigma)^{-1}R(\sigma,\tau))^s \times$$

$$\times\, |g'(\tau)|\,\hat{u}(\tau) \sim \sum_{|\gamma|=0}^{\infty}\frac{1}{\gamma!}\mathrm{E}_{\omega\to\sigma}(\lambda+i(a+s-|\gamma|))^{-1} \times$$

$$\times\, \partial_{\tau}^{\gamma}(\omega g'(\sigma)^{-1}R(\sigma,\tau))^s\,|_{\tau=\sigma}D_{\omega}^{\gamma}\Phi((g'(\sigma)^{-1})^{*}\omega)\mathrm{E}_{\tau\to\omega}(\lambda)\,|g'(\tau)|\,\hat{u}(\tau).$$

Combining the last relation and (8.8), the proof is finished. In the exponential series expansion for the expression $\partial_{\tau}^{\gamma}e^{ih(\omega,\sigma,\tau)}\,|_{\tau=\sigma}$ for fixed γ, only finitely many nonzero terms occur. ■

Chapter 4

Pseudodifferential operators with discontinuous symbols on manifolds with conical singularities

In this Chapter we consider various algebras of pseudodifferential operators. The first paragraph is devoted to the study of pseudodifferential operators on the spaces $H^s_\beta(\mathbf{R}^n)$. Symbols of 'canonical' pseudodifferential operators of order a will be functions $(\mathbf{R}^n \setminus 0) \times (\mathbf{R}^n \setminus 0) \ni (x, \xi) \mapsto \Phi(x, \xi)$, homogeneous of degree a both in x and in ξ. A 'general' pseudodifferential operator $A: H^s_\beta(\mathbf{R}^n) \to H^{s-\operatorname{Re} a}_\beta(\mathbf{R}^n)$ of order a can be expressed in terms of a meromorphic pseudodifferential operator $\mathfrak{U}$ by formula (1.1) (compare with (2.3.6) and (2.3.7)). We derive from the results of Chapter 3 concerning meromorphic pseudodifferential operators a symbolic calculus for pseudodifferential operators on $\mathbf{R}^n$ (modulo operators acting from $H^s_\beta(\mathbf{R}^n)$ to $H^{s-\operatorname{Re} a+N}_{\beta+N}(\mathbf{R}^n)$ for arbitrary N). Similar questions are considered for pseudodifferential operators with 'periodic' (in the sense of §7, Chapter 3) symbols.

In §2 we introduce and study pseudodifferential operators on a cone. This allows us to define, in §3, pseudodifferential operators on a manifold with a finite set of conical points (including, in particular, the points at which the symbols of the operators have a discontinuity 'of the first kind'). Necessary and sufficient conditions for a pseudodifferential operator of arbitrary order to be Fredholm are elucidated.

Algebras generated by pseudodifferential operators of order zero will be considered in §4. The symbols of the operators are allowed to have isolated singularities. The set of singular points of a symbol depends on the operator, and by closing the algebras, symbols with an everywhere dense set of discontinuity points arise. We describe algebras of 'operator symbols', which are isomorphic to quotient algebras of pseudodifferential operators by the ideal of compact operators. Operator symbols are functions whose values are meromorphic pseudodifferential operators or periodic meromorphic pseudodifferential operators. We consider, in particular, an algebra of singular integral operators on a composite contour.

§1. Pseudodifferential operators on $\mathbb{R}^n$

1. Definition of the operators. Let $\mathfrak{U}$ be the meromorphic pseudodifferential operator of order $a = a_0$ with symbol $\Sigma_{j=0}^{\infty}\Phi_j(\phi,\theta)$ (cf. definition 3.3.1). For $u \in C_0^\infty(\mathbb{R}^n\setminus 0)$ we put

$$(Au)(x) = \tag{1.1}$$

$$= \frac{1}{\sqrt{2\pi}}\int_{\operatorname{Im}\lambda=\tau} r^{i(\lambda+ia+in/2)}\mathfrak{U}_{\psi\to\phi}(\lambda)\tilde{u}(\lambda+in/2,\psi)d\lambda,$$

where, as before, $r = |x|$, $\phi = x/|x|$, $\tilde{u}$ is the Mellin transform of the function u, and the number τ is such that the line $\operatorname{Im}\lambda = \tau$ does not contain poles of the operator-function $\mathfrak{U}$.

P r o p o s i t i o n 1.1. *For* $\tau = \beta - s$ *the operator* (1.1) *realizes a continuous map* $A\colon H_\beta^s(\mathbb{R}^n)\to H_\beta^{s-\operatorname{Re}a}(\mathbb{R}^n)$.

P r o o f. Multiply both sides of (1.1) by r^a and then apply the Mellin transform $M_{r\to\lambda+in/2}$. As the result we obtain $\widetilde{(Au)}(\lambda+ia+in/2,\phi) = \mathfrak{U}_{\psi\to\phi}(\lambda)\times$ $\times\tilde{u}(\lambda+in/2,\psi)$. Definition 3.3.1 and proposition 3.1.3 imply that the estimate

$$\|\widetilde{(Au)}(\lambda+ia+in/2,\cdot);H^{s-\operatorname{Re}a}(\lambda,S^{n-1})\|^2 \leqslant$$

$$\leqslant c_\tau\|\tilde{u}(\lambda+in/2,\cdot);H^s(\lambda,S^{n-1})\|^2$$

holds on the line $\operatorname{Im}\lambda = \tau$. After integration of this inequality we have

$$\int_{\operatorname{Im}\lambda=\beta-s+\operatorname{Re}a} \|\widetilde{(Au)}(\lambda+in/2,\cdot);H^{s-\operatorname{Re}a}(\lambda,S^{n-1})\|^2 d\lambda \leqslant$$

$$\leqslant c\int_{\operatorname{Im}\lambda=\beta-s} \|\tilde{u}(\lambda+in/2,\cdot);H^s(\lambda,S^{n-1})\|^2 d\lambda.$$

Hence, $\|Au;H_\beta^{s-\operatorname{Re}a}(\mathbb{R}^n)\| \leqslant c\|u;H_\beta^s(\mathbb{R}^n)\|$. ■

D e f i n i t i o n 1.2. The operator (1.1) will be called a *pseudodifferential operator of order a* on $H_\beta^s(\mathbb{R}^n)$. The formal series

$$\sigma(x,\theta) = r^{-a}\sum_{j=0}^{\infty}\Phi_j(x,\theta) \tag{1.2}$$

is called the *(complete) symbol* of A, and the term with index $j=0$ is called the *principal symbol*; recall that the function $(x,\theta)\mapsto\Phi_j(x,\theta)$ is homogeneous of degree a_j both in x and in θ, $\mathrm{Re}\,a_0>\mathrm{Re}\,a_1\geqslant\mathrm{Re}\,a_2\geqslant\cdots$, $\mathrm{Re}\,a_j\to-\infty$. If a canonical meromorphic pseudodifferential operator of order a is taken in (1.1) instead of $\mathfrak{U}$, the operator (1.1) is called a *canonical pseudodifferential operator* on $H^s_\beta(\mathbb{R}^n)$.

E x a m p l e. An expression of the form $\Sigma_{j=0}^k r^{-j}L_j(\phi,D_x)$, where $L_j(\phi,D_x)$ is a homogeneous differential operator of order $k-j$ with coefficients depending on ϕ, is a pseudodifferential operator of order k on any space $H^s_\beta(\mathbb{R}^n)$ (cf. (3.1.15)). Extend the function $(\phi,\theta)\mapsto L_j(\phi,\theta)$ as a homogeneous function of degree $k-j$ both in ϕ and θ. The symbol of the operator $\Sigma_{j=0}^k r^{-j}L_j(\phi,D_x)$ equals $r^{-k}\Sigma_{j=0}^k L_j(x,\theta)=\Sigma_{j=0}^k r^{-j}L_j(\phi,\theta)$.

Associate to the series (1.2) the formal operator series

$$\sum_{j=0}^{\infty} r^{a_j-a}\mathrm{A}_j, \tag{1.3}$$

where $a=a_0$ and the operator A_j is defined by (1.1) with $a=a_j$, $\mathfrak{U}(\lambda)=\mathfrak{U}_j(\lambda)=\mathrm{E}(\lambda+ia_j)^{-1}\Phi_j(\phi,\theta)\mathrm{E}(\lambda)$ (i.e. the A_j are canonical pseudodifferential operators of orders a_j). Proposition 1.2 implies that if there are no poles of $\mathfrak{U}_j$ on the line $\mathrm{Im}\lambda=\tau=\beta-s$, the term in (1.3) with index j is a bounded operator from $H^s_\beta(\mathbb{R}^n)$ to $H^{s-\alpha_j}_{\beta+\alpha_0-\alpha_j}(\mathbb{R}^n)\subset H^{s-\alpha_0}_\beta(\mathbb{R}^n)$, $\alpha_j=\mathrm{Re}\,a_j$. Assume that there are no poles of the operators $\mathfrak{U}$, $\mathfrak{U}_j$, $j=0,\cdots,N$, on the line $\mathrm{Im}\lambda=\beta-s$. Then, using (3.3.1) and by a reasoning as in the proof of proposition 1.1 we obtain the estimate

$$\|\mathrm{A}-S_N;H^s_\beta(\mathbb{R}^n)\to H^{s-\alpha_{N+1}}_{\beta+\alpha_0-\alpha_{N+1}}(\mathbb{R}^n)\|\leqslant c_N<\infty; \tag{1.4}$$

here S_N denotes the N-th partial sum of the series (1.3).

E.g., if the α_j ($j=0,1,\cdots$) are integers, then there are no poles of the operators $\mathfrak{U}$, $\mathfrak{U}_j$, $j=0,1,\cdots$, on the line $\mathrm{Im}\lambda=\tau$, for all $\tau\in\mathbb{R}$ except possibly for some set of isolated values. Hence (1.4) holds for any $N=0,1,\cdots$.

Theorem 3.3 immediately implies

T h e o r e m 1.3. *For any given series* (1.2) *there is a pseudodifferential operator* A *of order a on $H^s_\beta(\mathbb{R}^n)$ having it as symbol. If a meromorphic pseudodifferential* *operator* $\mathfrak{U}$ *is fixed, then* A *is a pseudodifferential operator on* $H^s_\beta(\mathbf{R}^n)$ *for all* $\tau = \beta - s$, *except possibly for a set of isolated values.*

2. The algebra of pseudodifferential operators.

T h e o r e m 1.4. *Let* $A: H^s_\beta(\mathbf{R}^n) \to H^{s-\mathrm{Re}\,a}_\beta(\mathbf{R}^n)$, $B: H^{s+\mathrm{Re}\,b}_\beta(\mathbf{R}^n) \to H^s_\beta(\mathbf{R}^n)$ *be two pseudodifferential operators, of orders a and b respectively, and let*

$$\sigma_A(x,\theta) = r^{-a} \sum_{j=0}^{\infty} \Phi_j(x,\theta), \quad \sigma_B(x,\theta) = r^{-b} \sum_{k=0}^{\infty} \Psi_k(x,\theta)$$

be their symbols. (Here $a = a_0$, $b = b_0$, Φ_j *and* Ψ_k *are homogeneous functions of degrees* a_j *and* b_k *(in both arguments* x *and* θ*).) Then the operator* $AB: H^{s+\mathrm{Re}\,b}_\beta(\mathbf{R}^n) \to H^{s-\mathrm{Re}\,a}_\beta(\mathbf{R}^n)$ *is a pseudodifferential operator of order* $a+b$ *(in the sense of definition 1.2), and its symbol is defined by*

$$\sigma_{AB}(x,\theta) = \sum_{|\gamma| \geq 0} \frac{1}{\gamma!} D^\gamma_\theta \sigma_A(x,\theta) \cdot \partial^\gamma_x \sigma_B(x,\theta),$$

where the series σ_A *and* σ_B *are to be differentiated and multiplied termwise.*

P r o o f. By definition 1.2,

$$Au = \frac{1}{\sqrt{2\pi}} \int_{\mathrm{Im}\lambda = \beta - s} r^{i(\lambda + ia + in/2)} \mathfrak{U}(\lambda) \tilde{u}(\lambda + in/2, \cdot) d\lambda,$$

$$Bu = \frac{1}{\sqrt{2\pi}} \int_{\mathrm{Im}\mu = \beta - s - \mathrm{Re}\,b} r^{i(\mu + ib + in/2)} \mathfrak{B}(\mu) \tilde{v}(\mu + in/2, \cdot) d\mu,$$

where $\mathfrak{U}$ and $\mathfrak{B}$ are meromorphic pseudodifferential operators of orders a and b. After multiplication of the first equation by r^a and application of the Mellin transform $M_{r \to \lambda + in/2}$ we obtain the relation $\widetilde{(Au)}(\lambda + ia + in/2, \cdot) = \mathfrak{U}(\lambda)\tilde{u}(\lambda + in/2, \cdot)$ on the line $\mathrm{Im}\lambda = \beta - s$. Similarly, $\widetilde{(Bv)}(\mu + ib + in/2, \cdot) = \mathfrak{B}(\mu)\tilde{v}(\mu + in/2, \cdot)$ on the line $\mathrm{Im}\mu = \beta - s - \mathrm{Re}\,b$. Hence

$$ABv = \frac{1}{\sqrt{2\pi}} \int_{\mathrm{Im}\mu = \beta - s - \mathrm{Re}\,b} r^{i(\mu + i(a+b) + in/2)} \times$$

$$\times\, \mathfrak{U}(\mu + ib) \mathfrak{B}(\mu) \tilde{v}(\mu + in/2, \cdot) d\mu.$$

It remains to prove that $\mathfrak{U}(\mu + ib)\mathfrak{B}(\mu)$ is a meromorphic pseudodifferential operator of order $a+b$ whose symbol equals

$$r^{a+b}\sum_{|\gamma|\geq 0}\frac{1}{\gamma!}D_\theta^\gamma\sigma_A(x,\theta)\cdot\partial_x^\gamma\sigma_B(x,\theta). \tag{1.5}$$

By proposition 3.2.2 the symbol of the meromorphic pseudodifferential operator $\mathfrak{U}(\mu+ib)$ of order a equals

$$r^b\sum_{|\gamma|,j\geq 0}\frac{1}{\gamma!}\partial_x^\gamma r^{-b}D_\theta^\gamma\Phi_j(x,\theta).$$

Theorem 3.3.4 implies that $\mathfrak{U}(\mu+ib)\mathfrak{B}(\mu)$ is a meromorphic pseudodifferential operator of order $a+b$, and that its symbol has the form

$$r^b\sum_{|\gamma|,|\delta|,j,k\geq 0}\frac{1}{\gamma!\delta!}\partial_x^\gamma r^{-b}D_\theta^{\gamma+\delta}\Phi_j(x,\theta)\partial_x^\gamma\Psi_k(k,\theta)=$$

$$=r^b\sum_{|\kappa|\geq 0}\frac{1}{\kappa!}D_\theta^\kappa\Phi_j(x,\theta)\partial_x^\kappa(r^{-b}\Psi_k(x,\theta))$$

(we have used Leibniz' formula). Obviously, the righthand side coincides with (1.5). ■

Theorem 1.5. *Let* $A:H_\beta^s(\mathbb{R}^n)\to H^{s-\mathrm{Re}\,a}(\mathbb{R}^n)$ *be a pseudodifferential operator of order* a *and let* $A^*:H^{-s+\mathrm{Re}\,a}(\mathbb{R}^n)\to H_{-\beta}^{-s}(\mathbb{R}^n)$ *be the operator adjoint to* A *with respect to the extension of the scalar product of* $H_0^0(\mathbb{R}^n)=L_2(\mathbb{R}^n)$. *Then* A^* *is a pseudodifferential operator of order* $\bar{a}$, *and its symbol* σ_{A^*} *can be expressed in terms of the symbol* σ_A *of* A *by the formula*

$$\sigma_{A^*}(x,\theta)=\sum_{|\gamma|\geq 0}\frac{1}{\gamma!}\partial_x^\gamma D_\theta^\gamma\overline{\sigma_A(x,\theta)}.$$

Proof. Note that for $w\in H_\beta^{s-\mathrm{Re}\,a}(\mathbb{R}^n)$, $v\in H_{-\beta}^{-s+\mathrm{Re}\,a}(\mathbb{R}^n)=(H_\beta^{s-\mathrm{Re}\,a}(\mathbb{R}^n))^*$ the equality

$$(w,v)=\int_{\mathrm{Im}\lambda=\beta-s+\mathrm{Re}\,a}<\tilde{w}(\lambda+in/2,\cdot),\ \tilde{v}(\bar{\lambda}+in/2,\cdot)>d\lambda \tag{1.6}$$

holds, where (w,v) and $<f,g>$ are the extensions of the scalar products in $L_2(\mathbb{R}^n)$ and $L_2(S^{n-1})$, respectively. Put $w=Au$. Then $\tilde{w}(\lambda+ia+in/2,\cdot)=\mathfrak{U}(\lambda)\tilde{u}(\lambda+in/2,\cdot)$ on the line $\mathrm{Im}\lambda=\beta-s$. Here $\mathfrak{U}$ is the meromorphic pseudodifferential operator of order a whose symbol is $r^a\sigma_A(x,\theta) = \Sigma_{j=0}^{\infty}\Phi_j(x,\theta)$. This and (1.6) imply

$$(Av) =$$

$$= \int_{\mathrm{Im}\lambda=\beta-s\mathrm{Re}\,a} <\mathfrak{U}(\lambda-ia)u(\lambda+in/2-ia,\cdot),v(\lambda+in/2,\cdot)>d\lambda=$$

$$= \int_{\mathrm{Im}\lambda=\beta-s} \mathfrak{U}(\lambda)\tilde{u}(\lambda+in/2,\cdot),\tilde{v}(\bar{\lambda}+i(n/2-\bar{a}),\cdot)>d\lambda =$$

$$= \int_{\mathrm{Im}\lambda=\beta-s} <\tilde{u}(\lambda+in/2,\cdot),\ \mathfrak{U}^*(\bar{\lambda})\tilde{v}(\bar{\lambda}+i(n/2-a),\cdot)>d\lambda.$$

Put $\tilde{h}(\bar{\lambda}+in/2,\cdot) = \mathfrak{U}^*(\bar{\lambda})\tilde{v}(\bar{\lambda}+i(n/2-a),\cdot)$ on the line $\mathrm{Im}\lambda = \beta-s$. Rewrite this as $\tilde{h}(\mu+ia+in/2,\cdot) = \mathfrak{U}^*(\mu+i\bar{a})\tilde{v}(\mu+in/2,\cdot)$, where $\mathrm{Im}\mu = s-\beta-\mathrm{Re}\,a$. Thus,

$$h(r,\cdot) = \frac{1}{\sqrt{2\pi}}\int_{\mathrm{Im}\mu=s-\beta-\mathrm{Re}\,a} r^{i(\mu+i\bar{a}+in/2)}\mathfrak{U}^*(\mu+i\bar{a})\tilde{v}(\mu+in/2,\cdot)d\mu.$$

For arbitrary functions $u,v \in C_0^{\infty}(\mathbf{R}^n\setminus 0)$ we have $(Au,v) = (u,h)$. Hence $h = A^*v$. It remains to prove that the symbol of the meromorphic pseudodifferential operator $\mathfrak{U}^*(\mu+i\bar{a})$ equals

$$r^{\bar{a}}\sum_{|\gamma|\geq 0}\frac{1}{\gamma!}\partial_x^{\gamma}D_{\theta}^{\gamma}\overline{\sigma_A(x,\theta)}. \tag{1.7}$$

Theorem 3.3.5 implies the asymptotic expansion

$$\mathfrak{U}^*(\mu+i\bar{a}) \sim \sum_{j,|\gamma|\geq 0}\frac{1}{\gamma!}\mathrm{E}(\mu+i(\bar{a}+\bar{a}_j-|\gamma|))^{-1}\partial_{\phi}^{\gamma}D_{\theta}^{\gamma}\Phi_j(\phi,\theta)\mathrm{E}(\mu+i\bar{a}).$$

Apply proposition 3.2.2 to each term. Then

$$\mathfrak{U}^*(\mu+i\bar{a}) \sim \sum_{j,|\gamma|,|\delta|\geq 0}\frac{1}{\gamma!\delta!}\delta_x^{\delta}r^{-\bar{a}}|_{|x|=1}\times$$

$$\times\ \mathrm{E}(\mu+i(\bar{a}_j-|\gamma|-|\delta|))^{-1}\partial_x^{\gamma}D_{\theta}^{\gamma+\delta}\Phi_j(x,\theta)|_{|x|=1}\mathrm{E}(\mu) =$$

$$= \sum_{|\kappa|\geq 0}\frac{1}{\kappa!}\mathrm{E}(\mu+i(\bar{a}-|\kappa|))^{-1}\partial_x^{\kappa}D_{\theta}^{\kappa}(r^{-\bar{a}}\Phi_j(x,\theta))|_{|x|=1}\mathrm{E}(\mu).$$

This means that the symbol of $\mathfrak{U}^*(\mu+i\bar{a})$ coincides with (1.7). ■

We now give some information necessary in the proof of theorem 1.7. Let H_1, H_2 be Hilbert spaces. A bounded linear operator $\mathrm{B}: H_1 \to H_2$ is called Fredholm if its range, $\mathrm{Im}\,\mathrm{B}$, is closed and if the spaces $\ker \mathrm{B} = \{x \in H_1 : \mathrm{B}x = 0\}$ and $\mathrm{coker}\,\mathrm{B} = H_2/\mathrm{Im}\,\mathrm{B}$ are finite-dimensional. It is well-known (e.g. see [40]) that if $\mathrm{K}: H_1 \to H_2$ is a compact operator and B is a Fredholm operator, then $\mathrm{B}+\mathrm{K}$ is also a Fredholm operator. The following assertion is a particular instance of theorem 3.3 given in [21].

P r o p o s i t i o n 1.6. *Let* $\lambda \mapsto \mathrm{B}(\lambda): H_1 \to H_2$ *be an operator-function that is holomorphic in a domain* Ω *and whose values are Fredholm operators. If at some point* $\lambda_0 \in \Omega$ *the operator* $\mathrm{B}(\lambda_0): H_1 \to H_2$ *is an isomorphism, then all the* $\mathrm{B}(\lambda)$ *are isomorphisms, except possibly for a countable set whose limit points can lie on the boundary of* Ω *only.*

Let A be an operator of the form (1.1). A number τ is called admissible for A if for all $u \in C^\infty(S^{n-1})$ the inequality

$$c_1 \|u; H^s(\lambda, S^{n-1})\| \leqslant \|\mathfrak{U}(\lambda)u; H^{s-\mathrm{Re}\,a}(\lambda, S^{n-1})\| \leqslant$$
$$\leqslant c_2 \|u; H^s(\lambda, S^{n-1})\|$$

holds on the line $\mathrm{Im}\,\lambda = \tau$. In other words, for admissible τ the line $\mathrm{Im}\,\lambda = \tau$ does not contain poles of the operator-function $\lambda \mapsto \mathfrak{U}(\lambda)$, the inverse operator $\mathfrak{U}(\lambda)^{-1}$ exists and the estimate

$$\|\mathfrak{U}(\lambda)^{-1}; H^{s-\mathrm{Re}\,a}(\lambda, S^{n-1}) \to H^s(\lambda, S^{n-1})\| \leqslant c_\tau$$

holds. It is clear that only for admissible $\tau = \beta - s$ the operator (1.1) realizes an isomorphism $\mathrm{A}: H^s_\beta(\mathbb{R}^n) \to H^{s-\mathrm{Re}\,a}_\beta(\mathbb{R}^n)$.

A pseudodifferential operator A is called *elliptic* if its principal symbol Φ_0 does not vanish on $S^{n-1} \times S^{n-1}$.

T h e o r e m 1.7. *Let* A *be an elliptic pseudodifferential operator of order a of the form* (1.1) *and suppose there are no poles of the meromorphic pseudodifferential operator* $\mathfrak{U}$ *in the strip* $\tau_1 \leqslant \mathrm{Im}\,\lambda \leqslant \tau_2$. *Then every number* $\tau \in [\tau_1, \tau_2]$ *(with the possible exception of a finite set of values) is admissible for* A.

P r o o f. The pseudodifferential operator A and the meromorphic pseudodifferential operator $\mathfrak{U}$ in (1.1) are simultaneously elliptic. Hence the conditions and theorem 3.3.8 imply the existence of a meromorphic pseudodifferential operator $\mathfrak{R}$ of order $-a$ such that the difference $\mathfrak{J} = \mathfrak{R}\mathfrak{U} - \mathrm{I}$ is a meromorphic pseudodifferential operator of order $-\infty$. This means, in particular, that for arbitrary $h > 0$ the following inequality holds on the set $\{\lambda : |\mathrm{Im}\lambda| \leqslant h, |\mathrm{Re}\lambda| \geqslant c(h)\}$:

$$\|\mathfrak{J}(\lambda): H^s(\lambda, S^{n-1}) \to H^s(\lambda, S^{n-1})\| \leqslant c_p(1+|\lambda|)^{-p},$$

whatever p will be. Hence, for sufficiently large $|\lambda|$ there exists on this set an inverse operator $(\mathrm{I}+\mathfrak{J}(\lambda))^{-1}$, and with it also the operator $\mathfrak{U}(\lambda)^{-1} = (\mathrm{I}+\mathfrak{J}(\lambda))^{-1}\mathfrak{R}(\lambda)$. In view of proposition 3.2.2 and the compactness of the imbedding $H^{s-\mathrm{Re}a+1}(\lambda, S^{n-1}) \subset H^{s-\mathrm{Re}a}(\lambda, S^{n-1})$, the operator $\mathfrak{U}(\lambda) - \mathfrak{U}(\mu): H^s(\lambda, S^{n-1}) \to H^{s-\mathrm{Re}a}(\lambda, S^{n-1})$ is compact for any μ. Together with the invertibility of $\mathfrak{U}(\lambda)$ for certain λ this leads to the conclusion that the values of the function $\lambda \mapsto \mathfrak{U}(\lambda)$ are Fredholm operators. By applying proposition 1.6 we find that $\mathfrak{U}(\lambda)$ is an invertible operator on the whole λ-plane, except at isolated points which can only condense at poles of $\mathfrak{U}$. Hence a strip $\tau_1 \leqslant \mathrm{Im}\lambda \leqslant \tau_2$ contains at most finitely many points at which the operator $\mathfrak{U}(\lambda)$ does not have a bounded inverse. ∎

R e m a r k 1.8. In theorem 3.3 of [21] proposition 1.6 is extended to finite-meromorphic functions; in this case the points at which the inverse operator $\mathfrak{U}(\lambda)^{-1}$ does not exist does not condense at poles. However, canonical meromorphic pseudodifferential operators are, in general, not finite-meromorphic (cf. §1 of Chapter 3). Hence the question of whether the singularities of $\mathfrak{U}^{-1}$ can condense at points at which the operator-function $\lambda \mapsto \mathfrak{U}(\lambda)$ is not finite-meromorphic remains open.

R e m a r k 1.9. The isolated singular points at which the operator $\mathfrak{U}(\lambda)^{-1}$ does not exist are poles of the operator-function $\lambda \mapsto \mathfrak{U}(\lambda)^{-1}$ and are 'normal eigenvalues' of the function $\lambda \mapsto \mathfrak{U}(\lambda)$ (here λ plays the role of spectral parameter); cf. [21]. The proof of theorem 1.7 implies that the poles of the functions $\lambda \mapsto \mathfrak{U}(\lambda)^{\pm 1}$ which are located in a strip $|\mathrm{Im}\lambda| < h$ cannot condense at infinity. Hence a line $\mathrm{Im}\lambda = \tau$ is admissible for the operator A if and only if it does not contain poles of the operators $\mathfrak{U}^{\pm 1}$.

R e m a r k 1.10. Let $\mathfrak{R}$ be the meromorphic pseudodifferential operator considered in the proof of theorem 1.7. Clearly, $\mathfrak{R}(\lambda) = \mathfrak{R}(\lambda)\mathfrak{U}(\lambda)\mathfrak{U}(\lambda)^{-1} = (I+\mathfrak{J}(\lambda))\mathfrak{U}(\lambda)^{-1}$, i.e. $\mathfrak{U}^{-1}$ differs from $\mathfrak{R}$ by a term of order $-\infty$ (it is, however, not excluded that the singularities of this term may condense to poles of $\mathfrak{U}$).

3. The quotient norm of a pseudodifferential operator. We introduce in $H^s_\beta(\mathbf{R}^n)$ the equivalent norm

$$|u;H^s_\beta(\mathbf{R}^n)|_e = \left[\int_{\operatorname{Im}\lambda=\beta-s} \|\mathrm{E}(\sigma-is)\tilde{u}(\lambda+in/2,\cdot);H^0(S^{n-1})\|^2 d\lambda\right]^{1/2},$$

where $\sigma = \operatorname{Re}\lambda$ if $|\operatorname{Re}\lambda| \geqslant 1$, and $\sigma = 1$ if $|\operatorname{Re}\lambda| < 1$ (cf. Remark 3.6.3). Since for $\operatorname{Im}\lambda = 0$ the operator $\mathrm{E}(\lambda):H^0(S^{n-1})\to H^0(S^{n-1})$ is unitary (corollary 1.5.6), we have $|u;H^0_\beta(\mathbf{R}^n)|_e = \|u;H^0_\beta(\mathbf{R}^n)\|$.

P r o p o s i t i o n 1.11. *Let* A *be a canonical pseudodifferential operator of the form* (1.1), *where* $\mathfrak{U}(\lambda) = \mathrm{E}(\lambda+ia)^{-1}\Phi(\phi,\theta)\mathrm{E}(\lambda)$, *a is a real number,* $\tau = \beta-a$, *and* $\beta-a \neq k+n/2$, $\beta \neq -k-n/2$, $k = 0,1,\cdots$. *Then there is for any positive* ϵ *a continuous operator* $\mathrm{T}:H^a_\beta(\mathbf{R}^n)\to H^1_{\beta+1}(\mathbf{R}^n)$ *such that the inequality*

$$|(\mathrm{A}-\mathrm{T})v;H^0_\beta(\mathbf{R}^n)|_e \leqslant (K+\epsilon)|v;H^a_\beta(\mathbf{R}^n)|_e \tag{1.8}$$

holds, where $K = \|\Phi;C(S^{n-1}\times S^{n-1})\|$ *is an exact constant.*

P r o o f. By theorem 3.6.5 there is an operator-function $\lambda\mapsto\mathfrak{J}(\lambda)$ on the line $\operatorname{Im}\lambda = \beta-a$ such that

$$\|\mathfrak{J}(\lambda)\tilde{v}(\lambda+in/2,\cdot);H^1(\lambda,S^{n-1})\| \leqslant$$
$$\leqslant c\|\tilde{v}(\lambda+in/2,\cdot);H^a(\lambda,S^{n-1})\|$$

and

$$\|(\mathfrak{U}(\lambda)-\mathfrak{J}(\lambda))\tilde{v}(\lambda+in/2,\cdot);H^0(S^{n-1})\| \leqslant \tag{1.9}$$
$$\leqslant (K+\epsilon)\|\mathrm{E}(\sigma-ia)\tilde{v}(\lambda+in/2,\cdot);H^0(S^{n-1})\|.$$

Introduce an operator T, coinciding with the righthand side of (1.1) after replacing $\mathfrak{U}(\lambda)$ by $\mathfrak{J}(\lambda)$. It is easy to see that the map $T:H^a_\beta(\mathbb{R}^n)\to H^1_{\beta+1}(\mathbb{R}^n)$ is continuous. The estimate (1.9) implies (1.8).

It remains to prove that K is exact. Let $\{u_j\}$, $\{\lambda_j\}$ be the sequences of functions and numbers used in proposition 3.6.4, let U_j be intervals on $\operatorname{Im}\lambda = \beta - a$ with centers at the points λ_j such that the length of U_j tends to zero as $j\to\infty$. Let also ζ_j be a nonnegative smooth functions with supports in U_j, and equal to one near λ_j. Put $\tilde{v}_j(\lambda+in/2,\phi) = \zeta_j(\lambda)u_j(\phi)$. We have $|Tv_j;H^0_\beta(\mathbb{R}^n)|_e \leq \delta_j|v_j;H^a_\beta(\mathbb{R}^n)|_e$, where $\delta_j\to 0$. It is obvious that $\|\mathfrak{U}(\lambda)\tilde{v}_j(\lambda+in/2,\cdot)\| \leq \|(\mathfrak{U}(\lambda)-\mathfrak{U}(\lambda_j))\tilde{v}_j(\lambda+in/2,\cdot)\| + \|\mathfrak{U}(\lambda_j)\tilde{v}_j(\lambda+in/2,\cdot)\|$ (the norms are taken in $H^0(S^{n-1})$). The first term does not exceed $\gamma_j\|\tilde{v}_j(\lambda+in/2,\cdot);H^a(\lambda,S^{n-1})\|$, where $\gamma_j\to 0$, while the second term can, by proposition 3.6.4, be written as $c_0\|\mathrm{E}(\sigma_j-ia)\tilde{v}_j(\lambda+in/2,\cdot);H^0(S^{n-1})\| + \epsilon_j\|\tilde{v}_j(\lambda+in/2,\cdot);\ H^0(S^{n-1})\|$, with $c_0 = |\Phi(\phi_0,\theta_0)|$, $\epsilon_j\to 0$. This means that $|(\mathrm{A}-\mathrm{T})v_j;\ H^0_\beta(\mathbb{R}^n)|_e = c_0|v_j;\ H^a_\beta(\mathbb{R}^n)|_e + \epsilon_j|v_j;\ H^a_\beta(\mathbb{R}^n)|_e$, $\epsilon_j\to 0$. This implies that the constant K is exact (since the point $(\phi_0,\theta_0)\in S^{n-1}\times S^{n-1}$ can be chosen arbitrarily). ■

4. Periodic pseudodifferential operators on $\mathbb{R}^n$. In **1**, starting with meromorphic pseudodifferential operators, we arrived at pseudodifferential operators on $\mathbb{R}^n$ on the spaces $H^s_\beta(\mathbb{R}^n)$. Similarly, the periodic meromorphic pseudodifferential operators obtained in §7, Chapter 3, lead to periodic pseudodifferential operators on $\mathbb{R}^n$.

A canonical periodic pseudodifferential operator has the form (3.7.3). The representation (3.7.5) is valid for such an operator. Let **A** be the general periodic meromorphic pseudodifferential operator of order a defined in Chapter 3, §3.3. For $u\in C_0^\infty(\mathbb{R}^n\setminus 0)$ we put

$$\mathrm{A}u = (\mathrm{G}_{r\to(r,\mu+ia+in/2)})^{-1}\mathbf{A}(r;\mu)\mathrm{G}_{r\to(r,\mu+ia+in/2)}u, \tag{1.10}$$

where $\operatorname{Im}\mu = \tau$ and τ is such that the interval $\mathbf{I}_\tau = \{\mu\in\mathbb{C}\colon 0\leq \operatorname{Re}\mu\leq 2\pi,\ \operatorname{Im}\mu = \tau\}$ does not contain poles of **A**. The norm $\|u;H^s_\beta(\mathbb{R}^n)\|$ is equivalent to the norm

$$\left[\int_{\mathbf{I}_\tau}\|\mathrm{G}u(\mu+in/2,\cdot);H^s(\Pi)\|^2 d\mu\right]^{1/2},\quad \tau = \beta - s.$$

(This can be verified using (3.7.3).) Since $\mathbf{A}(\cdot,\mu):H^s(\Pi)\to H^{s-\mathrm{Re}\,a}(\Pi)$ is bounded, uniformly with respect to $\mu\in\mathbf{L}_\tau$, for $\tau=\beta-s$ the operator (1.10) realizes a continuous map $\mathbf{A}:H^s_\beta(\mathbb{R}^n)\to H^{s-\mathrm{Re}\,a}_\beta(\mathbb{R}^n)$.

For $\tau=\beta-s$ the operator (1.10) is called a (general) periodic pseudodifferential operator of order a in $H^s_\beta(\mathbb{R}^n)$. If $\Sigma\Phi_j(r,\phi,\theta)$ is the symbol of A, then $r^{-a}\Sigma\Phi_j(r,x,\theta)$ is called the symbol of A; recall that the function $(r,x,\theta)\mapsto\Phi_j(r,x,\theta)$ is regarded as being homogeneous of degree a_j both in x and in θ, and that $\Phi_j(re,\cdot)=\Phi_j(r,\cdot)$, $r=|x|$. The operator (1.10) has an asymptotic series expansion $\Sigma r^{a_j-a}A_j$, where the A_j are canonical periodic pseudodifferential operators of orders a_j (compare with (1.4)). Theorem 3.7.3 implies that there is a periodic pseudodifferential operator with arbitrary, pre-given, symbol.

T h e o r e m 1.12. *Let* $\mathbf{A}:H^s_\beta(\mathbb{R}^n)\to H^{s-\mathrm{Re}\,a_0}_\beta(\mathbb{R}^n)$, $\mathbf{B}:H^{s+\mathrm{Re}\,b_0}_\beta(\mathbb{R}^n)\to H^s_\beta(\mathbb{R}^n)$ *be periodic pseudodifferential operators, and let* $\sigma_{\mathbf{A}}(r,\phi,\theta)=\Sigma_{j=0}^{\infty}r^{a_j-a_0}\Phi_j(r,\phi,\theta)$, $\sigma_{\mathbf{B}}(r,\phi,\theta)=\Sigma_{k=0}^{\infty}r^{b_k-b_0}\Psi_k(r,\phi,\theta)$ *be their symbols. Then the operator* $\mathbf{AB}:H^{s+\mathrm{Re}\,b_0}_\beta(\mathbb{R}^n)\to H^{s-\mathrm{Re}\,a_0}_\beta(\mathbb{R}^n)$ *is a periodic pseudodifferential operator, and its symbol is*

$$\sum_{j,k,\gamma} r^{a_j+b_k-|\gamma|-a_0-b_0}\times \tag{1.11}$$

$$\times\left\{\frac{1}{\gamma!}r^{|\gamma|+b_0-b_k}D^\gamma_\theta\Phi_j(r,\phi,\theta)\partial^\gamma_x(r^{-b_0}\Psi_k(r,x,\theta))\right\}.$$

The expression between brackets, $S_{jk\gamma}$, *is periodic in* r: $S_{jk\gamma}(re,\cdot)=S_{jk\gamma}(r,\cdot)$. *The principal symbol of* AB *is, according to* (1.11), *equal to* $\Phi_0(r,\phi,\theta)\Psi_0(r,\phi,\theta)$.

P r o o f. The asymptotic series for A and B can be written as

$$\mathbf{A}\sim(\mathbf{G}_{r\to(r,\mu+ia_0+in/2)})^{-1}\sum_j\mathbf{A}_j(r,\mu)\mathbf{G}_{r\to(r,\mu+in/2)},$$

$$\mathbf{B}\sim(\mathbf{G}_{r\to(r,\mu+in/2)})^{-1}\sum_k\mathbf{B}_k(r,\mu-ib_0)\mathbf{G}_{r\to(r,\mu-ib_0+in/2)},$$

where $\mathbf{A}_j$, $\mathbf{B}_k$ are periodic meromorphic pseudodifferential operators with symbols $\Phi^{(j)}$ and $\Psi^{(k)}$, $\mathrm{Im}\,\mu=\beta-s$. Hence

$$\mathbf{AB}\sim\sum_{j,k}(\mathbf{G}_{r\to(r,\nu+i(a_0+b_0)+in/2)})^{-1}\mathbf{A}_j(r,\nu+ib_0)\times$$

$$\times \mathbf{B}_k(r,\nu)\mathbf{G}_{r\to(r,\nu+in/2)},$$

moreover, $\operatorname{Im}\nu = \beta - s - \operatorname{Re} b_0$. Further,

$$\mathbf{A}_j(r,\nu+ib_0)U = \sum_{k=-\infty}^{+\infty} r^{2\pi ik}\mathfrak{U}_j(r,\nu+ib_0+2k\pi)U_k.$$

By proposition 3.2.2, the symbol of the meromorphic pseudodifferential operator $\mathfrak{U}_j(r,\nu+ib_0)$ is equal to $r^{b_0}\Sigma_{|\gamma|\geq 0}(1/\gamma!)\partial_x^\gamma r^{-b_0}D_\theta^\gamma\Phi_j(r,x,\theta)$. By theorem 3.7.4, $\mathbf{A}_j(r,\nu+ib_0)\mathbf{B}_k(r,\nu)$ is a periodic meromorphic pseudodifferential operator, with symbol

$$\sum_{\gamma,\kappa}\frac{1}{\gamma!\kappa!}r^{b_0}\partial_x^\gamma r^{-b_0}D_\theta^{\gamma+\kappa}\Phi_j(r,\phi,\theta)r^{|\kappa|-b_k}\partial_x^\kappa\Psi_k(r,x,\theta).$$

So, the operator AB has the asymptotic series expansion

$$(\mathbf{G}_{r\to(r,\mu+i(a_0+b_0)+n/2)})^{-1}\sum_{j,k,\gamma}\mathbf{A}_{jk\gamma}(r,\mu)\mathbf{G}_{r\to(r,\mu+in/2)},$$

where $\mathbf{A}_{jk\gamma}$ is a periodic meromorphic pseudodifferential operator with symbol

$$\frac{1}{\gamma!}r^{|\gamma|+b_0-b_k}D_\theta^\gamma\Phi_j(r,\phi,\theta)\partial_x^\gamma(r^{-b_0}\Psi_k(r,x,\theta)).$$

This series can be rewritten as

$$\sum_{j,k,\gamma} r^{a_j+b_k-|\gamma|-a_0-b_0}\times$$

$$\times(\mathbf{G}_{r\to(r,\mu+i(a_j+b_k-|\gamma|)+in/2)})^{-1}\mathbf{A}_{jk\gamma}\mathbf{G}_{r\to(r,\mu+in/2)}.$$

This immediately implies the assertion of the theorem. ■

In order to describe the adjoint operator we need

Proposition 1.13. *Suppose a function* $(\mu,r,\phi)\mapsto V(\mu,r,\phi)$ *is 'periodic' in* $r: V(\mu,re,\phi) = V(\mu,r,\phi)$, $\mu\in\mathbf{C},\phi\in S^{n-1}$. *Suppose also that* $u\in C^\infty(\mathbf{R}^n\setminus 0)$. *Then*

$$\int_{\mathbf{R}^n}(\mathbf{G}_{r\to(r,\mu+in/2)})^{-1}V\cdot\bar{u}\,dx =$$

$$= \int_{\mathbf{I}_\tau}d\mu\int_1^e\frac{dr}{r}\int_{S^{n-1}}V\overline{(\mathbf{G}_{r\to(r,\bar\mu+in/2)}u)}\,d\phi.$$

P r o o f. By the inversion formula (3.7.2) we have

$$\int_{\mathbb{R}^n} (G_{r\to(r,\mu+in/2)})^{-1} V\cdot\bar{u}dx =$$

$$= \frac{1}{\sqrt{2\pi}} \int_0^\infty \int_{S^{n-1}} \overline{u(r,\phi) r^{n-1} dr d\phi \int_{L} r^{i(\mu+in/2)} V d\mu} =$$

$$= \sum_{j=-\infty}^{+\infty} \frac{1}{\sqrt{2\pi}} \int_{e^j}^{e^{j+1}} \overline{u(r,\phi) r^{n-1} dr d\phi \int_{L} r^{i(\mu+in/2)} V d\mu}.$$

Using the periodicity of V, the last expression can be rewritten as

$$\sum_{j=-\infty}^{+\infty} \frac{1}{\sqrt{2\pi}} \int_{S^{n-1}} d\phi \int_1^e \overline{u(re^j,\phi)(re^j)^n \frac{dr}{r}} \times$$

$$\times \int_{L} (re^j)^{i(\mu+in/2)} V(\mu,r,\phi) d\mu =$$

$$= \int_{L} d\mu \int_1^e \frac{dr}{r} \int_{S^{n-1}} V(\mu,r,\phi) \frac{1}{\sqrt{2\pi}} \sum_{j=-\infty}^{+\infty} \overline{(re^j)^{-i(\bar{\mu}+in/2)} u(re^j,\phi) d\phi}. \quad \blacksquare$$

The verification of the two following theorems is left to the reader.

T h e o r e m 1.14. *Let* $A: H^s_\beta(\mathbb{R}^n) \to H^{s-\operatorname{Re} a_0}_\beta(\mathbb{R}^n)$ *be a periodic pseudodifferential operator of order* a_0, *and let* $\sum r^{a_j - a_0} \Phi_j(r,\phi,\theta)$ *be its symbol. Then the operator* $A^*: H^{\operatorname{Re} a_0 - s}_{-\beta}(\mathbb{R}^n) \to H^{-s}_{-\beta}(\mathbb{R}^n)$ *adjoint to* A *with respect to the scalar product in* $L_2(\mathbb{R}^n)$ *is also a periodic pseudodifferential operator. The order of* A^* *is* $\bar{a}_0$, *and its symbol is*

$$\sum_{j,|\gamma|=0}^{\infty} r^{\bar{a}_j - |\gamma| - \bar{a}_0} \left\{ \frac{1}{\gamma!} r^{|\gamma| + \bar{a}_0 - \bar{a}_j} \partial^\gamma_x D^\gamma_\xi \overline{(r^{-a_0} \Phi_j(r,x,\theta))} \right\}. \tag{1.12}$$

The expression between brackets, $S_{j\gamma}(r,\phi,\theta)$, *satisfies the condition* $S_{j\gamma}(re,\cdot) = S_{j\gamma}(r,\cdot)$. *The principal symbol of* A^* *is, by (1.12), equal to* $\overline{\Phi_0(r,\phi,\theta)}$.

T h e o r e m 1.15. *Let* A *be an elliptic periodic pseudodifferential operator of order* a_0 *(i.e. the principal symbol* $\Phi_0(r,\phi,\theta)$ *of this operator does not vanish on* $[1,e] \times S^{n-1} \times S^{n-1}$*). Then there is a periodic pseudodifferential operator* Q *of order* $-a_0$ *such that the symbols of* AQ *and QA are the identity function.*

§2. Pseudodifferential operators on a conic manifold

1. The conic manifold X. A set $K \subset \mathbb{R}^n \setminus 0$ is called *conic* if $x \in K$ implies $tx \in K$ for all $t > 0$.

Let X be a Hausdorff space and Θ a point of it. Suppose that the following conditions hold: 1) there is a finite cover $\{U_j\}$ of the subspace $X \setminus \Theta$ by open sets for each of which there is defined a homeomorphism κ_j onto a conic subset of $\mathbb{R}^n \setminus 0$; and 2) if $U_j \cap U_i \neq \varnothing$, then the map $\kappa_j \circ \kappa_i^{-1} : \kappa_i(U_i \cap U_j) \to \kappa_j(U_i \cap U_j)$ is homogeneous of degree one and infinitely differentiable. In this situation we can define, as usual, a differentiable structure, which will be called a *conic structure* of class $\check{C}^\infty$ on X. The space X endowed with this structure will be called an *n-dimensional conic manifold* (a *cone*) of class $\check{C}^\infty$ with vertex at Θ.

Multiplication of a point x of X by a positive number t is defined by $tx = \kappa_j^{-1}(t\kappa_j(x))$, if x belongs to the coordinate neighborhood U_j. The fact that this is a well-defined definition follows from the homogeneity of the transformation $\kappa_j \circ \kappa_i^{-1}$. By the same token, $\kappa_j(tx) = t\kappa_j(x)$, i.e. the homeomorphism κ_j is a homogeneous map of the first degree.

We denote by $H^s(\lambda, \mathbb{R}^n)$ the space of homogeneous functions on $\mathbb{R}^n \setminus 0$ of degree $i\lambda - n/2$, with norm $\|u; H^s(\lambda, \mathbb{R}^n)\| = \|u|_{S^{n-1}}; H^s(\lambda, S^{n-1})\|$, where $u|_{S^{n-1}}$ is the trace of u on S^{n-1}. Let also $\{\zeta\}$ be a partition of unity of $X \setminus \Theta$ subordinate to the conic atlas $\{U, \kappa\}$ and consisting of infinitely differentiable homogeneous functions of degree zero (for simplicity, the indices are sometimes omitted). Introduce the space $H^s(\lambda, X)$ of homogeneous functions on $X \setminus \Theta$ of degree $i\lambda - n/2$, endowed with the norm

$$\|u; H^s(\lambda, X)\| = \sum_U \|\zeta_\kappa u; H^s(\lambda, \mathbb{R}^n)\|,$$

where $\zeta_\kappa u = (\zeta u) \circ \kappa^{-1}$ on $\kappa(U)$ and $\zeta_\kappa u = 0$ outside $\kappa(U)$. A distinct equivalent atlas and a distinct partition of unity lead to an equivalent norm.

2. Meromorphic pseudodifferential operators on X. We will assume that canonical meromorphic pseudodifferential operators, written using the exterior form (cf. Chapter 3, §1.4), act on homogeneous functions on $\mathbb{R}^n \setminus 0$ of degree $i\lambda - n/2$. Taking this understanding into account, definition 3.3.1. for general meromorphic pseudodifferential operators remains valid; we only need replace the $H^s(\lambda, S^{n-1})$

by the spaces $H^s(\lambda,\mathbb{R}^n)$ of homogeneous functions. Such operators will be called *meromorphic pseudodifferential operators on* $\mathbb{R}^n$.

We further consider operator-functions $\lambda\mapsto\mathfrak{U}(\lambda)$ whose values are operators given on $H^s(\lambda,X)$. We may free ourselves from the dependence on λ of the domains of definition of these operators. For this we fix a submanifold S of $X\setminus\mathcal{O}$ intersecting every ray $\{y\in X: y=tx_0, x_0\in X\setminus\mathcal{O}, t>0\}$ at a single point, and replace the homogeneous functions by their traces on S. Thus, we may speak of an operator-function $\lambda\mapsto\mathfrak{U}(\lambda)$ being meromorphic, and use general theorems concerning meromorphic operator-functions when applicable.

D e f i n i t i o n 2.1. A meromorphic operator-function $\lambda\mapsto\mathfrak{U}(\lambda)$ is called a *meromorphic pseudodifferential operator of order a on X* if the following conditions hold:

a) every strip $|\operatorname{Im}\lambda|<h$ contains only finitely many poles of $\mathfrak{U}$;

b) for two arbitrary infinitely differentiable functions ζ and η, homogeneous of degree zero and with intersecting supports (on $X\setminus\mathcal{O}$), the following inequality holds:

$$\|\zeta\mathfrak{U}(\lambda)\eta;H^s(\lambda,X)\to H^{s+p}(\lambda,X)\|\leq c(\mathfrak{F},N,p)(1+|\lambda|)^{-N},$$

where $\mathfrak{F}$ is an arbitrary admissible set, $\lambda\in\mathfrak{F}$, p and N are arbitrary positive integers, and $c(\mathfrak{F},N,p)=\text{const}<\infty$;

c) if the supports of ζ and η belong to one coordinate neighborhood U, then

$$\zeta\mathfrak{U}(\lambda)\eta=\zeta^{\kappa}\mathfrak{U}_U(\lambda)\eta_{\kappa},$$

where $\mathfrak{U}_U$ is a meromorphic pseudodifferential operator of order a on $\mathbb{R}^n$, $\zeta^{\kappa}v=\zeta(v\circ\kappa)$ on U and $\zeta^{\kappa}v=0$ outside U, while $\eta_{\kappa}u$ is to be understood as in **1**.

The fact that this definition is well-defined follows from the rule of change of variables in a meromorphic pseudodifferential operator on $\mathbb{R}^n$ (cf. §8, Chapter 3). This rule allows us to bring every meromorphic pseudodifferential operator in correspondence with a (principal) symbol Φ - a function on the bundle $T_0^*(X)$ of nonzero cotangent vectors over $X\setminus\mathcal{O}$; the principal symbol Φ_U of the meromorphic pseudodifferential operator $\mathfrak{U}_U$ on $\mathbb{R}^n$ figuring in formula (2.1) is the representation of Φ in the coordinate neighborhood U, i.e. $\Phi_U=\Phi\circ(\kappa_*)^{-1}$, where

$\kappa_*: T_0^*(X)|_U \to (\mathbf{R}^n\setminus 0)\times(\mathbf{R}^n\setminus 0)$, where the part of $T_0^*(X)$ above U is regarded as the direct product $U\times(\mathbf{R}^n\setminus 0)$, $\kappa_*(x,\tau) = (\kappa(x),\kappa'(x)\tau)$, $(x,\tau)\in U\times(\mathbf{R}^n\setminus 0)$. Recall that the function $(y,z)\mapsto\Phi_U(y,z)$ is homogeneous of degree a both in y and in z. Therefore we have that if τ_x is a cotangent vector in the fiber $T_0^*(X)_x$ above a point x, then for $t>0$,

$$\Phi(t\tau_x) = \Phi(\tau_{tx}) = t^a\Phi(\tau_x). \tag{2.2}$$

P r o p o s i t i o n 2.2. *Whatever a function Φ, smooth on $T_0^*(X)$ and satisfying* (2.2), *there is a meromorphic pseudodifferential operator $\mathfrak{U}$ of order a on X whose (principal) symbol coincides with Φ.*

P r o o f. Let $\{U_j,\kappa_j\}$ be an atlas on X, and $\{\zeta_j\}$ a partition of unity subordinate to it and consisting of smooth homogeneous functions of degree zero. Let $\{\eta_j\}$ be a set of smooth homogeneous functions of degree zero such that $\operatorname{supp}\eta_j\subset U_j$, $\eta_j\zeta_j = \zeta_j$.

Put

$$\mathfrak{U}(\lambda) = \sum_j(\zeta_j)^{\kappa_j}\mathfrak{U}_{U_j}(\lambda)(\eta_j)_{\kappa_j}, \tag{2.3}$$

where $\mathfrak{U}_{U_j}(\lambda) = \mathrm{E}(\lambda+ia)^{-1}\Phi_{U_j}(y,z)\mathrm{E}(\lambda)$ are canonical meromorphic pseudodifferential operators on $\mathbf{R}^n$ and the notations ζ^κ,η_κ have the same meaning as in (2.1).

We will show that $\mathfrak{U}$ is the required meromorphic pseudodifferential operator on X. It is obvious that condition a) of definition 2.1 is fulfilled. Equation (2.3) and proposition 3.1.6 imply that $\mathfrak{U}$ also satisfies b). Let, finally, the supports of η and ζ lie in one coordinate neighborhood U_k. In view of the fact that

$$\zeta\mathfrak{U}(\lambda)\eta = \sum_j(\zeta\zeta_j)^{\kappa_j}\mathfrak{U}_{U_j}(\lambda)(\eta\eta_j)_{\kappa_j}, \tag{2.4}$$

the sum will only contain those term for which $\operatorname{supp}\zeta_j\cap U_k\neq\varnothing$. Write the righthand side of (2.4) in the local coordinates of U_k. By the rule for change of variables, (2.4) takes the form

$$\zeta\mathfrak{U}(\lambda)\eta = \sum_j(\zeta\zeta_j)^{\kappa_k}\mathfrak{B}(\lambda)(\eta\eta_j)_{\kappa_k},$$

where $\mathfrak{B}$ is a meromorphic pseudodifferential operator on $\mathbf{R}^n$ with principal symbol Φ_{U_k}. By applying proposition 3.2.1. we obtain that the righthand side equals

$$\sum_j (\zeta\zeta_j\eta_j)^{\kappa_k}\mathfrak{C}(\lambda)\eta_{\kappa_k} = \sum_j (\zeta\zeta_j)^{\kappa_k}\mathfrak{C}(\lambda)\eta_{\kappa_k} = \zeta^{\kappa_k}\mathfrak{C}(\lambda)\eta_{\kappa_k},$$

where $\mathfrak{C}$ is a meromorphic pseudodifferential operator on $\mathbf{R}^n$ with the same principal symbol Φ_{U_k}. ■

Using theorem 3.3.4 the following assertion is easily verified.

Proposition 2.3. *Let $\mathfrak{U}$ and $\mathfrak{B}$ be meromorphic pseudodifferential operators on X of orders a and b. Then their composite $\mathfrak{U}\mathfrak{B}$ is a meromorphic pseudodifferential operator of order $a+b$, and its symbol is the product of the symbols of the factors.*

3. Pseudodifferential operators on X. For a function $u \in C_0^\infty(X \setminus \mathcal{O})$ we define the Mellin transform by

$$\tilde{u}(\lambda,x) = \frac{1}{\sqrt{2\pi}} \int_0^\infty t^{-i\lambda-1}u(tx)dt, \tag{2.5}$$

where λ is a complex parameter. It is clear that $x \mapsto \tilde{u}(\lambda,x)$ is a homogeneous function of degree $i\lambda$. The inversion formula

$$u(x) = \frac{1}{\sqrt{2\pi}} \int_{\mathrm{Im}\lambda=\tau} \tilde{u}(\lambda,x)d\lambda \tag{2.6}$$

and Parseval's equality

$$\int_{\mathrm{Im}\lambda=\tau} |\tilde{u}(\lambda,x)|^2 d\lambda = \int_0^\infty t^{2\beta}|u(tx)|^2 dt, \quad \tau = \beta + 1/2,$$

hold.

For arbitrary real s and β we introduce the space $H_\beta^s(X)$ as the completion of $C_0^\infty(X \setminus \mathcal{O})$ with respect to the norm

$$\|u;H_\beta^s(X)\| = \left[\int_{\mathrm{Im}\lambda=\tau} \|\tilde{u}(\lambda+in/2,\cdot);H^s(\lambda,X)\|^2 d\lambda\right]^{1/2}. \tag{2.7}$$

Let S be a 'directing' submanifold of the

manifold $X\setminus\Theta$ intersecting every ray $\{y\in X: y=tx_0, x_0\in X\setminus\Theta, t>0\}$ at a single point. For every point $x\in X\setminus\Theta$ we denote by $\rho_S(x)$ the positive number for which $\rho_S(x)^{-1}x\in S$. Clearly, $\rho_S(tx) = t\rho_S(x)$ for $t>0$.

D e f i n i t i o n 2.4. A *pseudodifferential operator of order a on a space* $H^s_\beta(X)$ is an operator of the form

$$(\mathrm{A}u)(x) = \frac{1}{2\pi}\rho_S(x)^{-a}\int_{\mathrm{Im}\lambda=\tau}\mathfrak{U}_{y\to x}(\lambda)\tilde{u}(\lambda+in/2,y)d\lambda, \tag{2.8}$$

where $\mathfrak{U}$ is a meromorphic pseudodifferential operator of order a on X (cf. definition 2.1) without poles on the line $\mathrm{Im}\lambda = \tau = \beta - s$.

P r o p o s i t i o n 2.5.

1) *An operator* (2.8) *realizes a continuous map* $\mathrm{A}: H^s_\beta(X)\to H^{s-\mathrm{Re}\,a}_\beta(X)$.

2) *For two arbitrary infinitely differentiable functions* ζ *and* η *on* $X\setminus\Theta$, *homogeneous of degree zero and with intersecting supports, the following inequality holds:*

$$\|\zeta\mathrm{A}\eta; H^s_\beta(X)\to H^{s-\mathrm{Re}\,a+N}_{\beta+N}(X)\|\leqslant c_N,$$

where N *is an arbitrary number,* $c_N = \mathrm{const}<\infty$.

3) *If the supports of* ζ *and* η *belong to a coordinate neighborhood* U *on* X, *then* $\zeta\mathrm{A}\eta u = \zeta^\kappa\mathrm{A}_U\eta_\kappa u$, *where* A_U *is a pseudodifferential operator of order* a *on* $\mathbb{R}^n$ *and the notations* $\zeta^\kappa, \eta_\kappa$ *have the same meaning as in* (2.1).

P r o o f. Put $v = \mathrm{A}u$. Then

$$\widetilde{(\rho^a_S, v)}(\lambda+in/2,x) = \mathfrak{U}_{y\to x}(\lambda)\tilde{u}(\lambda+in/2,y). \tag{2.9}$$

On the other hand,

$$\widetilde{(\rho^a_S v)}(\lambda+in/2,x) = \int_0^\infty t^{-i(\lambda+in/2)-1}\rho_S(tx)^a v(tx)dt =$$

$$= \int_0^\infty t^{-i(\lambda+ia+in/2)-1}v(tx)dt\,\rho_S(x)^a.$$

Hence for $x\in S$ we have

$$\widetilde{(\rho_S^a v)}(\lambda + in/2, x) = \tilde{v}(\lambda + ia + in/2, x). \tag{2.10}$$

Further,

$$\|v; H_\beta^{s-\mathrm{Re}\,a}(X)\| =$$

$$= \left[\int_{\mathrm{Im}\lambda=\beta-s} \|\tilde{v}(\lambda+ia+in/2,\cdot); H^{s-\mathrm{Re}\,a}(\lambda+ia,X)\|^2 d\lambda\right]^{1/2}.$$

This, equation (2.10) and the definition of the norm in $H^t(\mu, X)$ imply that

$$\|v; H_\beta^{s-\mathrm{Re}\,a}(X)\| \leqslant$$

$$\leqslant c\left[\int_{\mathrm{Im}\lambda=\beta-s} \|\widetilde{(\rho_S^a v)}(\lambda+in/2,\cdot); H^{s-\mathrm{Re}\,a}(\lambda,X)\|^2 d\lambda\right]^{1/2}.$$

By (2.9) this means that

$$\|v; H_\beta^{s-\mathrm{Re}\,a}(X)\| \leqslant$$

$$\leqslant c\left[\int_{\mathrm{Im}\lambda=\beta-s} \|\widetilde{(\rho_S^a v)}(\lambda+in/2,\cdot); H^{s-\mathrm{Re}\,a}(\lambda,X)\|^2 d\lambda\right]^{1/2} \leqslant$$

$$\leqslant c\|u; H_\beta^s(X)\|.$$

Assertions 2) and 3) readily follow from definitions 2.4 and 2.1. ■

A meromorphic pseudodifferential operator $\mathfrak{U}$ on X is called *elliptic* if its symbol $T_0^*(X) \ni \tau_x \mapsto \Phi(\tau_x)$ does not take the value zero on $T_0^*(X)$. An operator A of the form (2.8) will be called *elliptic* only if the corresponding meromorphic pseudodifferential operator $\mathfrak{U}$ is elliptic.

As in the case of a Euclidean space, a number $\tau = \beta - s$ will be called admissible for an operator $\mathrm{A}: H_\beta^s(X) \to H_\beta^{s-\mathrm{Re}\,a}(X)$ if this operator realizes an isomorphism.

The following statement is a variant for conic manifolds of theorem 1.7.

T h e o r e m 2.6. *Let* A *be an elliptic pseudodifferential operator of order* a *on* X *of the form* (2.8), *and suppose there are no poles of the meromorphic pseudodifferential operator* $\mathfrak{U}$ *in a strip* $\tau_1 \leqslant \mathrm{Im}\lambda \leqslant \tau_2$. *Then every number* $\tau \in [\tau_1, \tau_2]$ *(except for an at most finite set) is admissible for* A.

The p r o o f is completely analogous to that of theorem 1.7. If Φ is the symbol of $\mathfrak{U}$, then for $\mathfrak{R}$ we must take a meromorphic pseudodifferential operator of order $-a$ whose symbol is Φ^{-1}; the existence of this operator is guaranteed by proposition 2.2. ■

Remark 1.9 remains true for operators on the cone X.

4. The one-dimensional case (singular integral operators on a system of rays). Let $R_j = \{z \in \mathbf{C}: z = \rho e^{i\theta_j}\}$ be a ray in the complex plane emanating from the coordinate origin, $j = 1, \cdots, N$, $0 \leqslant \theta_1 < \cdots < \theta_N < 2\pi$, $0 \leqslant \rho < \infty$, and let $X = \cup_{j=1}^{N} R_j$. Consider the singular integral

$$(\mathcal{K}u)(x) = \frac{1}{\pi i}\int_X \frac{u(y)}{y-x}dy, \quad x \in X,$$

where $u \in C_0^\infty(X \setminus \mathcal{O})$.

P r o p o s i t i o n 2.7. *The following representation holds for the operator* (2.11):

$$(\mathcal{K}u)(r,\theta_j) = \frac{1}{\sqrt{2\pi}}\int_{-\infty}^{+\infty} r^{i(\lambda+in/2)} \times \tag{2.12}$$

$$\times \sum_{k=1}^{N} \mathcal{T}_{jk}(\lambda)\tilde{u}(\lambda+i/2,\theta_k)d\lambda, \quad j = 1, \cdots, N;$$

here $u(r,\theta_k) \equiv u(re^{i\theta_k})$, $\tilde{u}(\cdot,\theta_k)$ *is the Mellin transform of the function* $r \mapsto u(r,\theta_k)$, *and* $\mathcal{T}_{jk}$ *is defined by*

$$\mathcal{T}_{jk}(\lambda) = \tag{2.13}$$

$$= \begin{cases} \dfrac{\exp\{\lambda(\theta_k-\theta_j-\pi)+i(\theta_k-\theta_j)/2\}}{\cosh \pi\lambda} & \text{if } k>j, \\ -\dfrac{\exp\{\lambda(\theta_k-\theta_j+\pi)+i(\theta_k-\theta_j)/2\}}{\cosh \pi\lambda} & \text{if } k<j, \\ -\tanh \pi\lambda & \text{if } k=j, \end{cases}$$

P r o o f. Apply the Melling transform $M_{r\to\lambda+i/2}$ to equation (2.11):

$$\widetilde{(\mathcal{K}u)}(\lambda+i/2,\theta_j) = \tag{2.14}$$

$$= \frac{1}{\pi i\sqrt{2\pi}} \sum_{k=1}^{N} \int_0^\infty r^{-i(\lambda+i/2)-1} dr \int_0^\infty \frac{u(\rho,\theta_k)e^{i\theta_k} d\rho}{\rho e^{i\theta_k} - \rho e^{i\theta_j}}.$$

First we consider the terms for which $j \neq k$. Changing the order of integration we obtain

$$\int_0^\infty r^{-i\lambda-1/2} dr \int_0^\infty \frac{u(\rho,\theta_k)e^{i\theta_k} d\rho}{\rho e^{i\theta_k} - re^{i\theta_j}} = \tag{2.15}$$

$$= e^{i(\theta_k-\theta_j)} \int_0^\infty u(\rho e^{i\theta_k})\rho^{-i\lambda-1/2} d\rho \int_0^\infty \frac{t^{-i\lambda-1/2} dt}{e^{i(\theta_k-\theta_j)} - t},$$

where the new variable t equals the quotient r/ρ. If a is a nonnegative number,

$$\int_0^\infty \frac{t^{-i\lambda-1/2}}{a-t} dt = \frac{2\pi i \exp\{-(i\lambda+1/2)\ln a\}}{\exp\{-2\pi i(i\lambda+1/2)\}-1}. \tag{2.16}$$

This and (2.15) imply that for $k > j$ the k-th term at the righthand side of (2.14) equals $\exp\{\lambda(\theta_k-\theta_j-\pi)+i(\theta_k-\theta_j)/2\}/\cosh \pi\lambda$, while for $k > j$ it equals $-\exp\{\lambda(\theta_k-\theta_j-\pi)+i(\theta_k-\theta_j)/2\}/\cosh \pi\lambda$.

Now consider the j-th term in (2.14). The inner integral is to be understood as a principal value integral. By the well-known formula of Sokhotskii (cf., e.g., [41]),

$$\frac{1}{\pi i} \int_0^\infty \frac{u(\rho,\theta_j)d\rho}{\rho - r} = \frac{1}{\pi i} \int_0^\infty \frac{u(\rho,\theta_j)d\rho}{\rho - r - i0} - u(r,\theta_j).$$

This, taking into account (2.16), implies that the j-th term in (2.14) equals $-\tilde{u}(\lambda,\theta)\tanh \pi\lambda$. It remains to use the inversion formula (1.2.6) for the Mellin transform. ■

R e m a r k 2.8. For the sake of being specific we have assumed that all rays are emanating. If the q-th ray is incoming*), then $\mathfrak{T}_{jq}$ must be replaced by $-\mathfrak{T}_{jq}$. In particular, the entries of the matrix $\|\mathfrak{T}_{jk}\|_{j,k=1}^2$ corresponding to the operator

$$\frac{1}{\pi i} \int_{-\infty}^\infty \frac{u(y)dy}{y-x}$$

*) i.e. if $-\infty < \rho \leqslant 0$ for the q-th ray. (Translator's note.)

are defined by $\mathfrak{T}_{11}(\lambda) = -\mathfrak{T}_{22}(\lambda) = -\tanh\pi\lambda$, $\mathfrak{T}_{12}(\lambda) = -\mathfrak{T}_{21}(\lambda) = i/\cosh\pi\lambda$ (which is in accordance with Chapter 3, §1.2).

Let a and b be complex-valued functions on $X\setminus\mathcal{O}$ which are constant on each ray R_j, and let a_j, b_j be the values of a,b on R_j. A singular integral operator on X is an operator $\mathrm{A} = a\mathrm{I}+b\mathcal{K}$, where $\mathcal{K}$ is the integral (2.11). Applying proposition 2.7 we rewrite A as

$$(\mathrm{A}u)(r,\cdot) = \frac{1}{\sqrt{2\pi}}\int_{-\infty}^{+\infty} r^{i(\lambda+i/2)}\mathfrak{U}(\lambda)\tilde{u}(\lambda+i/2,\cdot)d\lambda, \tag{2.17}$$

where $\mathfrak{U}(\lambda) = \|\mathfrak{U}_{jk}(\lambda)\|_{j,k=1}^{N}$, $\mathfrak{U}_{jk}(\lambda) = a_j\delta_{jk}+b_j\mathfrak{T}_{jk}(\lambda)$, $\tilde{u}(\lambda+i/2,\cdot)$ is the vector $(\tilde{u}(\lambda+i/2,\theta_1),\cdots,\tilde{u}(\lambda+i/2,\theta_N))$, $(\mathrm{A}u)(r,\cdot)$ is a similar vector, and δ_{jk} are the Kronecker symbols.

Denote by $L_{2,\beta}(X)$ the completion of $C_0^\infty(X\setminus\mathcal{O})$ with respect to the norm

$$\|u;L_{2,\beta}(X)\| = \left[\int_0^\infty \rho^{2\beta}\sum_{j=1}^{N}|u(\rho,\theta_j)|^2 d\rho\right]^{1/2}.$$

If $2\beta\in(-1,1)$, then the operator $\mathrm{A} = a\mathrm{I}+b\mathcal{K}$, defined on $C_0^\infty(X\setminus\mathcal{O})$, is bounded on $L_{2,\beta}(X)$. For the closure $\overline{\mathrm{A}}$, defined on the whole space $L_{2,\beta}(X)$, we have the formula

$$(\overline{\mathrm{A}}u)(r,\cdot) = \frac{1}{\sqrt{2\pi}}\int_{\mathrm{Im}\lambda=\beta} r^{i(\lambda+i/2)}\mathfrak{U}(\lambda)\tilde{u}(\lambda+i/2,\cdot)d\lambda. \tag{2.18}$$

In order to convince ourselves of this, we must replace in (2.17) the line of integration $\mathrm{Im}\lambda = 0$ by the line $\mathrm{Im}\lambda = \beta$ and take into account that the functions $\mathfrak{U}_{jk}$ are bounded on the line $\mathrm{Im}\lambda = \beta$. It remains to use Parseval's equality (1.2.7) for the Mellin transform. Note that the operator A is continuous on $L_{2,\beta}(X)$ for arbitrary β satisfying $2\beta\neq 2k+1$, $k = 0,\pm1,\cdots$. However, for $|2\beta|>1$ it is necessary to first define the operator on a subset of $C_0^\infty(X\setminus\mathcal{O})$ that is dense in $L_{2,\beta}(X)$ (compare with theorem 2.3.5). Formula (2.18) remains valid for $\overline{\mathrm{A}}$.

5. Pseudodifferential operators with variable symbol on a conic manifold. Let S and ρ_S be the directing manifold and function introduced above definition 2.4. To a point $x\in X\setminus\mathcal{O}$ we assign 'polar coordinates' (ρ,θ), where $\theta = \rho_S(x)^{-1}x$ and

$\rho = \rho_S(x) > 0$. Replacing homogeneous functions on $X \setminus \Theta$ by their traces on S, we will further assume that meromorphic pseudodifferential operators are defined on functions given on S. With this understanding the Mellin transform is defined by

$$\tilde{u}(\lambda,\theta) = \frac{1}{\sqrt{2\pi}} \int_0^\infty \rho^{-i\lambda-1} u(\rho,\theta)\,d\rho.$$

The notation $\|\tilde{u}(\lambda,\cdot);H^s(\mu,X)\|$ has a meaning also for $\lambda \neq \mu$.

Let $\mathfrak{U}(\rho,\lambda)$ be a meromorphic pseudodifferential operator of order a on X, depending on a positive parameter ρ. We will assume that the representation $\mathfrak{U}(\rho,\lambda) = \mathfrak{U}(\lambda) + \mathfrak{U}_1(\rho,\lambda)$ holds, where $\mathfrak{U}_0$ is a meromorphic pseudodifferential operator not depending on ρ, while for $\mathfrak{U}_1$ we have the estimate

$$\| \, |\ln\rho|^q (\rho D_\rho)^k \mathfrak{U}_1(\rho,\lambda); H^s(\lambda,X) \to H^{s-\mathrm{Re}\,a}(\lambda,X)\| < c(k,q,s,F), \quad (2.19)$$

uniformly in $\rho > 0$, for all nonnegative integers k and q and for real s, on a line $\mathrm{Im}\,\lambda = \tau$ which does not contain poles of $\mathfrak{U}_0$. This implies that on $\mathrm{Im}\,\lambda = \tau$,

$$\|\tilde{\mathfrak{U}}_1(\nu,\lambda);\ H^s(\lambda,X) \to H^{s-\mathrm{Re}\,a}(\lambda,X)\| \leqslant c(1+|\nu|)^{-N} \quad (2.20)$$

for all s and all positive N; here

$$\tilde{\mathfrak{U}}_1(\nu,\lambda) = \frac{1}{\sqrt{2\pi}} \int_0^\infty \rho^{-i\nu-1} \mathfrak{U}_1(\rho,\lambda)\,d\rho, \quad \mathrm{Im}\,\nu = 0.$$

Assume that for two arbitrary infinitely differentiable functions ζ and η, homogeneous of degree zero and with disjoint supports, the inequality

$$\|\zeta \tilde{\mathfrak{U}}_1(\nu,\lambda)\eta; H^s(\lambda,X) \to H^{s+p}(\lambda,X)\| \leqslant \quad (2.21)$$

$$\leqslant c(1+|\nu|+|\lambda|)^{-N}$$

holds, where N and p are arbitrary positive numbers, and $\mathrm{Im}\,\lambda = \tau$.

In the proof of proposition 3.2.2. we have derived, in essence, the inequality

$$\max_{\sigma=\lambda,\mu} \|(\mathfrak{U}_0(\lambda) - \mathfrak{U}_0(\mu))u; H^{s-\mathrm{Re}\,a+1}(\sigma,X)\| \leqslant \quad (2.22)$$

$$\leqslant c|\lambda-\mu|(\|u;H^s(\lambda,X)\| + \|u;H^s(\mu,X)\|),$$

where $\mathrm{Im}\,\lambda = \mathrm{Im}\,\mu = \tau$. Assume that for $\mathfrak{U}_1$ the estimate

$$\max_{\sigma=\lambda,\mu} \|(\tilde{\mathfrak{U}}_1(\nu,\lambda)-\tilde{\mathfrak{U}}_1(\nu,\mu))u;H^{s-\mathrm{Re}\,a+1}(\sigma,X)\| \leqslant \tag{2.23}$$

$$\leqslant c(s,N)\,|\lambda-\mu|\,(1+|\nu|)^{-N}(\|u;H^s(\lambda,X)\|+\|u;H^s(\mu,X)\|)$$

holds, where s,ν,N are arbitrary real numbers, and $\mathrm{Im}\,\lambda = \mathrm{Im}\,\mu = \tau$.

D e f i n i t i o n 2.9. If condition (2.19) holds (hence also conditions (2.20), (2.21) and (2.23) hold), the operator $\mathfrak{U}(\rho,\lambda) = \mathfrak{U}_0(\lambda)+\mathfrak{U}_1(\rho,\lambda)$ will be called a *meromorphic pseudodifferential operator of order a with variable symbol on X.*

E x a m p l e 2.10. Suppose a function $(r,\phi,\theta)\mapsto\Phi(r,\phi,\theta)$ $(r>0,\phi,\theta\in S^{n-1})$ has a representation $\Phi(r,\phi,\theta) = \Phi_0(\phi,\theta)+\Phi_1(r,\phi,\theta)$, where $\Phi_0 \in C^\infty(S^{n-1}\times S^{n-1})$, while the function Φ_1 is infinitely differentiable and satisfies the conditions

$$\|\,|\ln r|^q(rD_r)^k\Phi_1(r,\cdot);C^p(S^{n-1}\times S^{n-1})\| \leqslant c_{p,q,k},$$

for all nonnegative integers p,q and k. Then the operator $\mathfrak{U}(r,\lambda) = \mathrm{E}_{\theta\to\phi}(\lambda+ia)^{-1}\Phi(r,\phi,\theta)\mathrm{E}_{\psi\to\theta}(\lambda)$ is a meromorphic pseudodifferential operator of order a with variable symbol on $\mathbb{R}^n$. Starting with such operators we can construct meromorphic pseudodifferential operators with variable symbol on X (compare with the proof of proposition 2.2).

D e f i n i t i o n 2.11. A *pseudodifferential operator of order a with variable symbol on a space* $H^s_\beta(X)$ is an operator given on $C_0^\infty(X\setminus\mathcal{O})$ by

$$(\mathrm{A}u)(x) = \tag{2.24}$$

$$= \frac{1}{\sqrt{2\pi}}\int_{\mathrm{Im}\lambda=\tau} \rho^{i(\lambda+ia+in/2)}\mathfrak{U}_{\phi\to\theta}(\rho,\lambda)\tilde{u}(\lambda+in/2,\phi)d\lambda,$$

where $\mathfrak{U}_{\phi\to\theta}(\rho,\lambda)$ is a meromorphic pseudodifferential operator of order a with variable symbol on X, the line $\mathrm{Im}\lambda = \tau = \beta-s$ does not contain poles of the operator-function $\mathfrak{U}$, and (ρ,θ) are the polar coordinates of a point x.

P r o p o s i t i o n 2.12. *The operator* (2.24) *for* $\tau = \beta-s$ *realizes a continuous map* $\mathrm{A}:H^s_\beta(X)\to H^{s-\mathrm{Re}\,a}_\beta(X)$.

P r o o f. By definition we have $\mathfrak{U}(\rho,\lambda) = \mathfrak{U}_0(\lambda)+\mathfrak{U}_1(\rho,\lambda)$, and, correspondingly,

$A = A_0 + A_1$. The operator A_0 is continuous (cf. proposition 2.5), hence it suffices to establish the continuity of the operator

$$(A_1 u)(\rho,\cdot) = \tag{2.25}$$

$$= \frac{\rho^{-a}}{\sqrt{2\pi}} \int_{\operatorname{Im}\lambda=\beta-s} \rho^{i(\lambda+in/2)} \mathfrak{U}_1(\rho,\lambda)\tilde{u}(\lambda+in/2,\cdot)d\lambda.$$

Multiplying (2.25) by ρ^a and applying the Mellin transform we obtain

$$\tilde{v}(\mu+ia+in/2,\cdot) = \int_{\operatorname{Im}\lambda=\beta-s} \tilde{\mathfrak{U}}_1(\mu-\lambda,\lambda)\tilde{u}(\lambda+in/2,\cdot)d\lambda,$$

where $v = \rho^a A_1 u$, $\operatorname{Im}\mu = \beta - s$. The inequality

$$\left[\frac{1+m^2+|\mu|^2}{1+m^2+|\lambda|^2}\right]^{s-\operatorname{Re}a} \leqslant c(1+|\mu-\lambda|^2)^{|s-\operatorname{Re}a|} \tag{2.26}$$

implies the estimate

$$\|\tilde{v}(\mu+ia+in/2,\cdot);H^{s-\operatorname{Re}a}(\mu+ia,X)\| \leqslant$$

$$\leqslant c \int_{\operatorname{Im}\lambda=\beta-s} (1+|\mu-\lambda|)^{|s-\operatorname{Re}a|} \times$$

$$\times \|\tilde{\mathfrak{U}}_1(\mu-\lambda,\lambda)\tilde{u}(\lambda+in/2,\cdot);H^s(\lambda,X)\|d\lambda.$$

Taking into account formula (3.1) we have

$$\|\tilde{v}(\mu+ia+in/2);H^{s-\operatorname{Re}a}(\mu+ia,X)\| \leqslant \tag{2.27}$$

$$\leqslant c \int_{\operatorname{Im}\lambda=\beta-s} (1+|\mu-\lambda|)^{-N}\|\tilde{u}(\lambda+in/2,\cdot);H^s(\lambda,X)\|d\lambda,$$

where N is a sufficiently large number. Put $k(v) = (1+|v|)^{-N}$, $U(\lambda) = \|\tilde{u}(\lambda+in/2,\cdot);\ H^s(\lambda,X)\|$, and denote the righthand integral in (2.27) by $J(\mu)$. By Minkowski's inequality,

$$\left[\int_{\operatorname{Im}\mu=\tau} J(\mu)^2 d\mu\right]^{1/2} = \left[\int_{\operatorname{Im}\mu=\tau} d\mu \left[\int_{-\infty}^{+\infty} k(v)U(\mu-v)dv\right]^2\right]^{1/2} \leqslant$$

$$\leqslant \int k(v)dv \left[\int_{\operatorname{Im}\mu=\tau} U(\mu)^2 d\mu\right]^{1/2}.$$

This, (2.27) and (2.7) imply $\|Au;H_\beta^{s-\mathrm{Re}\,a}(X)\| \leqslant c\|u;H_\beta^s(X)\|$. ■

6. Composition of pseudodifferential operators with variable symbols. Let $A:H_\beta^s(X)\to H_\beta^{s-\mathrm{Re}\,a}(X)$ and $B:H_\beta^{s+\mathrm{Re}\,b}(X)\to H_\beta^s(X)$ be pseudodifferential operators of orders a and b, respectively, with variable symbols on X, where A is given by (2.24) for $\tau = \beta - s$, and B has the form

$$\mathrm{B}v = \frac{1}{\sqrt{2\pi}}\int_{\mathrm{Im}\lambda=\sigma} \rho^{i(\lambda+ib+in/2)}\mathfrak{B}(\rho,\lambda)\tilde{v}(\lambda+in/2,\cdot)d\lambda,$$

$$\sigma = \beta - s - \mathrm{Re}\,b.$$

Put

$$\mathrm{C}v = \frac{1}{\sqrt{2\pi}}\int_{\mathrm{Im}\lambda=\sigma} \rho^{i(\lambda+i(a+b)+in/2)}\times$$

$$\times\,\mathfrak{U}(\rho,\lambda+ib)\mathfrak{B}(\rho,\lambda)\tilde{v}(\lambda+in/2,\cdot)d\lambda.$$

Theorem 2.13. *The operator* $\mathrm{AB}-\mathrm{C}:H_\beta^{s+\mathrm{Re}\,b}(X)\to H_\beta^{s-\mathrm{Re}\,a}(X)$ *is compact.*

The proof of the theorem is preceded by several propositions. Represent each operator $\mathfrak{U}(\rho,\lambda)$, $\mathfrak{B}(\rho,\lambda)$, A, and B as a sum, $\mathfrak{U}(\rho,\lambda) = \mathfrak{U}_0(\lambda)+\mathfrak{U}_1(\rho,\lambda)$, etc. Introduce the operators

$$\mathrm{C}_{01}v = \frac{1}{\sqrt{2\pi}}\int_{\mathrm{Im}\lambda=\sigma} \rho^{i(\lambda+i(a+b)+in/2)}\times$$

$$\times\,\mathfrak{U}_0(\lambda+ib)\mathfrak{B}_1(\rho,\lambda)\tilde{v}(\lambda+in/2,\cdot)d\lambda,$$

$$\mathrm{C}_{11}v = \frac{1}{\sqrt{2\pi}}\int_{\mathrm{Im}\lambda=\sigma} \rho^{i(\lambda+i(a+b)+in/2)}\times$$

$$\times\,\mathfrak{U}_1(\rho,\lambda+ib)\mathfrak{B}_1(\rho,\lambda)\tilde{v}(\lambda+in/2,\cdot)d\lambda.$$

It is clear that $\mathrm{AB}-\mathrm{C} = \mathrm{T}_{01}+\mathrm{T}_{11}$, where $\mathrm{T}_{01} = \mathrm{A}_0\mathrm{B}_1-\mathrm{C}_{01}$, $\mathrm{T}_{11} = \mathrm{A}_1\mathrm{B}_1-\mathrm{C}_{11}$.

Proposition 2.14. *Each of the operators* $\mathrm{T}_{01},\mathrm{T}_{11}$ *realizes a continuous map from* $H_\beta^{s+\mathrm{Re}\,b}(X)$ *to* $H_{\beta+1}^{s-\mathrm{Re}\,a+1}(X)$.

Proof. We have

$$\mathrm{T}_{01}v = \frac{\rho^{-a-b}}{\sqrt{2\pi}}\int_{\mathrm{Im}\mu=\sigma} \rho^{i(\mu+in/2)}\times$$

$$\times\, \mathfrak{U}_0(\mu+ib)\left[\int\limits_{\operatorname{Im}\lambda=\sigma} \tilde{\mathfrak{B}}_1(\mu-\lambda,\lambda)\tilde{v}(\lambda+in/2,\cdot)d\lambda\, + \right.$$

$$\left. -\,\mathfrak{B}_1(\rho,\mu)\tilde{v}\,(\mu+in/2,\cdot)\right]d\mu.$$

Multiply this equation by ρ^{a+b}, and then apply the Mellin transform. Then

$$\widetilde{(\mathrm{T}_{01}v)}(\mu+i(a+b)+in/2,\cdot) =$$

$$= \int\limits_{\operatorname{Im}\lambda=\sigma} (\mathfrak{U}_0(\mu+ib)-\mathfrak{U}_0(\lambda+ib))\tilde{\mathfrak{B}}_1(\mu-\lambda,\lambda)\tilde{v}(\lambda+in/2,\cdot)d\lambda.$$

Inequality (2.22) implies

$$\|\widetilde{(\mathrm{T}_{01}v)}(\mu+i(a+b)+in/2,\cdot);H^{s-\operatorname{Re}a+1}(\mu,X)\| \leqslant \tag{2.28}$$

$$\leqslant c \int\limits_{\operatorname{Im}\lambda=\sigma} |\mu-\lambda|\,(\|\tilde{\mathfrak{B}}_1(\mu-\lambda,\lambda)\tilde{v}(\lambda+in/2,\cdot);H^s(\lambda,X)\| +$$

$$+\,\|\tilde{\mathfrak{B}}_1(\mu-\lambda,\lambda)\tilde{v}(\lambda+in/2,\cdot);H^s(\mu,X)\|)d\lambda.$$

Since, by (2.27),

$$\|w;H^s(\mu,X)\| \leqslant c(1+|\mu-\lambda|)^{|s|}\,\|w;H^s(\lambda,X)\|, \tag{2.29}$$

and

$$\|\tilde{\mathfrak{B}}_1(\mu-\lambda,\lambda)\tilde{v}(\lambda+in/2,\cdot);H^s(\lambda,X)\| \leqslant \tag{2.30}$$

$$\leqslant c(1+|\mu-\lambda|)^{-N}\|\tilde{v}(\lambda+in/2,\cdot);H^{s+\operatorname{Re}b}(\lambda,X)\|,$$

(2.28) implies that T_{01} is a continuous operator (compare with the proof of proposition 2.12).

We now consider the operator

$$\mathrm{T}_{11}v = \frac{\rho^{-a-b}}{\sqrt{2\pi}} \int\limits_{\operatorname{Im}\lambda=\sigma} \rho^{i(\mu+in/2)}\mathfrak{U}_1(\rho,\mu+ib)\,\times$$

$$\times\left[\int\limits_{\operatorname{Im}\lambda=\sigma} \tilde{\mathfrak{B}}_1(\mu-\lambda,\lambda)\tilde{v}(\lambda+in/2,\cdot)d\lambda-\mathfrak{B}_1(\rho,\mu)\tilde{v}\,(\mu+in/2,\cdot)\right]d\mu.$$

After Mellin transformation and change of integration variable we find

$$\widetilde{(\mathrm{T}_{11}v)}(\nu+i(a+b)+in/2,\cdot) =$$

$$= \int_{\mathrm{Im}\mu=\sigma} d\mu \int_{\mathrm{Im}\lambda=\sigma} (\tilde{\mathfrak{U}}_1(\nu-\mu,\mu+ib)-\tilde{\mathfrak{U}}_1(\nu-\mu,\lambda+ib))\times$$

$$\times\tilde{\mathfrak{B}}(\mu-\lambda,\lambda)\tilde{v}(\lambda+in/2,\cdot)d\lambda.$$

Inequalities (2.23), (2.20) and (2.30) imply the estimate

$$\|\widetilde{(\mathrm{T}_{11}v)}(\nu+i(a+b)+in/2,\cdot);H^{s-\mathrm{Re}a+1}(\nu,X)\|\leqslant$$

$$\leqslant c\int_{\mathrm{Im}\mu=\sigma}(1+|\nu-\mu|)^{-N}d\mu\times$$

$$\times\int_{\mathrm{Im}\lambda=\sigma}(1+|\mu-\lambda|)^{-N}\|\tilde{v}(\lambda+in/2,\cdot);H^{s+\mathrm{Re}b}(\lambda,X)\|d\lambda.$$

Hence $\|\mathrm{T}_{11}v;H_{\beta+1}^{s-\mathrm{Re}a+1}(X)\|\leqslant c\|v;H_\beta^{s+\mathrm{Re}b}(X)\|$. ■

Proposition 2.15. *Let* T *be any of the operators* $\mathrm{A}_0\mathrm{B}_1$, C_{01}, $\mathrm{A}_1\mathrm{B}_2$, *or* C_{11}. *Then for all functions* $\tilde{u}$ *from the set* $\{\tilde{u}\colon \tilde{u}(\lambda+i(a+b)+in/2,\cdot)=\widetilde{(\mathrm{T}v)}(\lambda+i(a+b)+in/2,\cdot),\quad \|v;H_\beta^{s+\mathrm{Re}a}(X)\|\leqslant 1\}$, *defined on the line* $\mathrm{Im}\lambda=\beta-s$ *the following estimate holds:*

$$\Big[\int_{\mathrm{Im}\lambda=\beta-s}\|\tilde{u}(\lambda+i(a+b)+in/2+h,\cdot)+ \tag{2.31}$$

$$-\tilde{u}(\lambda+i(a+b)+in/2,\cdot);H^{s-\mathrm{Re}a}(\lambda,X)\|^2d\lambda\Big]^{1/2}\leqslant\chi(h),$$

where χ *is a continuous function,* $\chi(0)=0$, *independent, moreover, of the choice of an* $\tilde{u}$ *from the set indicated.*

Proof. Suppose, e.g., $\mathrm{T}=\mathrm{A}_0\mathrm{B}_1$ and $u=\mathrm{T}v$. For $\sigma=\beta-s-\mathrm{Re}b$,

$$\tilde{u}(\mu+i(a+b)+in/2+h,\cdot)-\tilde{u}(\mu+i(a+b)+in/2,\cdot)=$$

$$=\int_{\mathrm{Im}\lambda=\sigma}[\mathfrak{U}_0(\mu+ib+h)-\mathfrak{U}_0(\mu+ib)]\times$$

$$\times\tilde{\mathfrak{B}}_1(\mu+h-\lambda,\lambda)\tilde{v}(\lambda+in/2,\cdot)d\lambda+$$

$$+\int_{\mathrm{Im}\lambda=\sigma}\mathfrak{U}_0(\mu+ib)(\tilde{\mathfrak{B}}_1(\mu+h-\lambda,\lambda)-\tilde{\mathfrak{B}}_1(\mu-\lambda,\lambda))\tilde{v}(\lambda+in/2,\cdot)d\lambda.$$

The first of these integrals can be estimated using inequalities (2.22) and (2.30), and the second - by using the inequality

$$\|\tilde{\mathfrak{B}}_1(\mu+h-\lambda,\lambda)-\mathfrak{B}_1(\mu-\lambda,\lambda);H^{s+\mathrm{Re}\,b}(\lambda,X)\to H^s(\lambda,X)\|\leqslant$$

$$\leqslant\sigma_N(h)(1+|\mu-\lambda|)^{-N},$$

where σ_N is a continuous function, $\sigma_N(0)=0$. As the result we find that

$$\|\tilde{u}(\mu+i(a+b)+in/2+h,\cdot)+$$

$$-\tilde{u}(\mu+i(a+b)+in/2,\cdot);\ H^{s-\mathrm{Re}\,a}(\mu,X)\|\leqslant$$

$$\leqslant c\sigma_N(h)\int_{\mathrm{Im}\lambda=\sigma}(1+|\mu-\lambda|)^{-N}\|\tilde{v}(\lambda+in/2,\cdot);H^{s+\mathrm{Re}\,b}(\lambda,X)\|d\lambda$$

Using Minkowski's inequality (as in the proof of proposition 2.12) we are led to an estimate of the type (2.31). The other operators C_{01}, A_1B_1, and C_{11} are considered analogously. ■

P r o o f o f t h e o r e m 2.13. In view of the fact that $\mathrm{AB}-\mathrm{C}=\mathrm{T}_{01}+\mathrm{T}_{11}$, it suffices to verify that each operator $\mathrm{T}_{01},\mathrm{T}_{11}$ is compact from $H_\beta^{s+\mathrm{Re}\,b}(X)$ to $H_\beta^{s-\mathrm{Re}\,a}(X)$. Let v run over the unit sphere in $H_\beta^{s+\mathrm{Re}\,b}(X)$ and let $u=\mathrm{T}v$, where T is any of the operators $\mathrm{T}_{01},\mathrm{T}_{11}$. Let also $\{U,\kappa\}$ be an atlas on $X\setminus\mathfrak{O}$, and let $\{\zeta\}$ be a partition of unity subordinate to it. We convince ourselves of the fact that for any local chart the set $\{\zeta_\kappa u{:}u=\mathrm{T}v\}$ is compact in $H_\beta^{s-\mathrm{Re}\,a}(\mathbf{R}^n)$. For simplicity reasons we replace the notation $\zeta_\kappa u$ by u. Proposition 2.12 and the definition of the norms in the spaces $H_\beta^t(\mathbf{R}^n)$ implies that to an arbitrary positive number ϵ correspond numbers M and N such that

$$\|u;H_\beta^{s-\mathrm{Re}\,a}(\mathbf{R}^n)\|\leqslant$$

$$\leqslant\left[\sum_{m=0}^{M}\sum_{k=1}^{k_m}\int_{-N+i\tau}^{N+i\tau}(1+m^2+|\lambda|^2)^{s-\mathrm{Re}\,a}\times\right.$$

$$\left.\times|\tilde{u}_{mk}(\lambda+in/2,\cdot)|^2d\lambda\right]^{1/2}+\epsilon$$

for $\tau=\beta-s+\mathrm{Re}\,a$ (uniformly in v). Applying proposition 2.15 and a well-known compactness criterion in the space L_2 of square-integrable functions on a bounded interval (cf., e.g., [30]) we find that for any $\epsilon>0$ there is a finite ϵ-net for the set $\{\zeta_\kappa u{:}\ u=\mathrm{T}v\}$. ■

7. The adjoint operator. We denote by $d\sigma$ the volume element on the directing submanifold $S \subset X \setminus \Theta$ and introduce a volume element on the conic manifold X by $dv = \rho^{n-1} d\rho d\sigma$, where $\rho = \rho_S$. Let $\mathrm{A}: H^s_\beta(X) \to H^{s-\mathrm{Re}\,a}_\beta(X)$ be a pseudodifferential operator of order a with variable symbol on X of the form (2.24), with, moreover, $\tau = \beta - s$. We denote by $\mathrm{A}^*: H^{-s+\mathrm{Re}\,a}_{-\beta}(X) \to H^{-s}_{-\beta}(X)$ the operator adjoint to A with respect to the extension of the scalar product in $L_2(X,dv)$, and by $\mathfrak{U}(\rho,\lambda)^*$ the operator adjoint to $\mathfrak{U}(\rho,\lambda)$ with respect to the extension of the scalar product in $L_2(S,d\sigma)$.

T h e o r e m 2.16. *The following equation holds:*

$$\mathrm{A}^* v = \frac{1}{\sqrt{2\pi}} \int_{\mathrm{Im}\,\lambda = \tau^*} \rho^{i(\lambda + i\bar{a} + in/2)} \times \tag{2.32}$$

$$\times\, \mathfrak{U}^*(\rho, \lambda + i\bar{a}) \tilde{v}(\lambda + in/2, \cdot) d\lambda + \mathrm{T}v,$$

where $\tau^* = s - \mathrm{Re}\,a - \beta$, $\mathfrak{U}^*(\rho,\lambda) = \mathfrak{U}(\rho,\bar{\lambda})^*$, *and* $\mathrm{T}: H^{-s+\mathrm{Re}\,a}_{-\beta}(X) \to H^{-s}_{-\beta}(X)$ *is a compact operator.*

P r o o f. We have

$$(\mathrm{A}u, v) = \tag{2.33}$$

$$= \int_{\mathrm{Im}\,\mu = \beta - s + \mathrm{Re}\,a} <\widetilde{(\mathrm{A}v)}(\mu + in/2, \cdot), \tilde{v}(\bar{\mu} + in/2, \cdot)> d\mu =$$

$$= \int_{\mathrm{Im}\,\mu = \beta - s} <\widetilde{(\mathrm{A}u)}(\mu + ia + in/2, \cdot), \tilde{v}(\bar{\mu} + i(n/2 - \bar{a}), \cdot)> d\mu,$$

where (w,v) and $<f,g>$ are the extensions of the scalar products in $L_2(X,dv)$ and $L_2(S,d\sigma)$, respectively. Since

$$\widetilde{(\mathrm{A}u)}(\mu + ia + in/2, \cdot) = \mathfrak{U}_0(\mu) \tilde{u}(\mu + in/2, \cdot) +$$

$$+ \int_{\mathrm{Im}\,\mu = \beta - s} \tilde{\mathfrak{U}}_1(\mu - \lambda, \lambda) \tilde{u}(\lambda + in/2, \cdot) d\lambda,$$

(2.25) can be rewritten as

$$(\mathrm{A}u, v) = \tag{2.34}$$

$$= \int_{\mathrm{Im}\,\lambda = \tau} <\tilde{u}(\lambda + in/2, \cdot), \mathfrak{U}_0(\lambda)^* \tilde{v}(\bar{\lambda} + i(n/2 - \bar{a}), \cdot)> d\lambda +$$

$$+\int_{\mathrm{Im}\lambda=\tau} <\tilde{u}(\lambda+in/2,\cdot), \int_{\mathrm{Im}\mu=\tau} (\tilde{\mathfrak{U}}_1(\mu-\lambda,\lambda))^* \tilde{v}(\overline{\mu}+i(n/2-\overline{a}),\cdot)d\mu> d\lambda,$$

where $\tau = \beta - s$. Note that

$$(\tilde{\mathfrak{U}}_1(\mu-\lambda,\lambda))^* = \widetilde{(\mathfrak{U}_1^*(\overline{\lambda}-\overline{\mu},\overline{\lambda}))}$$

(the operator at the righthand side is obtained from $\mathfrak{U}(\rho,\lambda)$ by transition to the adjoint and subsequently applying the Mellin transform with respect to the first argument). Since $\mathfrak{U}_0(\lambda)^* = \mathfrak{U}_0^*(\lambda)$, (2.34) implies $(Au,v) = (u,h)$, where

$$\tilde{h}(\overline{\lambda}+in/2,\cdot) = \mathfrak{U}_0^*(\overline{\lambda})\tilde{v}(\overline{\lambda}+i(n/2-\overline{a}),\cdot)+$$

$$+\int_{\mathrm{Im}\mu=\tau} \mathfrak{U}_1^*(\overline{\lambda}-\overline{\mu},\overline{\lambda})\tilde{v}(\overline{\mu}+i(n/2-\overline{a}),\cdot)d\mu.$$

The latter equation can be rewritten as

$$\tilde{h}(\overline{\lambda}+in/2,\cdot) = \mathfrak{U}_0^*(\lambda+i\overline{a})\tilde{v}(\lambda+in/2,\cdot)+$$

$$+\int_{\mathrm{Im}\mu=\tau^*} \tilde{\mathfrak{U}}_1(\lambda-\mu,\lambda+i\overline{a})\tilde{v}(\mu+in/2),\cdot)d\mu.$$

Taking into account that $h = A^* v$ leads to (2.32), in which

$$\mathrm{T}v = \frac{1}{\sqrt{2\pi}}\int_{\mathrm{Im}\lambda=\tau^*} \rho^{i(\lambda+i\overline{a})}d\lambda \times \tag{2.35}$$

$$\times \int_{\mathrm{Im}\mu=\tau^*} \tilde{\mathfrak{U}}_1^*(\lambda-\mu,\lambda+i\overline{a})\tilde{v}(\mu+in/2,\cdot)d\mu+$$

$$-\frac{1}{\sqrt{2\pi}}\int_{\mathrm{Im}\lambda=\tau^*} \rho^{i(\lambda+i\overline{a})}\mathfrak{U}_1^*(\rho,\lambda+i\overline{a})\tilde{v}(\lambda+in/2,\cdot)d\lambda.$$

It remains to verify that the operator $\mathrm{T}:H_{-\beta}^{-s+\mathrm{Re}\,a}(X)\to H_{-\beta}^{-s}(X)$ defined by (2.35) is compact. This can be done by a reasoning similar to the one given in the proof of theorem 2.13. ■

§3. Pseudodifferential operators on manifolds with conical points

1. Manifolds with conical points. Let, as in §2, X be a conic manifold with vertex 0, and let S be a directing submanifold of it (having only one point in common with each ray). A neighborhood of 0 is a set on X containing a subset of the form $\{x=(\rho,\theta):\rho<\epsilon,\theta\in S\}$, where (ρ,θ) are the polar coordinates on X defined by using

S (cf. §2.5), and ϵ is an arbitrary positive number. If X, X' are conic manifolds, and $\kappa: X \to X'$ is a homogeneous map of degree one, infinitely differentiable, together with its inverse, everywhere except at the vertices, then κ is called a diffeomorphism from X to X'.

We denote by $\mathfrak{M}$ a Hausdorff space having the following properties: 1) there is a finite subset $Q = \{x^{(1)}, \cdots, x^{(l)}\}$ such that $\mathfrak{M} \setminus Q$ is an n-dimensional C^∞ manifold; 2) for every point $x^{(j)} \in Q$ there are given an open set $U_j \subset \mathfrak{M}$ containing $x^{(j)}$ and a homeomorphism κ_j of this set onto a neighborhood of the vertex O_j of an n-dimensional conic manifold X_j, $\kappa_j(x^{(j)}) = O_j$; and 3) there is a family $V_{l+1}, \cdots, V_p$ of coordinate neighborhoods on $\mathfrak{M} \setminus Q$, $\overline{V}_j \subset \mathfrak{M} \setminus Q$, forming together with the sets $U_1, \cdots, U_l$ a finite open cover of the manifold $\mathfrak{M}$.

Let $x \in U_j$ and let (ρ, θ) be the polar coordinates of its image $\kappa_j(x)$ in X_j. Coordinates of the form $(\rho, \theta_1, \cdots, \theta_{n-1})$ where $\theta_1, \cdots, \theta_{n-1}$ are certain local coordinates of the point θ on a directing submanifold S_j of the cone X_j will be called local coordinates of x in U_j.

We will assume that the local charts $\{U_j, \kappa_j\}_{j=1}^{l}$ and $\{V_k, \kappa_k\}_{k=l+1}^{p}$ form a $\check{C}^\infty$ atlas, i.e. in intersections of neighborhoods the local coordinates on one chart are infinitely differentiable functions of the local coordinates on the other chart. The notion of equivalent atlases is naturally introduced (in particular, if two charts (U_j, κ_j) and (U_j', κ_j') belong to equivalent atlases, then the map $\kappa_j' \circ \kappa_j^{-1}: \kappa_j(U_j \cap U_j') \to \kappa_j'(U_j \cap U_j')$ can be extended to a diffeomorphism of X_j to X_j'). The space $\mathfrak{M}$ endowed with this structure is called a $\check{C}^\infty$ *manifold with conical points*.

We introduce the space $H^s_\beta(\mathfrak{M})$

of functions on $\mathfrak{M}$, where s is a real number and $\beta = (\beta_1, \cdots, \beta_l)$ is a vector with real components. Let $\{\zeta_j\}_{j=1}^{p}$ be a partition of unity subordinate to the cover $U_1, \cdots, U_l, V_{l+1}, \cdots, V_p$. Put

$$\|u; H^s_\beta(\mathfrak{M})\| = \sum_{j=1}^{l} \|(\zeta_j u) \circ \kappa_j^{-1}; H^s_{\beta_j}(X_j)\| +$$

$$+ \sum_{j=l+1}^{p} \|(\zeta_j u) \circ \kappa_j^{-1}; H^s(\mathbf{R}^n)\|$$

(as usual, the function $(\zeta_j u) \circ \kappa_j^{-1}$ is extended by zero onto X_j $(j = 1, \cdots, l)$ or onto $\mathbf{R}^n$ ($j=l+1, \cdots, p$)). The space $H^s_\beta(\mathfrak{M})$ is defined as the completion of the set $C_0^\infty(\mathfrak{M}\setminus Q)$ with respect to the norm $\|\cdot;H^s_\beta(\mathfrak{M})\|$.

We will say that a function u given on $\mathfrak{M}\setminus Q$ belongs to the class (is of class) $\check{C}^\infty(\mathfrak{M})$ if it is infinitely differentiable outside Q, while near every conical point $x^{(j)} \in Q$ it is such that the functions $\rho \mapsto (\rho D_\rho)^q(u\circ\kappa_j^{-1})(\rho,\cdot) \in C^\infty(C_j)$ are continuous on an interval $[0,\delta]$. Here $\delta > 0$, S_j is a directing submanifold of the cone X_j, and (ρ,θ) are polar coordinates on X_j.

2. Pseudodifferential operators on a manifold with conical points. Let $(x,\xi)\mapsto\Phi(x,\xi)$ be a function of class $C^\infty(\mathbf{R}^n\times(\mathbf{R}^n\setminus 0))$, positively homogeneous of degree a in ξ, and independent of x outside a ball $|x| < r_0$. Let also $\chi \in C^\infty(\mathbf{R}^n)$, with $\chi(\xi) = 1$ if $|\xi| > 1$, and $\chi(\xi) = 0$ if $|\xi| < 1/2$.

The operator

$$(\mathrm{A}u)(x) = \int_{\mathbf{R}^n} e^{ix\xi}\chi(\xi)\Phi(x,\xi)\hat{u}(\xi)d\xi, \tag{3.1}$$

where $u \in C_0^\infty(\mathbf{R}^n)$ and $\hat{u}$ is the Fourier transform of u, is called a standard pseudodifferential operator of order a, and Φ is called its symbol. It is well-known (cf., e.g., [32]) that the operator (3.1) realizes a continuous map $\mathrm{A}:H^s(\mathbf{R}^n)\to H^{s-m}(\mathbf{R}^n)$.

We now define pseudodifferential operators on a manifold $\mathfrak{M}$ with conical points.

D e f i n i t i o n 3.1. A linear operator $\mathrm{A}:H^s_\beta(\mathfrak{M})\to H^{s-\mathrm{Re}\,a}_\beta(\mathfrak{M})$ is called a *pseudodifferential operator of order a on the space* $H^s_\beta(\mathfrak{M})$ if the following conditions are fulfilled.

A) For functions ζ,η from $\check{C}^\infty(\mathfrak{M})$ with supports in a neighborhood U of a conical point $x^{(j)}$ we have

$$\eta\mathrm{A}\zeta = \eta^\kappa \mathrm{A}_U\zeta_\kappa + \eta\mathrm{T}\zeta,$$

where $\mathrm{A}_U:H^s_{\beta_j}(X_j)\to H^{s-\mathrm{Re}\,a}_{\beta_j}(X_j)$ is a pseudodifferential operator of order a with variable symbol on the cone X_j, $\kappa:U\to X_j$ is the 'coordinate' homeomorphism, and $\mathrm{T}:H^s_\beta(\mathfrak{M})\to H^{s-\mathrm{Re}\,a}_\beta(\mathfrak{M})$ is a compact operator.

B) For functions $\eta,\zeta \in \check{C}^\infty(\mathfrak{M})$ with supports in a coordinate neighborhood

$V, \overline{V} \subset \mathfrak{M} \setminus Q,$

$$\eta A \zeta = \eta^{\kappa} A_V \zeta_{\kappa} + \eta T \zeta,$$

where A_V is a standard pseudodifferential operator of order a on $\mathbb{R}^n$ (of the form (3.1)) and T is, as before, a compact operator from $H^s_{\beta}(\mathfrak{M})$ to $H^{s-\operatorname{Re} a}_{\beta}(\mathfrak{M})$.

C) If $\eta, \zeta \in \overset{\vee}{C}{}^{\infty}(\mathfrak{M})$ and $\operatorname{supp}\zeta \cap \operatorname{supp}\eta = \emptyset$, then the operator $\eta A \zeta : H^s_{\beta}(\mathfrak{M}) \to H^{s-\operatorname{Re} a}_{\beta}(\mathfrak{M})$ is compact.

We denote by $\mathfrak{M}_Q$ the (C^1) manifold with boundary (a compactification of $\mathfrak{M} \setminus Q$) obtained by gluing to $\mathfrak{M} \setminus Q$ the 'origin' of all generators of the tangent cones X_j (distinct generators having distinct origins), and let $T_0^*(\mathfrak{M}_Q)$ be the bundle of nonzero cotangent vectors over $\mathfrak{M}_Q$. The boundary $\partial \mathfrak{M}_Q$ consists of l components, corresponding to the conical points $x^{(1)}, \cdots, x^{(l)}$. A generator of X_j with direction ϕ corresponds to the tangent (cotangent) space containing the vector ϕ (the dual vector ϕ^*).

For a pseudodifferential operator A on $\mathfrak{M}$ we can define its symbol Φ as a function on $T_0^*(\mathfrak{M}_Q)$. In a coordinate neighborhood V such that $\overline{V} \subset \mathfrak{M} \setminus Q$ the representative of Φ is the symbol Φ_V of the standard pseudodifferential operator A_V on $\mathbb{R}^n$. If U is a neighborhood of a conical point $x^{(j)}$, then the operator A_U (cf. A) of definition 3.1) has a representation (2.24). The corresponding meromorphic pseudodifferential operator $\mathfrak{U}^{(j)}(\rho,\lambda)$ with variable symbol is equal to a sum of operators $\mathfrak{U}_0^{(j)}(\lambda) + \mathfrak{U}_1^{(j)}(\rho,\lambda)$ (definition 2.9). In the fiber above a point in the boundary $\partial \mathfrak{M}_Q$ the symbol Φ of A coincides with the principal symbol of the meromorphic pseudodifferential operator $\mathfrak{U}_0^{(j)}(\lambda)$. So, the symbol Φ is a homogeneous function on the fiber of $T_0^*(\mathfrak{M}_Q)$ above each point $x \in \mathfrak{M}_Q$.

P r o p o s i t i o n 3.2. *For a given such function there is on $H^s_{\beta}(\mathfrak{M})$ a pseudodifferential operator of order a having this function as symbol.*

The p r o o f is similar to the verification of proposition 2.2. We only indicate the necessary changes. Let $U_1, \cdots, U_l, V_{l+1}, \cdots, V_p$ be the cover of $\mathfrak{M}$ and let $\{\zeta_j\}_{j=1}^p$ be a partition of unity subordinate to it. Denote by η_j, $j = 1, \cdots, p$, smooth functions for which $\zeta_j \eta_j = \zeta_j$, $\operatorname{supp}\eta_j \subset U_j$, $j = 1, \cdots, l$, and $\operatorname{supp}\eta_j \subset V_j$ for $j = l+1, \cdots, p$. Suppose that $\mathfrak{U}^{(j)}(\rho,\lambda)$ are meromorphic pseudodifferential operators on X_j, $j = 1, \cdots, l$, whose symbols coincide with the given functions (cf. proposition 2.2 and example 2.10). Define the operator A_{U_j} by (2.24), in which $\mathfrak{U}$ is replaced by $\mathfrak{U}^{(j)}$. Let also A_{V_j} be standard pseudodifferential operators on $\mathbb{R}^n$ with the given symbols. Put

$$A = \sum_{j=1}^{l} (\zeta_j)^{\kappa_j} A_{U_j}(\eta_j)_{\kappa_j} + \sum_{j=l+1}^{p} (\zeta_j)^{\kappa_j} A_{V_j}(\eta_j)_{\kappa_j}.$$

In order to prove that A is the operator looked for, we may reason as in the proof of proposition 2.2. Instead of proposition 3.2.2 we must use theorem 2.13 in this situation, and take into account, moreover, the following fact. By (2.3.4), for arbitrary functions $\zeta,\eta \in C_0^\infty(\mathbb{R}^n \setminus 0)$ we have

$$\zeta F^{-1}_{\xi\to x}\Phi(x,\xi)F_{y\to\xi}(\eta u)(y) =$$

$$= \frac{1}{\sqrt{2\pi}}\zeta(x) \int_{\operatorname{Im}\lambda=\tau} r^{i(\lambda+ia+in/2)} E_{\omega\to\phi}(\lambda+ia)^{-1} \times$$

$$\times \Phi(r,\phi,\omega)E_{\theta\to\phi}(\lambda)\widetilde{(\eta u)}(\lambda+in/2,\theta)d\lambda + Ku,$$

where $\tau = \beta - s$, K is an operator acting from every space $H^s_\beta(\mathbb{R}^n)$ to $C_0^\infty(\mathbb{R}^n)$, and the lefthand side is to be understood as the analytic extension with respect to the parameter a - the degree of homogeneity of the function $\xi \mapsto \Phi(\cdot,\xi)$. (Clearly, $\zeta F^{-1}\Phi(x,\xi)F\eta = \zeta F^{-1}\Phi(x,\xi)\chi(\xi)F\eta + T$, where T maps $H^s_\beta(\mathbb{R}^n)$ to $C_0^\infty(\mathbb{R}^n \setminus 0)$.) ■

A pseudodifferential operator A will be called *elliptic* if its symbol does not vanish on $T_0^*(\mathfrak{M}_Q)$. Thus, if A is elliptic, then the meromorphic pseudodifferential operator $\mathfrak{U}^{(j)}$ is elliptic on the cone X_j, $j = 1, \cdots, l$.

T h e o r e m 3.3. *A pseudodifferential operator* $A: H^s_\beta(\mathfrak{M}) \to H^{s-\operatorname{Re} a}_\beta(\mathfrak{M})$ *is Fredholm (i.e. is an operator with closed range and finite-dimensional kernel and co-kernel) if and only if* A *is elliptic and the line* $\operatorname{Im}\lambda = \beta - s$ *does not contain poles of the operator-function* $\lambda \mapsto \mathfrak{U}_0^{(j)}(\lambda)^{-1}$, $j = 1, \cdots, l$ *).

The study of properties of pseudodifferential operators given in the preceding paragraphs allows us to use for the proof of this theorem the traditional scheme, related to the construction of regularizers (parametrices), which invert the

*) Every strip $\tau_1 \le \operatorname{Im}\lambda \le \tau_2$ not containing poles of the operator-function $\lambda \mapsto \mathfrak{U}_0^{(j)}(\lambda)$ can contain at most finitely many poles of the function $\lambda \mapsto \mathfrak{U}_0^{(j)}(\lambda)^{-1}$ (cf. remarks 1.8 and 1.9).

behavior of A up to compact summands (cf., e.g., [1]). By the way, the proof can be performed by repeating, with obvious changes, the proof of proposition 4.4 to be given in the sequel.

§4. Algebras generated by pseudodifferential operators of order zero

Everywhere in this paragraph we will consider pseudodifferential operator of order zero only. In the definition of a standard pseudodifferential operator, we can do without the cut-off function χ, i.e. we can write a standard pseudodifferential operator on $\mathbb{R}^n$ as

$$(\mathrm{A}u)(x) = \int e^{ix\xi}\Phi(x,\xi)\hat{u}(\xi)d\xi.$$

1. The algebra generated by pseudodifferential operators on smooth manifolds with discontinuities 'of the first kind' in the symbols. We assume that the set of conical points is empty, and that $\mathfrak{M}$ is a compact C^∞ manifold without boundary. By proposition 3.1, the operator A_V in B) of definition 3.1 can be rewritten as

$$(\mathrm{A}_V u)(x) = \frac{1}{\sqrt{2\pi}} \int_{\mathrm{Im}\lambda=0} r^{i(\lambda+in/2)} \times \tag{4.1}$$

$$\times\, \mathrm{E}_{\theta\to\phi}(\lambda)^{-1}\Phi_V(r,\phi,\theta)\mathrm{E}_{\psi\to\theta}(\lambda)\tilde{u}(\lambda+in/2,\psi)d\lambda,$$

where (r,ϕ) are local spherical coordinates with origin at a point $x^{(0)} \in \kappa(V)$, $0 \leqslant r < \infty$, $\phi \in \mathbb{R}^n$, $|\phi| = 1$. Suppose that in every small coordinate neighborhood V, except perhaps at $x^{(0)}$, the symbol Φ_V is a smooth function of the local coordinates $x = (r,\phi)$ and $\theta \in S^{n-1}$. Suppose also that $\Phi_V(r,\phi,\theta) = \Phi_V^{(0)}(\phi,\theta) + \Phi_V^{(1)}(r,\phi,\theta)$, and let for arbitrary nonnegative integers p,q, and k,

$$\| \, |\ln r|^q (rD_r)^k \Phi_1(r,\cdot); C^p(S^{n-1}\times S^{n-1})\| \leqslant c_{p,q,k}. \tag{4.2}$$

Then (4.1) is a pseudodifferential operator of order zero with variable symbol on $H_0^0(\mathbb{R}^n) = L_2(\mathbb{R}^n)$ (in the sense of definition 2.11).

We will assume that a Riemannian metric is given on $\mathfrak{M}$, as well as a positive measure μ induced by it. By $L_2(\mathfrak{M})$ we denote the space of scalar-valued functions on $\mathfrak{M}$, endowed with the norm

$$\|u\| = \left[\int_{\mathfrak{M}} |u(x)|^2 d\mu\right]^{1/2}.$$

Let $S(\mathfrak{M})$ and $S^*(\mathfrak{M})$ be the bundles of tangent and cotangent unit vectors. The symbol Φ of a pseudodifferential operator A on $L_2(\mathfrak{M})$ is a complex-valued function on the Whitney sum $S(\mathfrak{M})\oplus S^*(\mathfrak{M})$. The conditions to which the function Φ_V is subjected imply that everywhere on $\mathfrak{M}$, except possibly at a finite point set $\operatorname{sing}\Phi$, the symbol Φ is a constant on a fiber $S(\mathfrak{M})_x$. (Near a point $x \in \operatorname{sing}\Phi$ the symbol satisfies inequalities (4.2).) Under these conditions we will say that at the points of $\operatorname{sing}\Phi$ the symbol has a discontinuity of the first kind. We stress that the set $\operatorname{sing}\Phi$ depends on the operator in question.

A pseudodifferential operator $A: L_2(\mathfrak{M}) \to L_2(\mathfrak{M})$ is put in correspondence with a family of maps

$$\mathfrak{U}(x,\lambda): L_2(S(\mathfrak{M})_x) \to L_2(S(\mathfrak{M})_x), \tag{4.3}$$

parametrized by points $\lambda \in \mathbf{R}$ and $x \in \mathfrak{M}$. The operator $\mathfrak{U}(x,\lambda)$ is defined by

$$\mathfrak{U}(x,\lambda) = \mathrm{E}_{\theta\to\phi}(\lambda)^{-1}\Phi(x,\phi,\theta)\mathrm{E}_{\psi\to\theta}(\lambda),$$

where $\phi,\psi \in S(\mathfrak{M})_x$, $\theta \in S^*(\mathfrak{M})_x$. The family (4.3) is called the *operator symbol* of the pseudodifferential operator A. Thus, pseudodifferential operators differing by compact terms only have the same operator symbol.

We denote by $\mathfrak{S}$ the involutive Banach algebra[*)] of operator-functions given on $\mathfrak{M}\times\mathbf{R}$, with pointwise multiplication, generated by the operator symbols and endowed with the norm

$$\|\mathfrak{U};\mathfrak{S}\| = \sup_{(x,\lambda)\in\mathfrak{M}\times\mathbf{R}} \|\mathfrak{U}(x,\lambda); L_2(S(\mathfrak{M})_x) \to L_2(S(\mathfrak{M})_x)\|.$$

Elements $\mathfrak{U}$ and $\mathfrak{U}^*$ adjoint in $\mathfrak{S}$ take values $\mathfrak{U}(x,\lambda)$ and $\mathfrak{U}(x,\lambda)^*$ adjoint with respect to the scalar product in $L_2(S(\mathfrak{M})_x)$. It is clear that $\mathfrak{S}$ is a C^*-algebra[*)].

In this paragraph, an algebra will always mean a C^*-algebra.

Let $\mathfrak{A}$ be the algebra generated by the compact operators and the pseudodifferential operators of order zero on $L_2(\mathfrak{M})$, and let $\mathcal{K}L_2(\mathfrak{M})$ be the ideal of compact operators. This section is dedicated to the proof of the following

[*)] Cf. the definitions in §1, Chapter 5.

assertion.

T h e o r e m 4.1. *The map* $\mathrm{A} \mapsto \mathfrak{U}$, *assigning to a pseudodifferential operator* A *its operator symbol* $\mathfrak{U} \in \mathfrak{S}$, *defines an isomorphism between the algebras* $\mathcal{A}/\mathcal{K}L_2(\mathfrak{M})$ *and* $\mathfrak{S}$. *In particular*[*)], *the norm of a residue class* $[A]$ *in the quotient algebra* $\mathcal{A}/\mathcal{K}L_2(\mathfrak{M})$ *is equal to the norm* $\|\mathfrak{U};\mathfrak{S}\|$ *of its operator symbol in the algebra* $\mathfrak{S}$.

The proof of the theorem is preceded by several propositions.

P r o p o s i t i o n 4.2. *Let* $\mathrm{A}^{(l,m)}$ *be a pseudodifferential operator or an operator adjoint to a pseudodifferential operator, let* $\mathfrak{U}^{(l,m)}$ *be its operator symbol,* $l = 1, \cdots, L$, $m = 1, \cdots, M$. *Then to the operators* $\mathrm{A} = \Sigma_l \Pi_m A^{(l,m)}$ *and* A^* *correspond the operator symbols* $\mathfrak{U} = \Sigma_l \Pi_m \mathfrak{U}^{(l,m)}$ *and* $\mathfrak{U}^*$.

The p r o o f of this proposition reduces to the verification of it for operators on $\mathbf{R}^n$, and hence follows from (4.1) and theorems 2.12 and 2.16. ■

R e m a r k 4.3. Let A and B be pseudodifferential operators of order zero on $L_2(\mathfrak{M})$ (in the sense of definition 3.1, in which A) is omitted), and let Φ, Ψ be their symbols (ordinary, not operator symbols). If the sets $\operatorname{sing}\Phi$ and $\operatorname{sing}\Psi$ of singular points are empty, then AB and A^* are, clearly, pseudodifferential operators of order zero; in the opposite case this need not hold, in general. Theorem 1.5 implies that for an $x \in \operatorname{sing}\Phi$ the operator symbol $\mathfrak{U}(x,\lambda)^*$ equals $\mathrm{E}(\lambda)^{-1}\overline{\Phi(x,\phi,\theta)}\mathrm{E}(\lambda) + \mathrm{T}(\lambda)$, with, moreover, $\|\mathrm{T}(\lambda); L_2(S^{n-1}) \to H^1(\lambda, S^{n-1})\| \leqslant \text{const}$ on the line $\operatorname{Im}\lambda = 0$. If $x \in \operatorname{sing}\Psi$, theorem 1.4 gives that $\mathfrak{U}(x,\lambda)\mathfrak{B}(x,\lambda) = \mathrm{E}(\lambda)^{-1}\Phi(x,\phi,\theta)\Psi(x,\phi,\theta)\mathrm{E}(\lambda) + \mathrm{S}(\lambda)$, while for $\mathrm{S}(\lambda)$ the same estimate holds as for $\mathrm{T}(\lambda)$. If the points of the sets $\operatorname{sing}\Phi$ and $\operatorname{sing}\Psi$ are regarded as being conical, then AB and A^* are pseudodifferential operators of order zero in the sense of definition 3.1 (but now taken into account requirement A).

P r o p o s i t i o n 4.4. *Let* $\mathrm{A}^{(l,m)}$ *be pseudodifferential operators, let* $\mathfrak{U}^{(l,m)}$, A *and* $\mathfrak{U}$ *be as proposition 4.2. Let also* $\Phi^{(l,m)}$ *be the ordinary (scalar) symbols of the* $\mathrm{A}^{(l,m)}$. *The operator* $\mathrm{A}: L_2(\mathfrak{M}) \to L_2(\mathfrak{M})$ *is Fredholm if and only if the following conditions*

[*)] An isomorphism of C^*-algebras is always an isometry, cf. §1, Chapter 5.

hold:

1) *the function* $\Phi = \Sigma_l \Pi_m \Phi^{(l,m)}$ *does not vanish on* $S(\mathfrak{M}) \oplus S^*(\mathfrak{M})$;

2) *at each point* $x \in \cup_{l,m} \operatorname{sing} \Phi^{(l,m)}$ *the operator-function* $\lambda \mapsto \mathfrak{U}(x,\lambda)^{-1}$ *does not have poles on the line* $\operatorname{Im}\lambda = 0$.

R e m a r k 4.5. At the points of the set $\mathfrak{M} \setminus \cup_{l,m} \operatorname{sing} \Phi^{(l,m)}$ condition 2) is satisfied in view of condition 1). Indeed, if $x \notin \cup \operatorname{sing} \Phi^{(l,m)}$ then $\Phi^{(l,m)}(x,\phi,\theta)$ is independent of ϕ, hence

$$\mathfrak{U}(x,\lambda) = \sum_l \prod_m E(\lambda)^{-1} \Phi^{(l,m)}(x,\theta) E(\lambda) =$$

$$= E(\lambda)^{-1} \sum_l \prod_m \Phi^{(l,m)}(x,\theta) E(\lambda).$$

By condition 1) the inverse operator $\mathfrak{U}(x,\lambda)^{-1} = E(\lambda)\Phi(x,\theta)^{-1}E(\lambda)$ exists on the line $\operatorname{Im}\lambda = 0$, which was required.

P r o o f o f p r o p o s i t i o n 4.4. Sufficiency of the conditions 1) and 2) is verified by constructing a left and a right regularizer (parametrix). Write A as

$$A = \sum \zeta^{\kappa} A_V \eta_{\kappa} + T,$$

where $\{\zeta\}$ is a partition of unity subordinate to the atlas $\{V,\kappa\}$, $\eta \in C^{\infty}(\mathfrak{M})$, $\operatorname{supp}\eta \subset V$, and $\zeta\eta = \zeta$; $T \in \mathcal{K}L_2(\mathfrak{M})$; A_V is the operator on $L_2(\mathbb{R}^n)$ given by

$$A_V u = \frac{1}{\sqrt{2\pi}} \int_{\operatorname{Im}\lambda=0} r^{i(\lambda+in/2)} \mathfrak{U}_V(r,\lambda) \tilde{u}(\lambda+in/2,\cdot)\,d\lambda,$$

and $\mathfrak{U}_V(r,\lambda)$ denotes the operator $\Sigma_l \Pi_m E(\lambda)^{-1} \Phi_V^{(l,m)}(r,\phi,\theta) E(\lambda)$.

Suppose that the neighborhood V contains a point of $\cup_{l,m} \operatorname{sing} \Phi^{(l,m)}$. Then in the local spherical coordinates with origin at this point we can represent $\Phi_V^{(l,m)}$ as $\Phi_V^{(l,m)}(r,\phi,\theta) = \Phi_0^{(l,m)}(\phi,\theta) + \Phi_1^{(l,m)}(r,\phi,\theta)$. Put $\mathfrak{U}_V(r,\lambda) = \mathfrak{U}_0(\lambda) + \mathfrak{U}_1(r,\lambda)$, where $\mathfrak{U}_0(\lambda) = \Sigma_l \Pi_m E(\lambda)^{-1}\, \Phi_0^{(l,m)}(\phi,\theta) E(\lambda)$. Condition (4.2) implies that for a fine partition of unity the norm of $\mathfrak{U}_1(r,\lambda): L_2(S^{n-1}) \to L_2(S^{n-1})$ can be taken arbitrarily small. By condition 2) the operator $\mathfrak{U}_0(\lambda)$ has an inverse on the line $\operatorname{Im}\lambda = 0$, and on this line $\|\mathfrak{U}_0(\lambda)^{-1}; L_2(S^{n-1}) \to L_2(S^{n-1})\| \leqslant \text{const}$ (a strip $|\operatorname{Im}\lambda| < \delta$, with δ a small positive number, contains only finitely many poles of the operator-function $\lambda \mapsto \mathfrak{U}_0(\lambda)^{-1}$; cf. remark 1.9). Hence the operator

$\mathfrak{U}_V(r,\lambda)^{-1}: L_2(S^{n-1}) \to L_2(S^{n-1})$ exists on the line $\operatorname{Im}\lambda = 0$. Put

$$\mathrm{R}_V f = \frac{1}{\sqrt{2\pi}} \int_{\operatorname{Im}\lambda=0} r^{i(\lambda+in/2)} \mathfrak{U}_V(r,\lambda)^{-1} \tilde{f}(\lambda+in/2,\cdot)d\lambda \tag{4.4}$$

If the neighborhood V does not contain points of $\cup_{l,m} \operatorname{sing} \Phi^{(l,m)}$, the existence of $\mathfrak{U}_V(r,\lambda)^{-1}$ is guaranteed by condition 1) already (cf. remark 4.5). The operator R_V is also in this case defined by (4.4).

Introduce the operator

$$\mathrm{R} = \sum \zeta^{\kappa} \mathrm{R}_V \eta_{\kappa} : L_2(\mathfrak{M}) \to L_2(\mathfrak{M}).$$

We show that $\mathrm{RA} = \mathrm{I}+\mathrm{T}$, where $\mathrm{T} \in \mathcal{K}L_2(\mathfrak{M})$. Restoring indices (the subscript of the local chart) and applying theorem 2.13 we have

$$\begin{aligned}\mathrm{RA} &= \sum_{j,k} (\zeta_j)^{\kappa_j} \mathrm{R}_{V_j} (\eta_j)_{\kappa_j} (\zeta_k)^{\kappa_k} \mathrm{A}_{V_k} (\eta_k)_{\kappa_k} + \mathrm{T} = \\ &= \sum_{j,k} (\zeta_j)^{\kappa_j} \mathrm{R}_{V_j} (\eta_j)_{\kappa_j} (\zeta_k)^{\kappa_j} \mathrm{A}_{V_j} (\eta_k)_{\kappa_j} + \mathrm{T} = \\ &= \sum_{j,k} (\zeta_j \zeta_k)^{\kappa_j} \mathrm{R}_{V_j} \mathrm{A}_{V_j} (\eta_k)_{\kappa_j} = \mathrm{I}+\mathrm{T},\end{aligned}$$

where the letter T denotes distinct compact operators. The equality $\mathrm{AR} = \mathrm{I}+\mathrm{T}$ is verified similarly. It is well-known (cf. [1], [3], [17]) that the operator A is Fredholm if and only if a left and a right regularizer exist. This means that conditions 1), 2) are sufficient.

We convince ourselves of necessity of these conditions. First suppose that condition 1) is violated at a point $x^{(0)} \notin \cup_{l,m} \operatorname{sing} \Phi^{(l,m)}$. Let (V,κ) be a local chart, $x^{(0)} \in V$, $\kappa(x^{(0)}) = 0$. For functions $\zeta,\eta \in C^{\infty}(\mathfrak{M})$ with supports in V we have

$$\zeta \mathrm{A} \eta = \zeta^{\kappa} \mathrm{A}_V \eta_{\kappa} + \mathrm{T}, \tag{4.5}$$

where $\mathrm{T} \in \mathcal{K}L_2(\mathfrak{M})$ and the symbol Φ_V of the standard operator A_V on $\mathbb{R}^n$ satisfies $\Phi_V(0,\theta_0) = 0$ for some $\theta_0 \in S^{n-1}$. Denote by $\{v_j\}$ a sequence of functions in $L_2(\mathbb{R}^n)$ such that $\|v_j\| = 1$, while $\operatorname{supp} v_j$ contract towards the coordinate origin. Clearly, the sequence $\{v_j\}$ converges weakly to zero in $L_2(\mathbb{R}^n)$. The sequence $w_j(y) = e^{-i\rho_j(\theta_0,y)} v_j(y)$ has the same properties, as $\rho_j \to +\infty$. Since $\Phi_V(0,\theta_0) = 0$, we have $\mathrm{A}_V w_j \to 0$ in $L_2(\mathbb{R}^n)$, and the weak convergence to zero of the w_j implies that $\|\mathrm{T}(w_j \circ \kappa)\| \to 0$ for any compact operator T. If A were

Fredholm, the estimate $\|u\| \leqslant (\|Au\| + \|Tu\|)$ would hold for some $T \in \mathcal{K}L_2(\mathfrak{M})$ (norms in $L_2(\mathfrak{M})$). This estimate is proved false on the sequence $\{w_j \circ \kappa\}$ by using (4.5).

Suppose now that condition 1) is violated at a point $x^{(0)} \in \cup_{l,m} \operatorname{sing} \Phi^{(l,m)}$, i.e. $\Phi(x^{(0)}, \phi_0, \theta_0) = 0$ for certain $\phi_0 \in S(\mathfrak{M})_{x^{(0)}}$, $\theta_0 \in S^*(\mathfrak{M})_{x^{(0)}}$. Put, as before, $\mathfrak{U}_0(\lambda) = \Sigma_l \Pi_m \mathrm{E}(\lambda)^{-1} \Phi^{(l,m)}(x^{(0)}, \phi, \theta) \mathrm{E}(\lambda)$. Proposition 3.6.4 implies that there are a sequence $\{u_j\} \subset C^\infty(S^{n-1})$, $\|u; L_2(S^{n-1})\| = 1$, and a sequence of numbers $\{\lambda_j\} \subset \mathbf{R}$, $\lambda_j \to \infty$, such that $\|\mathfrak{U}_0(\lambda_j) u_j; L_2(S^{n-1})\| \to 0$. Hence the sequence $\tilde{v}_j(\lambda + in/2, \psi) = \zeta_j(\lambda) u_j(\psi)$, where $\|v_j; L_2(\mathbf{R}^n)\| = 1$ and the support of ζ_j lies in an interval on the real axis with center at λ_j and of length $d_j \to 0$, satisfies

$$\|\mathrm{A}_0 v_j; L_2(\mathbf{R}^n)\| \to 0 \text{ as } j \to \infty; \tag{4.6}$$

here

$$\mathrm{A}_0 v = \frac{1}{\sqrt{2\pi}} \int_{\operatorname{Im}\lambda = 0} r^{i(\lambda + in/2)} \mathfrak{U}_0(\lambda) \tilde{v}(\lambda + in/2, \cdot) d\lambda. \tag{4.7}$$

The functions v_j in (4.6) may be replaced by the functions

$$w_j(r, \phi) = \frac{1}{\sqrt{2\pi}} r_j^{-n/2} \int_{\operatorname{Im}\lambda = 0} \left[\frac{r}{r_j} \right]^{i(\lambda + in/2)} \tilde{v}_j(\lambda + in/2, \phi) d\lambda,$$

where $r_j \to 0$. The supports of the v_j 'contract' towards the coordinate origin; more precisely, there is a sequence $\{\delta_j\}$, $\delta_j \to 0$, such that $\int_{|x| > \delta_j} |w_j(x)|^2 dx \to 0$. The sequence w_j converges weakly to zero in $L_2(\mathbf{R}^n)$.

Let $\mathfrak{U}_0(\lambda)$ and $\mathfrak{U}_1(r, \lambda)$ have the same meaning as in the first part of the proof, and let the operator A_1 be the righthand side of (4.7) after replacing $\mathfrak{U}(\lambda)$ by $\mathfrak{U}_1(r, \lambda)$. Let also $\chi \in C^\infty(\mathbf{R}^n)$, with $\chi(x) = 1$ if $|x| < \epsilon/2$, and $\chi(x) = 0$ if $|x| > \epsilon$. We have

$$\mathrm{A}_1 w_j = \chi \mathrm{A}_1 w_j + [\mathrm{A}_1, \chi] w_j + \mathrm{A}_1 (1 - \chi) w_j, \tag{4.8}$$

where $[\mathrm{A}_1, \chi] = \mathrm{A}_1 \chi - \chi \mathrm{A}_1$. Since the commutator $[A_1, \chi]$ is compact, the sequence $[A_1, \chi] w_j$ tends to zero in $L_2(\mathbf{R}^n)$. For small ϵ the term $\chi A_1 w_j$ remains small, and for large j the last term in (4.8) is small. The estimate $\|u\| \leqslant c(\|Au\| + \|Tu\|)$ is in this case not satisfied by the sequence $\{w_j\}$.

In order to prove necessity of condition 2) we can use a similar reasoning.

Recall that a pole λ_0 of the operator-function $\lambda \mapsto \mathfrak{U}_0(\lambda)^{-1}$ is an eigenvalue of $\mathfrak{U}_0$ (cf. remark 1.9). In other words, there is a nonzero solution $u \in L_2(S^{n-1})$ of the equation $\mathfrak{U}_0(\lambda_0)u = 0$. For u_j we may now take an eigenvector u of $\mathfrak{U}_0$ corresponding to an eigenvalue λ_0 on the line $\operatorname{Im}\lambda = 0$, and for $\{\zeta_j\}$ we may take a sequence of functions with supports contracting towards λ_0. ■

Theorem 2.16 and the proof of proposition 4.4 imply that proposition 4.4 remains valid if there are adjoints to pseudodifferential operators among the $A^{(l,m)}$. Therefore the following assertion is true.

P r o p o s i t i o n 4.6. *Let* $A, \mathfrak{U}$ *be as in proposition* 4.2, *and let, as before,* $\Phi = \Sigma_l \Pi_m \Phi^{(l,m)}$. *An operator* $A^* A - \mu I: L_2(\mathfrak{M}) \to L_2(\mathfrak{M})$ *is Fredholm if and only if the following conditions hold:* 1) *the function* $\overline{\Phi}\Phi - \mu$ *does not take the value zero on* $S(\mathfrak{M}) \oplus S^*(\mathfrak{M})$; *and* 2) *at each point* $x \in \cup_{l,m} \operatorname{sing} \Phi^{(l,m)}$ *the operator* $\mathfrak{U}(x,\lambda)^* \mathfrak{U}(x,\lambda) - \mu I$ *does not have eigenvalues on the line* $\operatorname{Im}\lambda = 0$.

P r o o f o f t h e o r e m 4.1. By proposition 4.2 it suffices to prove that

$$\inf_{T} \|A + T; L_2(\mathfrak{M}) \to L_2(\mathfrak{M})\| = \|\mathfrak{U}; \mathfrak{C}\|, \tag{4.9}$$

where T runs over the ideal $\mathcal{K}L_2(\mathfrak{M})$ of compact operators. Equation (4.9) follows from proposition 4.6 and the fact that for any bounded operator the relation $\inf_T \|B + T\|^2 = \inf_T \|B^* B + T\| = \sup\{\mu: \mu \in \sigma_{ess}(B^* B)\}$ holds; here $\sigma_{ess}(B^* B)$ denotes the spectrum of the residue class $[B^* B]$ in the quotient algebra $\mathfrak{L}/\mathcal{K}L_2(\mathfrak{M})$, where $\mathfrak{L}$ is the algebra of bounded operators in $L_2(\mathfrak{M})$ (cf. [24]). ■

2. The algebra generated by pseudodifferential operators with smooth symbols. Let $\mathfrak{A}_0$ be the subalgebra of the algebra $\mathfrak{A}$ generated by the pseudodifferential operators with smooth (scalar) symbols (i.e. $\operatorname{sing}\Phi = \varnothing$ for the symbol of every pseudodifferential operator in $\mathfrak{A}_0$). Let also $A^{(l,m)}$, $\mathfrak{U}^{(l,m)}$, A, and $\mathfrak{U}$ be as in proposition 4.2. Since the symbols $\Phi^{(l,m)}$ of the operators $A^{(l,m)}$ do not depend on a point $\phi \in S(\mathfrak{M})_x$, we have

$$\mathfrak{U}(x,\lambda) = \sum_{l=1}^{L} \prod_{m=1}^{M} E_{\theta\to\phi}(\lambda)^{-1} \Phi^{(l,m)}(x,\theta) E(\lambda) =$$

$$= E_{\theta\to\phi}(\lambda)^{-1} \Phi(x,\theta) E_{\psi\to\theta}(\lambda),$$

where $\Phi(x,\theta) = \Sigma_l \Pi_m \Phi^{(l,m)}(x,\theta)$. The operator $E(\lambda): L_2(S^{n-1}) \to L_2(S^{n-1})$ is unitary on the line $\operatorname{Im}\lambda = 0$ (corollary 1.5.6). Hence

$$\|\mathfrak{U}(x,\lambda); L_2(S^{n-1}) \to L_2(S^{n-1})\| = \max_{\theta} |\Phi(x,\theta)|.$$

Equation (4.9) for $A \in \mathfrak{A}_0$ takes the form

$$\inf_{T} \|A+T; L_2(\mathfrak{M}) \to L_2(\mathfrak{M})\| = \max_{S^*(\mathfrak{M})} |\Phi| = \|\Phi; C(S^*(\mathfrak{M})\|.$$

Together with theorem 4.1 this leads to the following well-known assertion.

T h e o r e m 4.7. *Let $\mathfrak{A}_0$ be the algebra generated by the pseudodifferential operators with smooth (scalar) symbols on a smooth compact manifold $\mathfrak{M}$ without boundary. Then the quotient algebra $\mathfrak{A}_0 / \mathfrak{K}L_2(\mathfrak{M})$ is isomorphic to the algebra $C(S^*(\mathfrak{M}))$ of continuous functions on the bundle $S^*(\mathfrak{M})$ of cotangent unit vectors. In particular, the norm of a residue class $[A]$ is equal to the norm $\|\Phi; C(S^*(\mathfrak{M}))\|$ of its symbol Φ.*

Theorems 4.1 and 4.7 can, in an obvious way, be generalized to pseudodifferential operators acting on sections of vector bundles. In this case, the norm $\|\Phi; C(S^*(\mathfrak{M}))\|$ figuring in theorem 4.7 must be understood to mean $\sup_{j, S^*(\mathfrak{M})} s_j$, where s_j^2 are the eigenvalues of the matrix $\Phi^*\Phi$, $s_j \geqslant 0$. In the sequel we will not give indications as to the possibility of similar obvious generalizations.

3. The algebra of pseudodifferential operators on a manifold with boundary. Let $\mathfrak{M}$ be a C^∞ manifold with boundary $\partial\mathfrak{M}$, and let $\mathfrak{A}$ be the algebra of operators on $L_2(\mathfrak{M})$ generated by the pseudodifferential operators whose symbols satisfy the same conditions as in section **1** (i.e. they are allowed to have singularities of the first kind).

Denote by $S_+(\mathfrak{M})_x$, $x \in \partial\mathfrak{M}$, the hemisphere corresponding to the tangent vectors directed to the side of the manifold, and by $P^+ : L_2(S(\mathfrak{M})_x) \to L_2^+(S(\mathfrak{M})_x)$ the operator of multiplication by the characteristic function of $S_+(\mathfrak{M})_x$.

Let A be a pseudodifferential operator with symbol Φ, and let $\mathfrak{U}(x,\lambda) = E_{\theta\to\phi}(\lambda)^{-1}\Phi(x,\phi,\theta)E_{\psi\to\theta}(\lambda)$, where $\phi,\psi \in S(\mathfrak{M})_x$, $\theta \in S^*(\mathfrak{M})_x$. Put $\mathfrak{B}(x,\lambda) = P^+\mathfrak{U}(x,\lambda): L_2^+(S(\mathfrak{M})_x) \to L_2^+(S(\mathfrak{M})_x)$ if $x \in \partial\mathfrak{M}$, and $\mathfrak{B}(x,\lambda) = \mathfrak{U}(x,\lambda)$ if $x \notin \partial\mathfrak{M}$. Introduce the algebra $\mathfrak{G}$ generated by the operator-functions $(x,\lambda) \mapsto \mathfrak{B}(x,\lambda)$, with pointwise multiplication, involution, and norm

$$\|\mathfrak{B};\mathfrak{C}\| = \sup_{(x,\lambda)\in\mathfrak{M}\times\mathbf{R}} \|\mathfrak{B}(x,\lambda)\|.$$

T h e o r e m 4.8. *The map* $A \mapsto \mathfrak{B}$ *defines an isomorphism of the algebras* $\mathfrak{A}/\mathfrak{K}L_2(S^{n-1})$ *and* $\mathfrak{G}$.

The p r o o f is a repetition, with obvious changes, of the proof of theorem 4.1. ■

4. The algebra of pseudodifferential operators on a manifold with conical points. Let $\mathfrak{M}$ be a $\check{C}^\infty$ manifold with conical points (cf. §3.1). Denote by $\mathfrak{A}$ the algebra generated by the pseudodifferential operators on $H^0_\beta(\mathfrak{M}) \equiv H_\beta(\mathfrak{M})$ (here β is a vector with real components $(\beta_1, \cdots, \beta_l)$ and l is the number of conical points $x^{(1)}, \cdots, x^{(l)}$). We will assume that discontinuities of the first kind are allowed in the symbols of the pseudodifferential operators. For a pseudodifferential operator A we introduce the operator symbol $\mathfrak{U}$, defined for $x \neq x^{(j)}$, $j = 1, \cdots, l$, as in section **1**, and for $x = x^{(j)}$ as the operator function $\lambda \mapsto \mathfrak{U}(x^{(j)},\lambda)$ given on the line $l_{x(j)} = \{\lambda \in \mathbf{C}\colon \operatorname{Im}\lambda = \beta_j\}$; here $\mathfrak{U}(x^{(j)},\cdot)$ denotes a meromorphic pseudodifferential operator on the cone X_j (cf. definition 2.1). Put also $l_x = \{\lambda \in \mathbf{C}\colon \operatorname{Im}\lambda = 0\}$ for $x \neq x^{(j)}$, $j = 1, \cdots, l$.

Let $\mathfrak{G}$ be the algebra generated by the operator symbols and endowed with the norm

$$\|\mathfrak{U}:\mathfrak{G}\| = \sup_{\substack{x\in\mathfrak{M}\\ \lambda\in l_x}} \|\mathfrak{U}(x,\lambda); L_2(S(\mathfrak{M})_x) \to L_2(S(\mathfrak{M})_x\|$$

(if $x = x^{(j)}$, then $S(\mathfrak{M})_x$ denotes a directing manifold of X_j).

T h e o r e m 4.9. *The map* $A \mapsto \mathfrak{U}$ *defines an isomorphism of the algebras* $\mathfrak{A}/\mathfrak{K}H_\beta(\mathfrak{M})$ *and* $\mathfrak{G}$.

This assertion can be proved similarly as theorem 4.1; the presence of conical points does not present additional difficulties. Theorem 4.9 can, in an obvious way, be generalized to manifolds with boundary (cf. theorem 4.8). Of course, theorems 4.1, 4.7, and 4.8 follow from the assertions formulated in this section.

5. The algebra of one-dimensional singular integral operators with discontinuous

coefficients on a composite contour. Let $\{\gamma_u\}$ be a finite set of smooth (Lyapunov) closed directed arcs in the complex plane. Suppose that two arcs γ_μ and γ_ν can intersect at finitely many points only; moreover, either $\gamma_\mu \cup \gamma_\nu$ is a smooth arc in a neighborhood of a common point x, or γ_μ and γ_ν do not have a common tangent at x. Under these conditions the union $\Gamma = \cup_\mu \gamma_\mu$ will be called a piecewise smooth contour. Those points of Γ at which there is no neighborhood on Γ that is a smooth open arc will be called *singular points*, or *cusps*, and the other points will be called *regular*. The singular points will be denoted by $z_1, \cdots, z_m$.

A function h defined on $\Gamma \setminus \cup_{\nu=1}^m \{z_\nu\}$ is called piecewise continuous if: 1) h is continuous everywhere except at finitely many points of discontinuity of the first kind; and 2) when passing along an arc to a cusp h has a finite limit, which depends, in general, on the arc. For the sake of being specific we will assume that h is 'left'-continuous.

Put $\rho(z) = \Pi_{\nu=1}^m |z - z_0|^{\beta_\nu}$, where $2\beta_\nu \in (-1,1)$, and introduce the space $L_2(\Gamma,\rho)$ with norm

$$\|u\| = \left[\int_\Gamma |u(z)|^2 \rho(z)^2 \, |dz|\right]^{1/2}.$$

A *singular integral operator* with piecewise continuous coefficients a and b is an operator of the form $\mathrm{A}u = au + b\mathcal{K}_\Gamma u$, where

$$(\mathcal{K}_\Gamma u)(z) = \frac{1}{\pi i}\int_\Gamma \frac{u(\zeta)}{\zeta - z} d\zeta.$$

The operator A is continuous in $L_2(\Gamma,\rho)$ (cf., e.g., [17]).

Let $t_j^\nu = \{z \in \mathbf{C}\colon z = z_\nu + r_\nu e^{i\theta_{j\nu}}\}$, where $0 \leqslant r < \infty$, $0 \leqslant \theta_{1\nu} < \cdots < \theta_{N_\nu \nu} < 2\pi$, be the tangent rays to the arcs containing the cusps z_ν. For a piecewise continuous function h, let $h_j(z_\nu)$ denote the limit of $h(z)$ as $z \to z_\nu$ along the arc tangent to the ray t_j^ν, and let, at a regular point z, $h_\pm(z)$ denote the limit values of h 'from the right' and 'from the left'.

At each regular point $z \in \Gamma$ we define a (2×2)-matrix-function $\lambda \mapsto \mathfrak{U}(z,\lambda)$ on the line $\mathrm{Im}\,\lambda = 0$ by

$$\mathfrak{U}(z,\lambda) = \tag{4.10}$$

$$= \begin{bmatrix} a_+(z) + b_+(z)\tanh \pi\lambda & ib_+(z)/\cosh \pi\lambda \\ -ib_-(z)/\cosh \pi\lambda & a_-(z) - b_-(z)\tanh \pi\lambda \end{bmatrix}.$$

At a cusp z_ν we define a matrix-function $\lambda \mapsto \mathfrak{U}(z_\nu,\lambda)$ of dimension $N_0 \times N_0$ on the line $\operatorname{Im}\lambda = \beta_\nu$. If all arcs containing z_ν 'leave' z_ν, then we put $\mathfrak{U}(z_\nu,\lambda) = \|\mathfrak{U}_{jk}(z_\nu,\lambda)\|_{j,k=1}^{N_0}$, where $\mathfrak{U}_{jk}(z_\nu,\lambda) = a_j(z_\nu)\delta_{jk} + b_j(z_\nu)\mathfrak{T}_{jk}^{(\nu)}(\lambda)$, with $\mathfrak{T}_{jk}^{(\nu)}(\lambda)$ coinciding with the righthand side of (2.13) after replacing N by N_ν, and θ_j by $\theta_{j\nu}$, $j = 1, \cdots, N_\nu$. If, however, the k-th arc enters, then the k-th column of the matrix $\mathfrak{U}$ must be defined by $\mathfrak{U}_{jk}(z_\nu,\lambda) = a_j(z_\nu,\lambda)\delta_{jk} - b_j(z_\nu)\mathfrak{T}_{jk}^{(\nu)}(\lambda)$.

It is obvious that the off-diagonal entries of $\mathfrak{U}(z,\lambda)$ decrease exponentially for $z \in \Gamma$ fixed and λ tending to ∞. The diagonal entries have finite limits as λ tends to an arbitrary 'end' at infinity along a corresponding line. We will assume that the matrix $\mathfrak{U}$ is defined at these 'ends' at infinity, and equals its limit values.

The matrix-function $\mathfrak{U}$ will be called the *(operator) symbol* of the singular integral operator $\mathrm{A} = a\mathrm{I} + b\mathcal{K}_\Gamma$.

Denote by $\mathfrak{A}$ the algebra generated by the singular integral operators on $L_2(\Gamma,\rho)$. We also introduce the algebra $\mathfrak{S}$ generated by the symbols of the singular integral operators. The norm of an element $\mathfrak{U} \in \mathfrak{S}$ is defined by $\|\mathfrak{U};\mathfrak{S}\| = \sup_{\lambda,z,j} s_j(z,\lambda)$, where $s_j(z,\lambda)^2$ are the eigenvalues of the matrix $\mathfrak{U}(z,\lambda)^*\mathfrak{U}(z,\lambda)$, $s_j \geqslant 0$, z runs over Γ, and the parameter λ runs over the line $\operatorname{Im}\lambda = 0$ if z is a regular point, and over the line $\operatorname{Im}\lambda = \beta_\nu$ if $z = z_\nu$.

Theorem 4.10. *The set $\mathcal{K}L_2(\Gamma,\rho)$ of compact operators on $L_2(\Gamma,\rho)$ is contained in the algebra $\mathfrak{A}$. The map 'singular integral operator $\to$ symbol' defines an isomorphism of the algebras $\mathfrak{A}/\mathcal{K}L_2(\Gamma,\rho)$ and $\mathfrak{S}$; the norm of a class in $\mathfrak{A}/\mathcal{K}L_2(\Gamma,\rho)$ equals the norm of its symbol.*

The proof of this theorem is simpler than that of its multidimensional version theorem 4.9. It is given in [56].

6. The algebra of pseudodifferential operators with oscillating symbols. Let $\mathfrak{M}$ be a C^∞ manifold without boundary, and let A be a pseudodifferential operator of order zero on $\mathfrak{M}$. We will assume that the symbol Φ of A is a smooth function everywhere on $S^*(\mathfrak{M})$ except at a set $S^*(\mathfrak{M})|\operatorname{sing}\Phi$, where $\operatorname{sing}\Phi$ is a finite subset of $\mathfrak{M}$*). Suppose that in a neighborhood of a point $x \in \operatorname{sing}\Phi$ the symbol Φ has

*) $S^*(\mathfrak{M})|\operatorname{sing}\Phi$ denotes the part of $S^*(\mathfrak{M})$ above $\operatorname{sing}\Phi$, cf. §2, Chapter 5. (Translator's note.)

in local spherical coordinates a representation $\Phi_0(r,\phi,\theta)+\Phi_1(r,\phi,\theta)$, where Φ_0 is a 'periodic' function of r, i.e. $\Phi_0(re,\cdot) = \Phi_0(r,\cdot)$, $\Phi_0 \in C^\infty(d\times S^{n-1}\times S^{n-1})$ (d denotes the interval $[1,e]$ with end points identified), while for Φ_1 the estimates $\|(rD_r)^k\Phi_1(r,\cdot);C^p(S^{n-1}\times S^{n-1})\| = O(r^\delta)$ hold, for arbitrary k,p, and a certain $\delta>0$. The set of functions Φ satisfying these requirements is denoted by $\mathfrak{N}$. We introduce the algebra $\mathcal{Q}$ generated by the pseudodifferential operators of order zero on $L_2(\mathfrak{M})$ whose symbols belong to $\mathfrak{N}$. For a pseudodifferential operator A with symbol $\Phi \in \mathfrak{N}$ we define its operator symbol $\mathbf{A}(x,\lambda)$, $x \in \mathfrak{M}$, $0\leqslant\lambda\leqslant 2\pi$, acting on functions $v \in C^\infty(d\times S(\mathfrak{M})_x)$, by

$$\mathbf{A}(x,\lambda)v = \sum_{j=-\infty}^{+\infty} r^{2\pi ij}\mathfrak{U}(x,\lambda+2\pi j)v_j;$$

here $\mathfrak{U}(x,\lambda) = \mathrm{E}_{\theta\to\phi}(\lambda)^{-1}\Phi(x,r,\phi,\theta)\mathrm{E}_{\psi\to\theta}(\lambda)$, $\phi,\psi \in S(\mathfrak{M})_x$, $\theta \in S^*(\mathfrak{M})_x$,

$$v_j(\phi) = \int_1^e v(r,\phi)r^{-2\pi ij-1}\,dr.$$

By $\mathfrak{D}$ we denote the algebra generated by the functions $(x,\lambda)\mapsto\mathbf{A}(x,\lambda)$, with pointwise operations and norm

$$\|\mathrm{A};\mathfrak{D}\| = \sup_{x,\lambda}\|\mathbf{A}(x,\lambda);\ L_2(\Pi)\to L_2(\Pi)\|,$$

$$\Pi = d\times S^{n-1}.$$

T h e o r e m 4.11. *The map* $\mathrm{A}\to\mathbf{A}$ *defines an isomorphism of the algebras* $\mathcal{Q}/\mathcal{K}L_2(\mathfrak{M})$ *and* $\mathfrak{D}$.

This theorem can be proved along the same lines as theorem 4.1.

Chapter 5

The spectrum of a C^*-algebra of pseudodifferential operators with discontinuous symbols on a closed manifold

The spectrum of an algebra is the set of equivalence classes of irreducible representations of this algebra, endowed with a special topology. The algebra $\mathcal{Q}$ generated by the pseudodifferential operators with smooth symbols on a smooth closed manifold $\mathfrak{M}$ becomes commutative after factorization by the ideal $\mathcal{K}L_2(\mathfrak{M})$ of compact operators. Hence all irreducible representations of the quotient algebra $\mathcal{Q}/\mathcal{K}L_2(\mathfrak{M})$ are one-dimensional, and its spectrum can be identified with the maximal ideal space, which is the bundle $S^*(\mathfrak{M})$ of cotangent unit vectors. If $A \in \mathcal{Q}$, $[A] \in \mathcal{Q}/\mathcal{K}L_2(\mathfrak{M})$ is its residue class, and π runs over the spectrum of $\mathcal{Q}/\mathcal{K}L_2(\mathfrak{M})$, then the map $\pi \mapsto \pi[A]$ coincides with the function $S^*(\mathfrak{M}) \ni \pi \mapsto \Phi(\pi) \in \mathbf{C}$, where Φ is the symbol of A.

In this Chapter we study the spectrum of an algebra of pseudodifferential operators with singularities either in the symbols or on the manifold. The quotient algebra $\mathcal{Q}/\mathcal{K}L_2(\mathfrak{M})$ remains commutative, among its irreducible representations there are infinite-dimensional ones. We compute all equivalence classes of irreducible representations, indicate a realization of these representations (i.e. describe a map $\pi \mapsto \pi[A]$), and elucidate a topology on the spectrum.

§1 contains an outline of the results used from the theory of C^*-algebras. It is supplemented by references to [24], where the corresponding proofs can be found. §2 is devoted to the statement of a theorem about the spectrum of the algebra $\mathcal{Q}$ generated by the pseudodifferential operators with discontinuities 'of the first kind' in the symbols; variants and consequences are discussed also. In §3, §4 we prepare the proof of this theorem, and study 'local' algebras (of values of operator symbols at a point or restrictions of them onto a line). We finish the proof of the theorem in §5. In it, necessary and sufficient conditions for an arbitrary element of $\mathcal{Q}$ to be Fredholm are given also. A description of certain important ideals in $\mathcal{Q}$ is contained in §6. In §7 we study the spectrum of the algebra of pseudodifferential operators on a manifold with conical points, and in §8 we study the spectrum of the algebra of pseudodifferential operators with oscillating symbols.

§1. Results from the theory of C^*-algebras

1. C^*-algebras. Let $\mathcal{A}$ be an algebra over the field of complex numbers. We say that a map $x \mapsto x^*$ of $\mathcal{A}$ into itself is an *involution* if the following conditions hold: $(x^*)^* = x$, $(x+y)^* = x^*+y^*$, $(\lambda x)^* = \bar{\lambda}x^*$, and $(xy)^* = y^*x^*$. An algebra endowed with an involution is called *involutive**).

An *involutive normed algebra* is a normed algebra $\mathcal{A}$ endowed with an involution such that $\|x^*\| = \|x\|$ for all $x \in \mathcal{A}$. A complete involutive normed algebra is called an *involutive Banach algebra.*

If the involution has the property $\|x^*\|^2 = \|x^*x\|$ for any element $x \in \mathcal{A}$, then the involutive Banach algebra $\mathcal{A}$ is called a *C^*-algebra.*

Let $\mathcal{A}$ be an involutive algebra. A subalgebra of $\mathcal{A}$ that is mapped into itself under the involution is, by definition, an involutive subalgebra. Every closed involutive subalgebra of a C^*-algebra is a C^*-algebra.

Denote by $\mathfrak{L}(H)$ the algebra of all bounded linear operators on a Hilbert space H. Involution in $\mathfrak{L}(H)$ means transition to the adjoint operator. Every closed involutive subalgebra of $\mathfrak{L}(H)$ is a C^*-algebra.

2. Morphisms of C^*-algebras. Let $\mathcal{A}$ and $\mathcal{B}$ be involutive algebras. A *morphism* $f:\mathcal{A}\to\mathcal{B}$ is a linear map satisfying $f(xy) = f(x)f(y)$, $f(x^*) = f(x)^*$ for all $x,y \in \mathcal{A}$. A bijective morphism is called an *isomorphism.*

It turns out that every morphism of C^*-algebras $\mathcal{A}$ and $\mathcal{B}$ (simply regarded as a morphism of involutive algebras) is continuous. If $f:\mathcal{A}\to\mathcal{B}$ is an injective morphism, then $\|f(x)\| = \|x\|$ for all $x \in \mathcal{A}$.

Every C^*-algebra is isomorphic to a closed involutive subalgebra of an algebra $\mathfrak{L}(H)$.

---

*) An involutive algebra is also called an algebra with involution. (Translator's note.)

3. Quotient algebras. Let $\mathfrak{A}$ be a C^*-algebra and J a closed two-sided ideal in $\mathfrak{A}$. Then J is self-adjoint (i.e. is preserved as a set under involution) and the quotient algebra $\mathfrak{A}/J$, endowed with the quotient norm and corresponding involution, is a C^*-algebra.

Proposition 1.1. *Let $\mathfrak{A}$ and $\mathfrak{B}$ be C^*-algebras, $\phi:\mathfrak{A}\to\mathfrak{B}$ a morphism, and I the kernel of ϕ. Then I is a closed ideal in $\mathfrak{A}$, and the image $\phi(\mathfrak{A})$ is closed in $\mathfrak{B}$.*

Proof. The map ϕ is continuous, since every morphism of C^*-algebras is continuous. Hence I is a closed ideal and $\mathfrak{A}/I$ is a C^*-algebra. The morphism $\mathfrak{A}/I\to\mathfrak{B}$ obtained by factoring through ϕ is injective, hence an isometry. Thus, the image $\phi(\mathfrak{A})$ is complete and closed in $\mathfrak{B}$. ■

4. Representations. A *representation* of a C^*-algebra $\mathfrak{A}$ in a Hilbert space H is a morphism $\pi:\mathfrak{A}\to\mathfrak{L}(H)$. The space H is called the *representation space* of π, and the (Hilbert) dimension of H is called the *dimension of the representation*.

Two representations π,π' of an algebra $\mathfrak{A}$ in spaces H,H' are equivalent, by definition, if there is a Hilbert space isomorphism $U:H\to H'$ such that $U\pi(x) = \pi'(x)U$ for all $x\in\mathfrak{A}$.

We say that a vector $\xi\in H$ is *cyclic* (or *totalizing*) for a representation π of an algebra $\mathfrak{A}$ in H if the closure of the set $\pi(\mathfrak{A})\xi$ coincides with H.

A representation π of an algebra $\mathfrak{A}$ in H is called *irreducible* if π satisfies either one of the following (equivalent) conditions:

1) the only closed vector subspaces in H that are invariant under $\pi(\mathfrak{A})$ are 0 and H;

2) the commutator of $\pi(\mathfrak{A})$ in $\mathfrak{L}(H)$ consists of scalars;

3) either every nonzero vector $\xi\in H$ is cyclic for π, or π is the null representation of dimension 1*).

A *two-sided primitive ideal* in a C^*-algebra $\mathfrak{A}$ is the kernel of an irreducible representation of $\mathfrak{A}$. Every closed two-sided ideal in $\mathfrak{A}$ is the intersection of the primitive ideals containing it.

*) This is called the nontrivial representation in [24]. (Translator's note.)

In the sequel, an algebra will always mean a C^*-algebra.

Proposition 1.2. *Let π be an irreducible representation of an algebra $\mathcal{A}$ in a space H. The following assertions are true:*

1) *if I is a two-sided ideal in $\mathcal{A}$ and $\pi(I) \neq 0$, then the restriction $\pi|_I$ of π to I is irreducible;*

2) *if I_1, I_2 are two-sided ideals in $\mathcal{A}$ and $\pi(I_1) \neq 0$, $\pi(I_2) \neq 0$, then $\pi(I_1 I_2) \neq 0$.*

Proof. 1) Put $E = \{x \in H : \pi(I)x = 0\}$. Since $\pi(\mathcal{A})E \subset E$, and by requirement $E \neq H$, we have $E = 0$. Hence $\pi(I)\xi \neq 0$ whatever the nonzero vector $\xi \in H$. The subspace $\pi(I)\xi$ is invariant under $\pi(\mathcal{A})$. Since π is irreducible we find $\pi(I)\xi = H$. Thus, every nonzero vector $\xi \in H$ is cyclic for $\pi|_I$.

2) As was proved in 1), $\pi(I_2)H = H$, $\pi(I_1)\pi(I_2)H = H$. Hence $\pi(I_1 I_2) \neq 0$. ■

Proposition 1.3. *Let I_1, I_2 be two-sided ideals in an algebra $\mathcal{A}$, and let I be a primitive ideal. If $I_1 I_2 \subset I$ (in particular, if $I_1 \cap I_2 \subset I$), then either $I_1 \subset I$ or $I_2 \subset I$.*

Proof. Assume the contrary, i.e. $I_1 \not\subset I$ and $I_2 \not\subset I$. Applying the second assertion of proposition 1.1. to an irreducible representation π of $\mathcal{A}$ with kernel I, we find $\pi(I_1 I_2) \neq 0$, whence $I_1 I_2 \not\subset I$. ■

Proposition 1.4. *Let $\mathcal{A} = \mathcal{K}(H)$ be the algebra of compact operators on a Hilbert space H. Then every nonnull irreducible representation of $\mathcal{A}$ is equivalent to the identity representation.*

Proof. See [24], 4.1.5. ■

Proposition 1.5. ([24], 4.1.10). *Let π be an irreducible representation of an algebra $\mathcal{A}$ in H. If $\pi(\mathcal{A}) \cap \mathcal{K}(H) \neq 0$, then $\mathcal{K}(H) \subset \pi(\mathcal{A})$ and every irreducible representation of $\mathcal{A}$ with the same kernel as π is equivalent to π.*

Proposition 1.6 ([24], 2.10.2) (extension of a representation). *Let $\mathcal{B}$ be a* *subalgebra of $\mathcal{A}$, and ρ a representation of $\mathcal{B}$ in a Hilbert space $\mathcal{K}$. Then there is a Hilbert space H, containing K as a subspace, and a representation π of $\mathcal{A}$ in H, such that $\rho(x) = \pi(x)|_K$ for all $x \in \mathcal{B}$. If ρ is irreducible, π can be chosen to be irreducible.*

5. The spectrum of a C^*-algebra. Let Prim $\mathcal{A}$ be the set of two-sided primitive ideals in $\mathcal{A}$, and suppose $T \subset \operatorname{Prim}\mathcal{A}$. Denote by $I(T)$ the intersection of the ideals in T. The set $I(T)$ is a two-sided ideal in $\mathcal{A}$. Let $\overline{T}$ be the set of primitive ideals containing $I(T)$. It turns out that there is a unique topology on Prim $\mathcal{A}$ such that for all $T \subset \operatorname{Prim}\mathcal{A}$ the set $\overline{T}$ is the closure of T in this topology. This topology is called the *Jacobson topology* on Prim $\mathcal{A}$.

Introduce the set $\hat{\mathcal{A}}$ of equivalence classes of nonnull irreducible representations of $\mathcal{A}$. The map $\pi \mapsto \ker \pi$ defines a canonical surjection $\hat{\mathcal{A}} \to \operatorname{Prim}\mathcal{A}$.

The *spectrum* of $\mathcal{A}$ is the set $\hat{\mathcal{A}}$ endowed with the topology that is the pre-image of the Jacobson topology under the canonical map $\hat{\mathcal{A}} \to \operatorname{Prim}\mathcal{A}$.

We say that a topological space is a T_0-space if, given two points of this space, there is a neighborhood of one point not containing the other point.

The following three conditions are equivalent ([24], 3.1.6):

1) $\hat{\mathcal{A}}$ is a T_0-space;

2) two irreducible representations of $\mathcal{A}$ with the same kernel are equivalent;

3) the canonical map $\hat{\mathcal{A}} \to \operatorname{Prim}\mathcal{A}$ is a homeomorphism.

Finally we given another description of the topology on the spectrum $\hat{\mathcal{A}}$ of an algebra $\mathcal{A}$. Let $\{x_i\}$ be a family of elements of $\mathcal{A}$ that is everywhere dense in $\mathcal{A}$ and let $Z_i = \{\pi \in \hat{\mathcal{A}}: \|\pi(x_i)\| > 1\}$. Then the Z_i form a base for the topology in $\hat{\mathcal{A}}$. Thus, if $\mathcal{A}$ is separable, then $\hat{\mathcal{A}}$ has a countable base.

Proposition 1.7. *Let I be a closed two-sided ideal of an algebra $\mathcal{A}$ and let $\hat{\mathcal{A}}^I = \{\pi \in \hat{\mathcal{A}}: \pi|_I \neq 0\}$. Then the map $\pi \mapsto \pi|_I$ is a bijection of $\hat{\mathcal{A}}^I$ onto $\hat{I}$.*

The proof is contained in [24], 2.11.2.

Let, as before, $\mathcal{K}(H)$ be the algebra of compact operators on a Hilbert space H, let X be a locally compact space, and let $\mathcal{K}_0(X,H)$ be the algebra of continuous functions on X, with values in $\mathcal{K}(H)$, and tending to zero (in the operator norm) at infinity. The norm in $\mathcal{K}_0(X,H)$ is defined by

$$\|v\| = \sup_{x\in X}\|v(x);H\to H\|.$$

The following proposition is an instance of assertion 10.4.4 from [24].

P r o p o s i t i o n 1.8. *For every $x \in X$ we define a representation $\pi(x)$ of the algebra $\mathcal{K}_0(X,H)$ by $\pi(x)v = v(x)$. Then $x \mapsto v(x)$ is a homeomorphism of X onto the spectrum $\widehat{\mathcal{K}_0(X,H)}$ of $\mathcal{K}_0(X,H)$.*

6. Rich subalgebras. A subalgebra $\mathcal{B}$ of an algebra $\mathcal{Q}$ is called rich if the following conditions hold:

1) for every irreducible representation π of $\mathcal{Q}$ the representation $\pi|_{\mathcal{B}}$ is irreducible;

2) if π_1, π_2 are inequivalent irreducible representations of $\mathcal{Q}$, then the restrictions $\pi_1|_{\mathcal{B}}$, $\pi_2|_{\mathcal{B}}$ are inequivalent.

P r o p o s i t i o n 1.9 (cf. 11.1.4 in [24]). *If $\mathcal{B}$ is a rich subalgebra of the algebra $\mathcal{K}_0(X,H)$, then $\mathcal{B} = \mathcal{K}_0(X,H)$.*

7. A criterion for invertibility of an element of an algebra.

P r o p o s i t i o n 1.10. *Let $\mathcal{Q}$ be an algebra with identity. An element $a \in \mathcal{Q}$ is invertible if and only if for every irreducible representation π of $\mathcal{Q}$ the operator $\pi(a)$ is invertible (on the representation space H_π).*

P r o o f. Necessity of the condition for $a \in \mathcal{Q}$ to be invertible is obvious. We convince ourselves of the sufficiency. Suppose a does not have a right (or left) inverse. Then the element $b = b^* = aa^*$ (respectively, a^*a) cannot be inverted. Introduce the commutative algebra $\mathcal{B}$ generated by b and the identity of $\mathcal{Q}$. It is well-known ([24], 1.5) that $\mathcal{B}$ is isomorphic to the algebra of continuous functions on the spectrum $\operatorname{Sp} b$. Since b is noninvertible, 0 belongs to $\operatorname{Sp} b$, and there is a one-dimensional representation ρ of $\mathcal{B}$ annihilating b. By proposition 1.6 there is a representation $\pi \in \hat{\mathcal{Q}}$ extending ρ in the sense that H_ρ is a one-dimensional subspace of H_π and $\pi(x)|_{H_\rho} = \rho(x)$ for all $x \in \mathcal{B}$. Since $\pi(b)|_{H_\rho} = 0$, the operator

$\pi(b)$ (hence also $\pi(a)$) is noninvertible. ∎

P r o p o s i t i o n 1.11 ([24], 1.3.10). *Let $\mathfrak{B}$ be a subalgebra of an algebra $\mathfrak{A}$ which has an identity which, moreover, belongs to $\mathfrak{B}$. Then for any $x \in \mathfrak{B}$ we have* $\mathrm{Sp}_{\mathfrak{A}}x = \mathrm{Sp}_{\mathfrak{B}}x$, *where* $\mathrm{Sp}_C y$ *denotes the spectrum of an element y in the algebra C.*

Thus, if an element $x \in \mathfrak{B}$ is invertible in $\mathfrak{A}$, then it is invertible in $\mathfrak{B}$ also.

§2. The spectrum of a C^*-algebra of pseudodifferential operators with discontinuities of the first kind in the symbols on a smooth closed manifold (statement of the main theorem)

Let $\mathfrak{M}$ be a closed Riemannian manifold of class C^∞, and let $S(\mathfrak{M})$ and $S^*(\mathfrak{M})$ be the bundles of tangent and cotangent unit vectors. Let $\mathfrak{M}$ be the set of functions Φ given and everywhere smooth on $S^*(\mathfrak{M})$ except at a set $S^*(\mathfrak{M}) \mid \mathrm{sing}\,\Phi$, where $\mathrm{sing}\,\Phi$ is a finite subset of $\mathfrak{M}$, depending on the function Φ, and $S^*(\mathfrak{M}) \mid \mathrm{sing}\,\Phi$ is the part of $S^*(\mathfrak{M})$ above $\mathrm{sing}\,\Phi$. At the points of $\mathrm{sing}\,\Phi$ the function Φ is allowed to have discontinuities of the first kind (for a description of such discontinuities see Chapter 4, §4.1). A function $\Phi \in \mathfrak{M}$ can be regarded as being given on the Whitney sum $S(\mathfrak{M}) \oplus S^*(\mathfrak{M})$, where, moreover, the map $(x, \phi, \theta) \mapsto \Phi(x, \phi, \theta)$ $(x \in \mathfrak{M},\ \phi \in S(\mathfrak{M})_x,\ \theta \in S^*(\mathfrak{M})_x)$ does not depend on ϕ if $x \notin \mathrm{sing}\,\Phi$.

We introduce the space $\hat{\mathfrak{G}}$, which, as will be proved in the sequel, is homeomorphic to the spectrum of the quotient algebra $\mathfrak{A}/\mathcal{K}L_2(\mathfrak{M})$, where $\mathfrak{A}$ is the algebra generated by the pseudodifferential operators of order zero and with symbols from $\mathfrak{M}$ (here and in the sequel, an algebra means a C^*-algebra).

Put $\hat{\mathfrak{G}} = (S(\mathfrak{M}) \oplus S^*(\mathfrak{M})) \cup \mathfrak{L}$, where $\mathfrak{L}$ is the disjoint union of a set of lines $\{l_x\}$ indexed by the points of the manifold $\mathfrak{M}$. A topology on $\hat{\mathfrak{G}}$ is defined as follows. The part of $S(\mathfrak{M}) \oplus S^*(\mathfrak{M})$ lying above a neighborhood U of a point x_0 is regarded as the set of triples (x, ϕ, θ), where $x \in U$, and ϕ, θ are tangent and cotangent unit vectors. Let $V(\phi_0)$ and $W(\theta_0)$ be neighborhoods of points ϕ_0, θ_0 on the unit spheres, and let $K(\phi_0)$ be a subset of U whose image under some coordinate map $\kappa: U \to \mathbb{R}^n$ is the intersection of an open cone with vertex at a point $\kappa(x_0)$ and containing all directions from $\kappa'(x_0)(V(\phi_0))$ and an open n-dimensional ball with center at $\kappa(x_0)$. A neighborhood in $\hat{\mathfrak{G}}$ of a point (x_0, ϕ_0, θ_0) in case

$\phi_0\theta_0 = 0$ is the union of sets $\{x_0\} \times V(\phi_0) \times W(\theta_0)$, $K(\phi_0) \times S^{n-1} \times W(\theta_0)$, of the set of lines l_x, $x \in K(\phi_0)$, and l_{x_0}. If $\phi_0\theta_0 \gtrless 0$, a neighborhood is the same such set, but with l_{x_0} replaced by a set $\{\lambda \in l_{x_0} : \lambda \gtrless N\}$ (every line l_x is regarded as the real axis). Finally, a neighborhood of a point $\lambda \in l_x$ is an open interval on l_x containing λ. These neighborhoods form a base for a topology on $\hat{\mathfrak{S}}$. This topology is not separable: any two neighborhoods of distinct points (x_0,ϕ_1,θ_1) and (x_0,ϕ_2,θ_2) have nonempty intersection. Clearly, $\hat{\mathfrak{S}}$ is a T_0-space, i.e. for each pair of points there is a neighborhood not containing the other point. The topology $\mathcal{T}(S,S^*)$ induced on $S(\mathfrak{M}) \oplus S^*(\mathfrak{M})$ is the weakest in which all functions on $\mathfrak{M}$ are continuous.

Let $\mathcal{Q}$ be the algebra obtained by closing, with respect to the operator norm in $L_2(\mathfrak{M})$, the set of compact operators and pseudodifferential operators of order zero whose symbols belong to $\mathfrak{M}$. The following fundamental theorem holds.

T h e o r e m 2.1. *Let* A *be a pseudodifferential operator of order zero,* $\phi \in \mathfrak{M}$ *its symbol, and* $\mathfrak{U}$ *the operator symbol (cf.* (4.4.3) *and further). The following assertions hold:*

1) *The maps* $\pi(x,\phi,\theta): \mathrm{A} \mapsto \phi(x,\phi,\theta)$ $(\phi \in S(\mathfrak{M})_x,\ \theta \in S^*(\mathfrak{M})_x)$ *and* $\pi(x,\lambda): \mathrm{A} \mapsto \mathfrak{U}(x,\lambda)$, $\lambda \in l_x$, *induce pairwise inequivalent irreducible representations of the quotient algebra* $\mathcal{Q}/\mathcal{K}L_2(\mathfrak{M})$; $\pi(x,\phi,\theta)$ *is a one-dimensional representation, and* $\pi(x,\lambda)$ *is a representation in the algebra of bounded operators on* $L_2(S(\mathfrak{M})_x)$.
2) *Every irreducible representation of the quotient algebra* $\mathcal{Q}/\mathcal{K}L_2(\mathfrak{M})$ *is equivalent to one of the representations mentioned in* 1).
3) *The correspondence* $\pi(x,\phi,\theta) \mapsto (x,\phi,\theta)$, $\pi(x,\lambda) \mapsto \lambda \in l_x$, *defines a bijection of the spectrum* $\widehat{(\mathcal{Q}/\mathcal{K}L_2(\mathfrak{M}))}$ *onto* $\hat{\mathfrak{S}}$. *The topology on* $\hat{\mathfrak{S}}$ *coincides with the Jacobson topology.*

Using theorem 2.1 we can obtain a description of the spectrum of the algebra $\mathcal{Q}$. In fact, the following holds.

C o r o l l a r y 2.2. *Every irreducible representation of* $\mathcal{Q}$ *is equivalent to one of the representations mentioned in* 1) *of theorem* 2.1, *or to the identity representation (which is irreducible). The spectrum is the disjoint union of* $\hat{\mathfrak{S}}$ *and a point* e *(corresponding to the identity representation). In the Jacobson topology* $\{e\}$ *is an open set whose closure coincides with* $\hat{\mathfrak{G}} \cup \{e\}$.

P r o o f o f c o r o l l a r y 2.2. We first note that, since the algebra $\mathcal{A}$ contains all compact operators on $L_2(\mathfrak{M})$, its identity representation is irreducible. Let π be an irreducible representation of $\mathcal{A}$ such that $\pi(\mathcal{K}L_2(\mathfrak{M})) \neq 0$. Then the restriction $\pi \mid_{\mathcal{K}L_2(\mathfrak{M})}$ is also irreducible (the first part of proposition 1.2). By proposition 1.4, $\pi \mid_{\mathcal{K}L_2(\mathfrak{M})}$ is equivalent to the identity representation. This implies that the initial representation is also equivalent to the identity representation. So, in order to obtain the complete list of representations of $\mathcal{A}$ we must add to the representations given in theorem 2.1 the identity representation. ■

We now consider the subalgebra $\mathcal{A}_0$ of $\mathcal{A}$ generated by the pseudodifferential operators whose symbols do not have a discontinuity on a fixed set $\mathfrak{M}_0 \subset \mathfrak{M}$, i.e. $\operatorname{sing} \Phi \cap \hat{\mathfrak{M}} \neq \varnothing$ for the symbol Φ of an arbitrary $A \in \mathcal{A}_0$. Discard from $\hat{\mathfrak{G}}$ the lines l_x corresponding to points of $\mathfrak{M}_0$. Identify points (x,ϕ,θ) and (x,ψ,θ), where $x \in \mathfrak{M}_0$, $\phi,\psi \in S(\mathfrak{M})_x$, $\theta \in S^*(\mathfrak{M})_x$. The new space, endowed with the quotient topology, will be denoted by $\hat{\mathfrak{G}}_0$.

T h e o r e m 2.3. *All assertions of theorem* 2.1 *remain valid if* $\mathcal{A}$, $\hat{\mathfrak{G}}$ *are replaced by* $\mathcal{A}_0$, $\hat{\mathfrak{G}}_0$, *a representation* $\pi(x,\phi,\theta)$ *for* $x \in \mathfrak{M}_0$ *by the representation* $\pi(x,\theta): A \mapsto \Phi(x,\theta)$, *and the map* $\pi(x,\phi,\theta) \mapsto (x,\phi,\theta)$ *(for* $x \in \mathfrak{M}_0$*) by the map* $\pi(x,\theta) \mapsto (x,\theta)$.

In particular, for $\mathfrak{M}_0 = \mathfrak{M}$ we are led to the following assertion (which, by the way, follows immediately from theorem 4.4.7).

C o r o l l a r y 2.4. *If* $\mathfrak{M}_0 = \mathfrak{M}$, *then the algebra* $\mathcal{A}_0 / \mathcal{K}L_0(\mathfrak{M})$ *is commutative; all its irreducible representations are one-dimensional. The maximal ideal space coincides with the bundle* $S^*(\mathfrak{M})$ *of cotangent unit vectors. A residue class* [A] *is mapped by the Gel'fand homomorphism to its symbol.*

Note that the topology of $\hat{\mathfrak{G}}_0$ does not have a countable base if $\mathfrak{M} \setminus \mathfrak{M}_0$ is uncountable. This and the assertions at the end of §1 imply that for $\mathfrak{M} \setminus \mathfrak{M}_0$ uncountable the algebra $\mathcal{A}_0 / \mathcal{K}L_2(\mathfrak{M})$ (and also $\mathcal{A}_0$) is not separable. It is easy to check that this algebra is separable if $\mathfrak{M} \setminus \mathfrak{M}_0$ is countable.

Finally, theorem 2.1 and corollary 2.2 imply that for any irreducible representation π of $\mathcal{A}$ in a space H the image $\pi(\mathcal{A})$ contains the ideal $\mathcal{K}(H)$. This need be verified only for the representation $\pi(x,\lambda)$ in $L_2(S(\mathfrak{M})_x)$. The set $\pi(x,\lambda)(\mathfrak{A})$ contains, e.g., the commutator $\mathfrak{U}(x,\lambda)f - f\mathfrak{U}(x,\lambda)$, where $\mathfrak{U}(x,\lambda) = \mathrm{E}(\lambda)^{-1}\Phi(x,\phi,\theta)\mathrm{E}(\lambda)$, $f \in C^\infty(S(\mathfrak{M})_x)$. This commutator is a compact operator on $L_2(S(\mathfrak{M})_x)$ (cf. proposition 3.2.1). But then, by proposition 1.4, continuity of the representation implies the inclusion $\mathcal{K}L_2(S(\mathfrak{M})_x) \subset \pi(x,\lambda)(\mathfrak{A})$. Thus $\mathfrak{A}$ is a type I algebra (cf. [24], 9.1).

Finalizing this paragraph we give a one-dimensional version of theorem 2.1. Let $\mathfrak{M}$ be a union of finitely many smooth simply closed oriented contours in the complex plane. Consider a singular integral operator

$$(\mathrm{A}u)(z) = a(z)u(z) + \frac{b(z)}{\pi i}\int_{\mathfrak{M}} \frac{v(\zeta)}{\zeta - z}d\zeta, \tag{2.1}$$

where a, b are functions with, possibly, finitely many singularities of the first kind and left - continuous. Introduce the space $\hat{\mathfrak{G}}$. The bundle $S(\mathfrak{M})\oplus S^*(\mathfrak{M})$ can be represented as the disjoint union of four copies $\mathfrak{M}_{++}$, $\mathfrak{M}_{+-}$, $\mathfrak{M}_{-+}$, $\mathfrak{M}_{--}$ of $\mathfrak{M}$, each oriented as $\mathfrak{M}$. Let $z \in \mathfrak{M}$, let z_l and z_r be arbitrary points of $\mathfrak{M}$ lying 'left' and 'right' of z (near z), and let $<$ be the order relation on an arc. Points of, e.g., $\mathfrak{M}_{+-}$ will be denoted by z^{+-}, etc. We define a topology $\mathfrak{T}(S,S^*)$ on $S(\mathfrak{M})\oplus S^*(\mathfrak{M})$. Neighborhoods forming a fundamental system of neighborhoods (an fsn) in the topology $\mathfrak{T}(S,S^*)$ are: for a point z^{++}, the sets of the form $\{x^{++}: z^{++} \leqslant x < z_r^{++}\} \cup \{x^{-+}: z^{-+} < x^{-+} < z_r^{-+}\}$; for a point z^{-+}, the sets $\{x^{-+}: z^{-+} \geqslant x^{-+} > z_l^{-+}\} \cup \{x^{++}: z^{++} > x^{++} > z_l^{++}\}$; etc. The space $\hat{\mathfrak{G}}$ is the union of the four contours $\mathfrak{M}_{\pm\pm}$ and the set of lines $\{l_x\}$, $x \in \mathfrak{M}$. Let $p: S(\mathfrak{M})\oplus S^*(\mathfrak{M}) \to \mathfrak{M}$ be projection, and V a neighborhood of z^{++} (z^{--}) on $S(\mathfrak{M})\oplus S^*(\mathfrak{M})$ in $\mathfrak{T}(S,S^*)$. Unions of sets V, $\{l_x\}$ $(x \in p(V)\setminus z)$ and $\{\lambda \in l_z: \lambda > N\}$ form an fsn on $\hat{\mathfrak{G}}$ of z^{++} (z^{--}). Similar unions with $\{\lambda \in l_z: \lambda > N\}$ replaced by $\{\lambda \in l_z: \lambda < N\}$ form a fundamental system of neighborhoods of z^+ (z^-). Finally, for an arbitrary point $\lambda \in l_z$ an fsn is formed by the open intervals on l_z containing z. Note that, in distinction to the one-dimensional case, for $\dim \mathfrak{M} \geqslant 2$ there are points (x,ϕ,θ) in $S(\mathfrak{M})\oplus S^*(\mathfrak{M})$ every neighborhood of which contains the line l_x (in case $\phi\theta = 0$). The topology on $\hat{\mathfrak{G}}$ is not separable.

T h e o r e m 2.5. *Let* A *be a singular integral operator* (2.1), *let* $\mathfrak{U}(z,\lambda)$ *be the matrix* (4.4.10), *and let* $\mathfrak{A}$ *be the algebra generated by the operators* (2.1). *Then*

$\mathcal{Q} \supset \mathcal{K}L_2(\mathfrak{M})$, and the following assertions hold:

1) *The maps* $\pi(z, \pm, +)$: $A \mapsto a_{\pm}(z) + b_{\pm}(z)$, $\pi(z, \pm, -)$: $A \mapsto a_{\pm}(z) - b_{\pm}(z)$, $\pi(z, \lambda)$: $A \mapsto \mathfrak{U}(z, \lambda)$, $\lambda \in l_z$, *generate pairwise inequivalent irreducible representations of the algebra* $\mathcal{Q}/\mathcal{K}L_2(\mathfrak{M})$.
2) *Every irreducible representation of* $\mathcal{Q}/\mathcal{K}L_2(\mathfrak{M})$ *is equivalent to one such representation.*
3) *The correspondences* $\pi(z, \pm, \pm) \mapsto z^{\pm\pm}$, $\pi(\lambda, z) \mapsto \lambda \in l_z$ *define a bijection of the spectrum of* $\mathcal{Q}/\mathcal{K}L_2(\mathfrak{M})$ *onto* $\hat{\mathfrak{G}}$. *The topology on* $\hat{\mathfrak{G}}$ *coincides with the Jacobson topology.*

The study of the spectrum of $\mathcal{Q}/\mathcal{K}L_2(\mathfrak{M})$ reduces, by theorem 4.4.1, to the study of the spectrum of the algebra $\mathfrak{G}$ of operator symbols. The proof of theorem 2.1 is prepared in §3, §4 in which the representations of the 'local' algebras $\mathfrak{G}(\lambda)$ and $\mathfrak{G}(l_x)$ are described. The algebra $\mathfrak{G}(\lambda)$ is generated by the values $\mathfrak{U}(x, \lambda)$ of operator symbols at $\lambda \in l_x$, and $\mathfrak{G}(l_x)$ is generated by the restrictions $\mathfrak{U}(x, \cdot)$ of the symbols $\mathfrak{U} \in \mathfrak{G}$ to the lines l_x. The proof of theorem 2.1 is finished in §5. The verification of the other theorems stated in this paragraph does not present any difficulties then. (Note that in view of theorem 4.4.8 the first assertion in theorem 2.5 is obvious; in particular, in order to convince ourselves of the inequivalence of representations $\pi(z, \lambda)$ with distinct λ it suffices to compare the traces of the matrices $\pi(z, \lambda)$A.)

§3. Representations of the algebra $\mathfrak{G}(\lambda)$ generated by the operators $E(\lambda)^{-1}\Phi(\phi, \omega)E(\lambda)$

P r o p o s i t i o n 3.1. *The algebra* $\mathfrak{G}(\lambda)$ *generated by the operators* $E_{\omega\to\phi}(\lambda)^{-1}\Phi(\phi, \omega)E_{\psi\to\omega}(\lambda)$ *in* $L_2(S^{n-1})$ *is irreducible, for every* $\lambda \in \mathbb{R}$.

P r o o f. The algebra $\mathfrak{G}(\lambda)$ contains all operators of multiplication by smooth functions (which correspond to symbols that are independent of ω). Hence, every invariant subspace of this algebra is contained in a subspace of the form $\chi_\Omega L_2(S^{n-1})$, where χ_Ω is the characteristic function of a set Ω, $0 < \operatorname{mes} \Omega < \operatorname{mes} S^{n-1}$. In fact, let $\mathcal{H}$ be an invariant subspace and let $P: L_2(S^{n-1}) \to \mathcal{H}$ be orthogonal projection. Then P commutes with all operators in $\mathfrak{G}(\lambda)$, in particular with operators of multiplication by functions f from

$C^\infty(S^{n-1})$. Therefore, $\mathrm{P}(f) = f\mathrm{P}(1)$, i.e. P acts as the operator of multiplication by $\mathrm{P}(1)$. Furthermore, $\mathrm{P}^2(f) = \mathrm{P}(f\mathrm{P}(1)) = f\mathrm{P}(\mathrm{P}(1)) = f\mathrm{P}(1)^2$. On the other hand, $\mathrm{P}^2(f) = \mathrm{P}(f) = f\mathrm{P}(1)$. Hence $\mathrm{P}(1)^2 = \mathrm{P}(1)$. Thus, $\mathrm{P}(1)$ takes only two values, 0 and 1, and thus is the characteristic function of a certain set Ω.

Suppose there is a nontrivial invariant subspace; let u be an element of it. We may assume that the support of u lies in an open hemisphere. Choose a point ω_0 such that the set $\{\psi\in S^{n-1}: |\psi\omega_0| < \epsilon\}$, where ϵ is a sufficiently small number, does not intersect $\operatorname{supp} u$. Denote by $\{G_m\}$ a sequence of smooth averaging kernels such that

$$\int_{S^{n-1}} G_m(\theta,\psi)u(\psi)d\psi \to u(\theta) \text{ in } L_2(S^{n-1}).$$

Then

$$\int_{S^{n-1}} G_m(\theta,\psi)(-\omega_0\psi+i0)^{-i\lambda-n/2}u(\psi)d\psi \to \tag{3.1}$$

$$\to(-\omega_0\theta+i0)^{-i\lambda-n/2}u(\theta)$$

for almost all θ. Fix a $\theta = \theta_0$ such that the limit (3.1) exists and is distinct from zero. For sufficiently large m,

$$\int_{S^{n-1}} G_m(\theta_0,\psi)(-\omega_0\psi+i0)^{-i\lambda-n/2}u(\psi)d\psi \neq 0. \tag{3.2}$$

Obviously, the function $v(\psi) = G_m(\theta_0,\psi)u(\psi)$ belongs to the invariant subspace.

Denote by $\{\Phi_k\}$ a δ-sequence of smooth functions such that the sets $\operatorname{supp}\Phi_k$ contract towards a point ω_0. Also, let a be a $C^\infty(S^{n-1})$ function equal to zero in a small neighborhood of the equator $\{\phi:\phi\omega_0=0\}$ and equal to one outside another small neighborhood of this equator.

Put $\mathfrak{U}_k(\lambda) = \mathrm{E}(\lambda)^{-1}a(\phi)\Phi_k(\omega)\mathrm{E}(\lambda)$. Then

$$\mathfrak{U}_k(\lambda)v \to C(\lambda)a(\phi)(\phi\omega_0+i0)^{i\lambda-n/2} \times \tag{3.3}$$

$$\times \int_{S^{n-1}} (-\omega_0\psi+i0)^{-i\lambda-n/2}v(\psi)d\psi,$$

where $C(\lambda) = (2\pi)^{-n}e^{i\pi n/2}\Gamma(-i\lambda+n/2)\Gamma(i\lambda+n/2)$ (cf. (1.2.1) and (1.4.1)). Formula (3.2) and the equation

$$(\phi\omega_0+i0)^{-i\lambda-n/2} =$$

$$= (\phi\omega_0)_+^{-i\lambda-n/2} + e^{i(i\lambda-n/2)\pi}(\phi\omega_0)_-^{i\lambda-n/2}$$

imply that the righthand side of (3.3) cannot vanish almost everywhere on $S^{n-1}\setminus\Omega$. We have obtained a contradiction. ■

P r o p o s i t i o n 3.2. *Every compact operator on* $L_2(S^{n-1})$ *belongs to the algebra* $\mathfrak{G}(\lambda)$.

P r o o f. Proposition 3.2.1 implies that $\mathfrak{G}(\lambda)$ contains certain nonzero compact operators (commutators). Since $\mathfrak{G}(\lambda)$ is irreducible (proposition 3.1), by proposition 1.5 every compact operators is an element of it. ■

Denote by $V(n,2)$ the manifold of pairs (ϕ,ω) of mutually orthogonal unit vectors of $\mathbf{R}^n$.

L e m m a 3.3. *Let* (ϕ_0,ω_0) *be an arbitrary point of* $V(n,2)$, *and let* $\mathfrak{U}(\lambda) = \mathrm{E}(\lambda)^{-1}\Phi(\phi,\omega)\mathrm{E}(\lambda)$, *where* $\Phi(\phi,\omega) = \Sigma a_k(\phi)\Phi_k(\omega)$ *(the sum is finite,* a_k, Φ_k *are smooth functions). Then there is a sequence* $\{w_q\}$ *of* $C^\infty(S^{n-1})$ *functions such that* $\|w_q;L_2(S^{n-1})\| = 1$, $\{w_q\}$ *converges weakly to zero in* $L_2(S^{n-1})$, *and*

$$\|\mathfrak{U}(\lambda)w_q;L_2(S^{n-1})\| \to |\Phi(\phi_0,\omega_0)| \quad \text{as} \quad q\to\infty.$$

P r o o f. Let v be a $C^\infty(S^{n-1})$ function with support concentrated in a small neighborhood of ϕ_0, and with $\|v;L_2(S^{n-1})\| = 1$. Denote by g a smooth function on the sphere, extended onto $\mathbf{R}^n\setminus 0$ as a homogeneous function of degree zero. We will assume that ∇g vanishes nowhere on the support of v and that $\nabla g(\phi_0) = \omega_0$. (Since $(\phi_0,\omega_0) \in V(n,2)$ the latter is possible.) Theorem 3.5.1 implies

$$e^{-i\mu g(\phi)}\mathfrak{U}_{\psi\to\phi}(\lambda)(e^{i\mu g(\psi)}v(\psi)) - \Phi(\phi,\mu\nabla g(\phi)+\sigma\phi)v(\phi) = \tag{3.4}$$
$$= O((|\mu|+|\sigma|)^{-1}) \text{ as } |\mu|+|\sigma|\to\infty,$$

where μ is a real number and $\sigma = \mathrm{Re}\,\lambda$. Using this relation we can prove that for any null sequence $\{\epsilon_q\}$ of positive numbers there are sequences of smooth functions, $\{v_q\}$, and numbers, $\{\mu_q\}$, with the following properties: $\mu_q \to +\infty$, $\|v_q;L_2(S^{n-1})\| = 1$, the supports of the v_q contract towards the point ϕ_0, and

$$|\mathfrak{U}_{\psi\to\phi}(\lambda)(e^{i\mu_q g(\psi)}v_q(\psi)) - \Phi(\phi,\mu_q\nabla g(\phi)+\sigma\phi)e^{i\mu_q g(\phi)}v_q(\phi)| < \epsilon_q.$$

Clearly, we can set $w_q = e^{i\mu_q g}v_q$. ■

C o r o l l a r y 3.4. *The following inequality holds:*

$$\inf_{T\in\mathcal{K}L_2(S^{n-1})} \|\mathfrak{U}(\lambda)+T\| \geqslant \max_{(\phi,\omega)\in V(n,2)} |\Phi(\phi,\omega)|. \tag{3.5}$$

L e m m a 3.5. *Let* $\mathrm{T} = \mathrm{E}(\lambda)^{-1}\Psi(\phi,\omega)\mathrm{E}(\lambda)$, *where the symbol* Ψ *is of class* $C^\infty(S^{n-1}\times S^{n-1})$ *and vanishes in a neighborhood of the manifold* $V(n,2)$. *Then the following inequality holds for any* s : $\|\mathrm{T}u;H^s(S^{n-1})\| \leqslant c_s\|u;L_2(S^{n-1})\|$.

The p r o o f follows from the fact that the kernel of the integral operator $v\mapsto Qv$, where $(Qv)(\phi) = \mathrm{E}_{\omega\to\phi}(\lambda)^{-1}\Psi(\phi,\omega)v(\omega)$, is a function of class $C^\infty(S^{n-1}\times S^{n-1})$.

Denote by $\sigma_{\mathrm{ess}}(\mathfrak{U}\mathfrak{U}^*)$ the essential spectrum of the operator $\mathfrak{U}\mathfrak{U}^*$, i.e. the spectrum of the residue class $[\mathfrak{U}\mathfrak{U}^*+\mathrm{T}]$ in the algebra $\mathfrak{L}(L_2(S^{n-1}))/\mathcal{K}L_2(S^{n-1})$.

L e m m a 3.6. *The inequality*

$$\sup\sigma_{\mathrm{ess}}(\mathfrak{U}(\lambda)\mathfrak{U}(\lambda)^*) \leqslant \max_{V(n,2)} |\Phi|^2 \tag{3.6}$$

holds, where $\mathfrak{U}(\lambda) = \mathrm{E}(\lambda)^{-1}\Phi(\phi,\omega)\mathrm{E}(\lambda)$.

P r o o f. It suffices to prove that if $\mu > \max_{V(n,2)}|\Phi|^2$, then the operator $\mu-\mathfrak{U}(\lambda)\mathfrak{U}(\lambda)^*$ is Fredholm. The principal symbol of $\mathfrak{U}(\lambda)\mathfrak{U}(\lambda)^*$ is $|\Phi|^2$ (cf. theorems 3.3.4 and 3.3.5). Hence the symbol $\mu-|\Phi|^2$ of $\mu-\mathfrak{U}(\lambda)\mathfrak{U}(\lambda)^*$ is positive in a neighborhood of $V(n,2)\subset S^{n-1}\times S^{n-1}$. Let T be an operator with vanishing symbol near $V(n,2)$. By lemma 3.5, T is compact. We may clearly assume that the symbol χ of $\mu-\mathfrak{U}(\lambda)\mathfrak{U}(\lambda)^*+\mathrm{T}$ is positive everywhere on $S^{n-1}\times S^{n-1}$. Now theorem 3.3.4 and the compactness of the imbedding $H^1(S^{n-1})\subset L_2(S^{n-1})$ imply that $\mathfrak{R}(\lambda) = \mathrm{E}(\lambda)^{-1}\chi(\phi,\omega)\mathrm{E}(\lambda)$ is a two-sided regularizer for $\mu-\mathfrak{U}(\lambda)\mathfrak{U}(\lambda)^*+\mathrm{T}$ (i.e. the product of these operators is zero, up to a compact term). This, as is well-known (cf. [1], [17], [40]), implies that $\mu-\mathfrak{U}(\lambda)\mathfrak{U}(\lambda)^*$ is a Fredholm operator. ■

C o r o l l a r y 3.7. *The following estimate holds:*

$$\inf_{\mathrm{T}\in\mathcal{K}L_2(S^{n-1})} \|\mathfrak{U}(\lambda)+\mathrm{T}\| \leqslant \max_{V(n,2)} |\Phi|. \tag{3.7}$$

P r o o f. Recall that

$$\inf_{\mathrm{T}\in\mathcal{KL}_2(S^{n-1})} \|\mathfrak{U}+\mathrm{T}\|^2 = \inf_{\mathrm{T}} \|\mathfrak{U}\mathfrak{U}^* + \mathrm{T}\| = \sup \sigma_{\mathrm{ess}}(\mathfrak{U}\mathfrak{U}^*)$$

(cf. [24]). This and (3.6) imply (3.7). ■

Let $\mathfrak{U}_{jk}(\lambda) = \mathrm{E}(\lambda)^{-1}\Phi^{(j,k)}(\phi,\omega)\mathrm{E}(\lambda)$, $\mathfrak{U}(\lambda) = \Sigma_{j=1}^{m}\Pi_{k=1}^{N}\mathfrak{U}_{jk}$, $M,N<\infty$. Theorems 3.3.4 and 3.3.5 imply that $\mathfrak{U}(\lambda) = \mathrm{E}(\lambda)^{-1}\Phi\mathrm{E}(\lambda)+\mathrm{T}_1$, $\mathfrak{U}(\lambda)^* = \mathrm{E}(\lambda)^{-1}\overline{\Phi}\mathrm{E}(\lambda)+\mathrm{T}_2$, where $\Phi = \Sigma_j\Pi_k\Phi^{(j,k)}$, $\mathrm{T}_1,\mathrm{T}_2 \in \mathcal{KL}_2(S^{n-1})$. By corollaries 3.4 and 3.7 the following relation is valid:

$$\inf_{\mathrm{T}\in\mathcal{KL}_2(S^{n-1})} \|\mathfrak{U}(\lambda)+\mathrm{T}\| = \max_{V(n,2)} |\Phi|.$$

Thus we have:

T h e o r e m 3.8. *The algebra* $\mathfrak{S}(\lambda)/\mathcal{KL}_2(S^{n-1})$ *is isomorphic to the algebra* $C(V(n,2))$ *of continuous functions on* $V(n,2)$.

Denote by $\hat{\mathfrak{S}}(\lambda)$ the union of $V(n,2)$ and the point λ. Let λ belong to an open set, and let a fundamental system of neighborhoods of a point $(\phi,\omega) \in V(n,2)$ consist of sets $\mathcal{W}(\phi,\omega) \cup \{\lambda\}$, where $\mathcal{W}(\phi,\omega)$ is a neighborhood of (ϕ,ω) in $V(n,2)$.

T h e o r e m 3.9. *Every irreducible representation of the algebra* $\mathfrak{S}(\lambda)$ *is either one-dimensional or the identity representation* $e(\lambda)$. *A one-dimensional representation can be realized as a map* $\pi(\phi,\omega):\mathfrak{U}\mapsto\Phi(\phi,\omega)$, *where* Φ *is the symbol of the class* $[\mathfrak{U}]$ *in* $\mathfrak{S}(\lambda)/\mathcal{KL}_2(S^{n-1})$, $(\phi,\omega) \in V(n,2)$. *The correspondence* $\pi(\phi,\omega)\mapsto(\phi,\omega)$, $e(\lambda)\mapsto\lambda$ *defines a bijection of the spectrum of* $\mathfrak{S}(\lambda)$ *onto* $\hat{\mathfrak{S}}(\lambda)$. *The topology on* $\hat{\mathfrak{S}}(\lambda)$ *coincides with the Jacobson topology.*

P r o o f. Let π be an irreducible representation of $\mathfrak{S}(\lambda)$. If $\pi(\mathcal{KL}_2(S^{n-1})) = 0$, then π induces an irreducible representation of the quotient algebra $\mathfrak{S}(\lambda)/\mathcal{KL}_2(S^{n-1})$. Theorem 3.8 implies that π can be realized as a map $\mathfrak{U}(\lambda)\mapsto\Phi(\phi,\omega)$, $(\phi,\omega) \in V(n,2)$. Every irreducible representation π such that $\pi(\mathcal{KL}_2(S^{n-1})) \neq 0$ is equivalent to the identity representation (cf. the proof of corollary 2.2). In order to finish the proof it suffices to recall the definition of the Jacobson topology on the spectrum of an algebra (at the end of §1.5). ■

§4. Representations of an algebra $\mathfrak{G}(l_x)$

Let $x \in \mathfrak{M}$ and let $\mathfrak{G}(l_x)$ be the algebra of restrictions to the line l_x of the elements of $\mathfrak{G}$. The norm in $\mathfrak{G}(l_x)$ of the restriction $\mathfrak{U}(x,\cdot)$ to l_x of an element $\mathfrak{U} \in \mathfrak{G}$ is defined by

$$\|\mathfrak{U}(x,\cdot);\mathfrak{G}(l_x)\| = \sup_{\lambda \in l_x}\|\mathfrak{U}(x,\lambda);L_2(S(\mathfrak{M})_x)\to L_2(S(\mathfrak{M})_x)\|.$$

P r o p o s i t i o n 4.1. *The algebra $\mathfrak{G}(l_x)$ is closed, hence a C^*-algebra.*

P r o o f. Let $\overline{\mathfrak{G}(l_x)}$ be the C^*-algebra generated by the restrictions $\mathfrak{U}(x,\cdot)$, $\mathfrak{U} \in \mathfrak{G}$. The image of the isomorphism $\mathfrak{G} \ni \mathfrak{U} \mapsto \mathfrak{U}(x,\cdot) \in \overline{\mathfrak{G}(l_x)}$ is closed (proposition 1.1). Hence $\mathfrak{G}(l_x) = \overline{\mathfrak{G}(l_x)}$. ■

P r o p o s i t i o n 4.2. *Let $\lambda_1,\lambda_2 \in l_x$, $\lambda_1 \neq \lambda_2$. The representations $\pi(x,\lambda_j):\mathfrak{U}(x,\cdot)\mapsto\mathfrak{U}(x,\lambda_j)$, $j = 1,2$, of $\mathfrak{G}(\lambda_x)$ in $L_2(S(\mathfrak{M})_x)$ are not equivalent.*

P r o o f. Let α be a function defined and smooth on $\mathfrak{M}\setminus x$, and having a finite limit $\beta(\phi)$ when approaching x in the direction ϕ, $\beta \in C^\infty(S(\mathfrak{M})_x)$. Clearly, the operator of multiplication by α belongs to $\mathfrak{G}$. Hence we may assume that the operators of multiplication by ϕ_j, where ϕ_j are the coordinates of the vector $\phi \in S(\mathfrak{M})_x$ (in a fixed local system), belong to $\mathfrak{G}(l_x)$.

Let $\Phi \in C^\infty(S(\mathfrak{M})_x \times S^*(\mathfrak{M})_x)$ and $\mathfrak{U}(\lambda) = \mathrm{E}(\lambda)^{-1}\Phi(\phi,\omega)\mathrm{E}(\lambda)$. Obviously, $\mathfrak{U} \in \mathfrak{G}(l_x)$. The formula

$$\mathfrak{U}(\lambda)\psi_j = \phi_j\mathfrak{U}(\lambda+i)+\mathrm{E}(\lambda)^{-1}D_{\omega_j}\Phi(\phi,\omega)\mathrm{E}(\lambda+i) \tag{4.1}$$

holds (cf. proposition 3.2.1). Replace in (4.1) $\Phi(\phi,\omega)$ by $\phi\omega\Phi(\phi,\omega)$. Taking into account (1.4.1), which defines $\mathrm{E}(\lambda)^{-1}$, and the relation $(\phi\omega+i0)^{i\mu}\phi\omega = (\phi\omega+i0)^{i(\mu-i)}$, we find that $\mathfrak{U}(\lambda+i) =$ $= \mathrm{E}(\lambda+i)^{-1}\phi\omega\Phi(\phi,\omega)\mathrm{E}(\lambda+i) = (\lambda+in/2)\mathrm{E}(\lambda)^{-1}\Phi(\phi,\omega)\mathrm{E}(\lambda+i)$. Now (4.1) takes the form

$$\begin{aligned}&\mathrm{E}_{\omega\to\phi}(\lambda)^{-1}\phi\omega\Phi(\phi,\omega)\mathrm{E}_{\psi\to\omega}\psi_j = \\ &= \phi_j(in/2+\lambda)\mathrm{E}_{\omega\to\phi}(\lambda)^{-1}\Phi(\phi,\omega)\mathrm{E}_{\psi\to\omega}(\lambda+i)+ \\ &+\mathrm{E}_{\omega\to\phi}(\lambda)^{-1}D_{\omega_j}(\phi\omega\Phi(\phi,\omega))\mathrm{E}_{\psi\to\omega}(\lambda+i).\end{aligned} \tag{4.2}$$

Definition (1.2.1), defining E(λ), and the relation $(-\omega\psi+i0)^{-i\mu}\omega\psi = -(-\omega\psi+i0)^{-i(\mu+i)}$ imply that

$$E(\lambda)^{-1}\Phi(\phi,\omega)E(\lambda)\omega\psi = \tag{4.3}$$

$$= -i(i\lambda-1+n/2)E(\lambda)^{-1}\Phi(\phi,\omega)E(\lambda+i).$$

Multiply (4.3) by ϕ_j and subtract the result from (4.2). Then

$$E(\lambda)^{-1}\phi\omega\Phi(\phi,\omega)E(\lambda)\psi_j-\phi_jE(\lambda)^{-1}\Phi(\phi,\omega)E(\lambda)\omega\psi = \tag{4.4}$$

$$= E(\lambda)^{-1}\Psi(\phi,\omega)E(\lambda+i),$$

where $\Psi(\phi,\omega) = i(n-1)\phi_j\Phi(\phi,\omega)+D_{\omega_j}(\phi\omega\Phi(\phi,\omega))$. It is obvious that the lefthand sides of (4.3) and (4.4) are elements of the algebra $\mathfrak{G}(l_x)$. Replace in (4.3) the function Φ by Ψ. We find that the operator-functions $\lambda\mapsto(i\lambda-1+n/2)E(\lambda)^{-1}\Psi(\phi,\omega)E(\lambda+i)$, $\lambda\mapsto E(\lambda)^{-1}\Psi(\phi,\omega)E(\lambda+i)$ belong to $\mathfrak{G}(l_x)$. Denote these functions by $\mathfrak{M}$ and $\mathfrak{N}$.

Suppose that for some distinct λ_1,λ_2 the representations $\pi(x,\lambda_1)$ and $\pi(x,\lambda_2)$ are equivalent. Then the norms of both the operators $\mathfrak{M}(\lambda_1)$ and $\mathfrak{M}(\lambda_2)$ and those of $\mathfrak{N}(\lambda_1)$ and $\mathfrak{N}(\lambda_2)$ must coincide. If $\lambda_1 \neq -\lambda_2$ this is impossible, since $|i\lambda_1-1-n/2| \neq |i\lambda_2-1-n/2|$. If however $\lambda_1 = -\lambda_2$, then the values of the operator-function $\mathfrak{M}(\lambda)-(i\lambda_1-1+n/2)\mathfrak{N}(\lambda) = i(\lambda-\lambda_1)\mathfrak{N}(\lambda)$ have distinct norms at λ_1 and λ_2. ■

Let $\mathcal{K}_0(l_x) = \mathcal{K}_0(l_x,L_2(S(\mathfrak{M})_x))$ be the algebra of continuous functions on l_x, with values in $\mathcal{K}L_2(S(\mathfrak{M})_x)$, which tend (in norm) to zero at infinity. In $\mathcal{K}_0(l_x)$ we introduce the norm

$$\|v;\mathcal{K}_0(l_x)\| = \sup_{\lambda\in l_x}\|v(\lambda);L_2(S(\mathfrak{M})_x)\to L_2(S(\mathfrak{M})_x)\|.$$

L e m m a 4.3. *The following inclusion holds:* $\mathcal{K}_0(l_x)\subset\mathfrak{G}(l_x)$.

P r o o f. Since representations $\mathfrak{U}(x,\cdot)\mapsto\mathfrak{U}(x,\lambda_1)$ and $\mathfrak{U}(x,\cdot)\mapsto\mathfrak{U}(x,\lambda_2)$ of $\mathfrak{G}(l_x)$ are irreducible, and inequivalent for distinct λ_1,λ_2 (propositions 3.1 and 4.2), their restrictions to $\mathcal{K}_0(l_x)\cap\mathfrak{G}(l_x)$ are also irreducible and inequivalent (proposition 1.7). Therefore $\mathcal{K}_0(l_x)\cap\mathfrak{G}(l_x)$ is a rich subalgebra of $\mathcal{K}_0(l_x)$. This and proposition 1.9 imply $\mathcal{K}_0(l_x)\cap\mathfrak{G}(l_x) = \mathcal{K}_0(l_x)$. ■

P r o p o s i t i o n 4.4. *The quotient algebra $\mathfrak{S}(l_x)/\mathcal{K}_0(l_x)$ and the algebra $C(S(\mathfrak{M})_x \times S^*(\mathfrak{M})_x)$ are isomorphic.*

P r o o f. Put $\mathfrak{U}(x,\lambda) = \Sigma_j \Pi_k \mathrm{E}(\lambda)^{-1}\Phi_{jk}\mathrm{E}(\lambda)$ (the subscripts j,k run over finite sets). By theorem 3.3.4, $\mathfrak{U}(x,\lambda) = \mathrm{E}(\lambda)^{-1}\Sigma_j\Pi_k\Phi_{jk}(\phi,\omega)\mathrm{E}(\lambda)+\mathrm{T}(\lambda)$, where $\mathrm{T} \in \mathcal{K}_0(l_x)$. This and corollary 3.4 imply

$$\inf_{T\in\mathcal{K}_0(l_x)} \sup_{\lambda\in l_x}\|\mathfrak{U}(x,\lambda)+\mathrm{T}(\lambda)\| \geqslant$$

$$\geqslant \left\|\sum_j\prod_k\Phi_{jk};C(S(\mathfrak{M})_x \times S^*(\mathfrak{M})_x)\right\|.$$

The reverse inequality is contained in theorem 3.6.5. Thus,

$$\inf_{T\in\mathcal{K}_0(l_x)} \|\mathfrak{U}(x,\cdot)+\mathrm{T};\mathfrak{S}(l_x)\| =$$

$$= \left\|\sum_j\prod_k\Phi_{jk};C(S(\mathfrak{M})_x \times S^*(\mathfrak{M})_x)\right\|.$$

From this we find that the map $\mathfrak{B}(\cdot)\mapsto\Psi$, where $\mathfrak{B}(\lambda) = \mathrm{E}(\lambda)^{-1}\Psi(\phi,\omega)\mathrm{E}(\lambda)$, defines an isomorphism between the quotient algebra $\mathfrak{S}(l_x)/\mathcal{K}_0(l_x)$ and the algebra $C(S(\mathfrak{M})_x \times S^*(\mathfrak{M})_x)$ of continuous functions. ■

C o r o l l a r y 4.5. *Let π be an irreducible representation of $\mathfrak{S}(l_x)$ such that $\pi(\mathcal{K}_0(l_x)) = 0$. Then π is equivalent to a representation of the form $\mathfrak{U}(x,\cdot)\mapsto\Phi(\phi_0,\omega_0)$, where Φ is the symbol of $\mathfrak{U}$ (i.e. the symbol of the corresponding class in $\mathfrak{S}(l_x)/\mathcal{K}_0(l_x)$) and $\phi_0 \in S(\mathfrak{M})_x$, $\omega_0 \in S^*(\mathfrak{M})_x$.*

Denote by $\hat{\mathfrak{S}}(l_x)$ the union of the product $S(\mathfrak{M})_x \times S^*(\mathfrak{M})_x$ and the line $l_x = \mathbf{R}$. We introduce a topology on $\hat{\mathfrak{S}}(l_x)$. If $(\phi,\omega) \in S(\mathfrak{M})_x \times S^*(\mathfrak{M})_x$, $\phi\omega \gtrless 0$, then a fundamental system of neighborhoods of (ϕ,ω) consists of sets $\mathcal{V}(\phi,\omega) \cup \{\lambda\in l_x:\lambda\gtrless N\}$, where $\mathcal{V}(\phi,\omega)$ is a neighborhood of (ϕ,ω) in $S(\mathfrak{M})_x \times S^*(\mathfrak{M})_x$, $N \in \mathbf{R}$. In case $\phi\omega = 0$ a fundamental system of neighborhoods is formed by the sets $\mathcal{V}(\phi,\omega) \cup l_x$. A neighborhood of $\lambda \in l_x$ is an interval on l_x.

T h e o r e m 4.6. *Let $\mathfrak{U}(x,\cdot)$ be an element of the algebra $\mathfrak{S}(l_x)$ and let $\Phi \in C(S(\mathfrak{M})_x \times S^*(\mathfrak{M})_x)$ be its symbol. Then:*

1) *the maps* $\pi(\phi,\omega):\mathfrak{U}(x,\cdot)\mapsto\Phi(\phi,\omega)$ *and* $\pi(x,\lambda):\mathfrak{U}(x,\cdot)\mapsto\mathfrak{U}(x,\lambda)$ *are pairwise inequivalent irreducible representations of* $\mathfrak{S}(l_x)$;

2) *every irreducible representation of* $\mathfrak{S}(l_x)$ *is equivalent to one of the representations in* 1);

3) *the correspondence* $\pi(\phi,\omega)\mapsto(\phi,\omega)$, $\pi(x,\lambda)\mapsto\lambda\in l_x$ *defines a bijection of the spectrum of* $\mathfrak{S}(l_x)$ *onto* $\hat{\mathfrak{S}}(l_x)$. *The topology on* $\hat{\mathfrak{S}}(l_x)$ *coincides with the Jacobson topology.*

The proof of this theorem is preceded by two lemmas.

L e m m a 4.7. *Let*

$$(\mathcal{E}_+(\lambda)v)(\phi) =$$

$$= \frac{1}{(2\pi)^{n/2}} e^{i(n/2-i\lambda)\pi/2}\Gamma(n/2-i\lambda)\int_{S^{n-1}}(\phi\omega)_+^{i\lambda-n/2}v(\omega)d\omega,$$

$$(\mathcal{E}_-(\lambda)v)(\phi) =$$

$$= \frac{1}{(2\pi)^{n/2}} e^{-\lambda\pi/2}\Gamma(n/2-i\lambda)\int_{S^{n-1}}(\phi\omega)_-^{i\lambda-n/2}v(\omega)d\omega,$$

where $d\omega$ *is the volume element on the unit sphere. On the line* $\mathrm{Im}\lambda = 0$ *we have the estimates*

$$\|\mathcal{E}_\pm(\lambda);L_2(S^{n-1})\to L_2(S^{n-1})\| \leqslant \text{const},$$

$$\|\mathcal{E}_\pm(\lambda);L_2(S^{n-1})\to L_2(S^{n-1})\| = O(e^{-\pi|\lambda|}) \quad as \quad \lambda\to\mp\infty.$$

P r o o f. Expand a function v in a series in spherical harmonics:

$$v(\phi) = \sum_{m=0}^{\infty}\sum_{k=1}^{k_m} v_{mk}Y_{mk}(\phi).$$

Consider, e.g., the operator $\mathcal{E}_+$. Using a reasoning similar to the proof of proposition 1.3.1, we are led to

$$\mathcal{E}_+(\lambda)v = \sum_{m,k}\mu_m(\lambda)v_{mk}Y_{mk},$$

where

$$\mu_m(\lambda) = c_n e^{\lambda\pi/2}\frac{\Gamma(n/2-i\lambda+m)\Gamma((1+i\lambda-m-n/2)/2)}{2^m\Gamma((n/2+i\lambda+m)/2)},$$

while, here and below, the c_n denote constants whose absolute values depend on the dimension of the space only.

In view of the formula $\Gamma(1-z)\Gamma(z) = \pi/\sin \pi z$, for $z = (-i\lambda+m+1+n/2)/2$ we have

$$\Gamma((1+i\lambda-m-n/2)/2) =$$
$$= \frac{\Gamma(1-(n/2-i\lambda+m+1)/2)\Gamma((n/2-i\lambda+m+1)/2)}{\Gamma((n/2-i\lambda+m+1)/2)} =$$
$$= \frac{\pi}{\Gamma((n/2-i\lambda+m+1)/2)\sin\{\pi(n/2-i\lambda+m+1)/2\}}.$$

Therefore

$$\mu_m(\lambda) = c_n e^{\lambda\pi/2} \times$$
$$\times \frac{\Gamma(n/2-i\lambda+m)}{2^m\Gamma((n/2-i\lambda+m+1)/2)\Gamma((n/2+i\lambda+m)/2)\sin\{\pi(n/2-i\lambda+m+1)/2\}}.$$

By the duplication formula $\Gamma(2z) = 2^{2z-1}\pi^{-1/2}\Gamma(z)\Gamma(z+1/2)$, for $z = (-i\lambda+m+n/2)/2$ we obtain

$$\frac{\Gamma(n/2-i\lambda+m)}{\Gamma((n/2-i\lambda+m)/2+1/2)} =$$
$$= 2^{n/2-i\lambda+m}\sqrt{\pi}\,\Gamma((n/2-i\lambda+m)/2).$$

Thus,

$$\mu_m(\lambda) = c_n 2^{-i\lambda} e^{\lambda\pi/2} \times$$
$$\times \frac{\Gamma((n/2-i\lambda+m)/2)}{\Gamma((n/2+i\lambda+m)/2)\sin\{\pi(n/2-i\lambda+m+1)/2\}}.$$

In order to derive from this the estimate for the norms of the operator $\mathcal{E}_+(\lambda)$ it remains to remark that for $\operatorname{Im}\lambda = 0$ the modulus of a quotient of Γ-functions is one. ■

L e m m a 4.8. *Let* $\mathfrak{U}(\lambda) = \mathrm{E}(\lambda)^{-1}\Phi(\phi,\omega)\mathrm{E}(\lambda)$, *where* $\Phi(\phi,\omega) = 0$ *for* $\phi\omega \leqslant 0$ *(respectively,* $\phi\omega \geqslant 0$*). Then* $\|\mathfrak{U}(\lambda)\| = O(e^{-\pi|\lambda|})$ *as* $\lambda \to -\infty$ *(respectively,* $\lambda \to +\infty$*) on the line* $\operatorname{Im}\lambda = 0$.

Proof. Clearly, $\mathfrak{U}(\lambda)$ can be written as $\mathfrak{U}(\lambda) = \mathcal{E}_+(\lambda)\Phi\mathrm{E}(\lambda)$ (respectively, $\mathfrak{U}(\lambda) = \mathcal{E}_-(\lambda)\Phi\mathrm{E}(\lambda)$). Now it suffices to expand the symbol Φ in a series:

$$\Phi(\phi,\omega) = \sum_{m,k} a_{mk}(\phi)Y_{mk}(\omega)$$

and use the preceding lemma (compare with the proof of proposition 1.3). ■

Proof of theorem 4.6. 1) The truth of this assertion follows from propositions 3.1 and 4.2.

2) Let π be an irreducible representation of $\mathfrak{G}(l_x)$ such that $\pi(\mathcal{K}_0(l_x)) = 0$. By Corollary 4.5, π is equivalent to a representation of the form $\mathfrak{U}(x,\cdot) \mapsto \Phi(\phi,\omega)$. If $\pi(\mathcal{K}_0(l_x)) \neq 0$, then the restriction $\pi|_{\mathcal{K}_0(l_x)}$ is an irreducible representation of $\mathcal{K}_0(l_x)$ (proposition 1.2). As is well known (proposition 1.8), every irreducible representation of $\mathcal{K}_0(l_x)$ is equivalent to a representation $\mathrm{T} \mapsto \mathrm{T}(\lambda)$, where $\mathrm{T} \in \mathcal{K}_0(l_x)$, $\lambda \in l_x$. In view of proposition 1.7 this means that the corresponding representation π of $\mathfrak{G}(l_x)$ is equivalent to a representation $\mathfrak{U}(x,\cdot) \mapsto \mathfrak{U}(x,\lambda)$.

3) It suffices to verify that the topology on $\hat{\mathfrak{G}}(l_x)$ coincides with the Jacobson topology, since the first part of the assertion follows from 1) and 2).

By §1.5 a base of the Jacobson topology on the spectrum of an algebra $\mathfrak{G}(l_x)$ is formed by the set of (equivalence classes of) representations π for which $\|\pi(\mathfrak{U}(x,\cdot))\| > 1$, where $\mathfrak{U}(x,\cdot)$ runs over a set that is dense in $\mathfrak{G}(l_x)$.

Let λ_0 be an arbitrary point of l_x and let $\mathfrak{U}(x,\cdot)$ be an element of $\mathfrak{G}(l_x)$ for which $\|\mathfrak{U}(x,\lambda_0)\| > 1$. The set $\{\pi:\|\pi\mathfrak{U}(x,\cdot)\| > 1\}$ is a neighborhood of λ_0 in the Jacobson topology. Since the operator-function $\lambda \mapsto \mathfrak{U}(x,\lambda)$ is norm-continuous, this set is also a neighborhood of λ_0 in the usual topology on the real line. In view of the fact that an operator-function $\lambda \mapsto c(\lambda)(\cdot,f)g$, $f,g \in L_2(S(\mathfrak{M})_x)$, where c is an arbitrary continuous function on the real line that vanishes at infinity, belongs to $\mathfrak{G}(l_x)$, every open interval on the real line is open also in the Jacobson topology.

Now we will consider neighborhoods of a point $(\phi_0,\omega_0) \in S(\mathfrak{M})_x \times S^*(\mathfrak{M})_x$, $\phi_0\omega_0 > 0$. Let $\Phi(\phi_0,\omega_0) > 1$ and $\mathfrak{U}(x,\lambda) = \mathrm{E}(\lambda)^{-1}\Phi(\phi,\omega)\mathrm{E}(\lambda)$. Since $\pi(\phi,\omega)\mathfrak{U} = \Phi(\phi,\omega)$, the inequality $\|\pi(\phi,\omega)\mathfrak{U}\| > 1$ holds for all (ϕ,ω) in a neighborhood of (ϕ_0,ω_0) on $S(\mathfrak{M})_x \times S^*(\mathfrak{M})_x$. By (3.4) the inequality $\|\pi(x,\lambda)\mathfrak{U}\| > 1$ holds for all $\lambda > N$, with N a large number. Hence a neighborhood of (ϕ_0,ω_0) in the

Jacobson topology is also neighborhood in the topology on $\hat{\mathfrak{S}}(l_x)$.

Let $\mathcal{V}(\phi_0,\omega_0) \cup \{\lambda \in l_x : \lambda > N\}$ be an arbitrary neighborhood of (ϕ_0,ω_0) in $\hat{\mathfrak{S}}(l_x)$. We show that it is a neighborhood of this point in the Jacobson topology. Choose a function $\Phi \in C^\infty(S(\mathfrak{M})_x \times S^*(\mathfrak{M})_x)$ such that $\Phi(\phi_0,\omega_0) > 1$ and $\Phi = 0$ outside $\mathcal{V}(\phi_0,\omega_0) \cup \{(\phi,\omega) : \phi\omega > 0\}$. Put $\mathfrak{U}(\lambda) = \mathrm{E}(\lambda)^{-1}\Phi\mathrm{E}(\lambda)$. Then $\{(\phi,\omega) : \|\pi(\phi,\omega)\mathfrak{U}\| > 1\} \subset \mathcal{V}(\phi_0,\omega_0)$. Now note that by lemma 4.8, $\pi(x,\lambda)\mathfrak{U} \to 0$ as $\lambda \to -\infty$. Moreover, lemma 3.5 implies that $\mathfrak{U}(x,\lambda)$ is, for every λ, a compact operator. Hence, by lemma 4.3, an operator-function $\lambda \mapsto c(\lambda)\mathfrak{U}(x,\lambda)$, where $c \in C_0^\infty(l_x)$, belongs to $\mathfrak{S}(l_x)$.

Choose the function c such that $0 \leqslant c \leqslant 1$ and $c(\lambda) = 1$ for $\lambda \in [-M,M]$, where M is a large number. Replacing, if necessary, $\mathfrak{U}$ by $\mathfrak{U} - c\mathfrak{U}$, we find an operator-function for which $\{\pi : \|\pi(\mathfrak{U})\| > 1\} \subset \mathcal{V}(\phi_0,\omega_0) \cup \{\lambda \in l_x : \lambda > N\}$.

So, it has been proved that the systems of neighborhoods of a point (ϕ_0,ω_0), $\phi_0\omega_0 > 0$, in the topology of $\hat{\mathfrak{S}}(l_x)$ and in the Jacobson topology coincide. This assertion is similarly proved for points (ϕ_0,ω_0) satisfying $\phi_0\omega_0 < 0$.

Let $\phi_0\omega_0 = 0$, $\Phi \in C^\infty(S(\mathfrak{M})_x \times S^*(\mathfrak{M})_x)$, $\Phi(\phi_0,\omega_0) > 1$, and $\mathfrak{U}(\lambda) = \mathrm{E}(\lambda)^{-1}\Phi\mathrm{E}(\lambda)$. In (3.4), g is understood to be an arbitrary homogeneous function of degree zero. Choose g such that the equation $\omega_0 = \nabla g(\phi_0)$ is fulfilled. Letting μ tend to infinity leads to the inequality $\|\mathfrak{U}(\lambda)\| = \|\pi(x,\lambda)\mathfrak{U}\| > 1$ for all $\lambda \in l_x$. Hence a neighborhood of (ϕ_0,ω_0) in the Jacobson topology is a neighborhood also in the topology of $\hat{\mathfrak{S}}(l_x)$. The converse assertion is obvious. ■

§5. Proof of theorem 2.1

Recall that $\mathfrak{S}$ denotes the algebra generated by the operator symbols defined just above theorem 4.4.1.

Let $\mathfrak{U}$ be an arbitrary element of $\mathfrak{S}$, and let $\{\mathfrak{U}_n\}$ be an approximating sequence for $\mathfrak{U}$ every term of which is a sum of finite product of generators:

$$\mathfrak{U}_n(x,\lambda) = \sum_j \prod_k \mathrm{E}(\lambda)^{-1}\Phi_{jk}^{(n)}\mathrm{E}(\lambda),$$

where $\Phi_{jk}^{(n)}(x,\cdot) \in C^\infty(S(\mathfrak{M})_x \times S^*(\mathfrak{M})_x)$. For all $x \in \mathfrak{M}$, except for a finite set, the maps $(\phi,\omega) \mapsto \Phi_{jk}^{(n)}(x,\phi,\omega)$, $(\phi,\omega) \in S(\mathfrak{M})_x \times S^*(\mathfrak{M})_x$, do not depend on ϕ. For such x,

$$\mathfrak{U}_n(x,\lambda) = \mathrm{E}(\lambda)^{-1}\sum_j\prod_k \Phi_{jk}^{(n)}(x,\cdot)\mathrm{E}(\lambda)$$

and

$$\sup_{\lambda \in l_x}\|\mathfrak{U}_n(x,\lambda);L_2(S(\mathfrak{M})_x)\to L_2(S(\mathfrak{M})_x)\| = \qquad (5.1)$$
$$= \|\Phi_n(x,\cdot);C(S^*(\mathfrak{M})_x)\|,$$

where $\Phi_n = \Sigma_j\Pi_k\Phi_{jk}^{(n)}$. We will denote the space $S(\mathfrak{M})\oplus S^*(\mathfrak{M})$ with the topology $\mathfrak{T}(S,S^*)$ induced by the topology on $\hat{\mathfrak{G}}$ (cf. §2) by $\hat{\mathfrak{M}}$, while $C(\hat{\mathfrak{M}})$ will denote the corresponding space of continuous functions. Formula (5.1) implies that $\|\Phi_n;C(\hat{\mathfrak{M}})\| \leq \|\mathfrak{U}_n;\mathfrak{G}\|$. Hence also for a limit element $\mathfrak{U} \in \mathfrak{G}$ there is defined its symbol $\Phi \in C(\hat{\mathfrak{M}})$. It is obvious that to the product of two elements $\mathfrak{U},\mathfrak{B}$ in $\mathfrak{G}$ corresponds the product $\Phi\Psi$ of their symbols.

Proposition 5.1. *Let $\mathfrak{U}$ be an element of $\mathfrak{G}$, with corresponding symbol $\Phi \in C(\hat{\mathfrak{M}})$. The correspondence $\pi(x,\phi,\theta):\mathfrak{U}\mapsto\Phi(x,\phi,\theta)$, $\phi \in S(\mathfrak{M})_x$, $\theta \in S^*(\mathfrak{M})_x$, defines a one-dimensional representation of the algebra $\mathfrak{G}$, while the correspondence $\pi(x,\lambda):\mathfrak{U}\mapsto\mathfrak{U}(x,\lambda)$, $x \in \mathfrak{M}$, $\lambda \in l_x$, defines an irreducible representation of $\mathfrak{G}$ in $L_2(S(\mathfrak{M})_x)$. These representations are pairwise inequivalent.*

Proof. The fact that $\pi(x,\phi,\theta)$ is a one-dimensional representation of $\mathfrak{G}$ follows from the assertions made before stating the proposition, while the fact that $\pi(x,\lambda)$ is a representation immediately follows from the definition of $\mathfrak{G}$. Irreducibility of $\pi(x,\lambda)$ is verified using proposition 3.1, while the proof of the inequivalence of distinct representations reduces to an application of proposition 4.2. ■

Our nearest goal is to prove that every irreducible representation of $\mathfrak{G}$ is equivalent to a representation listed in proposition 5.1. It is reached in proposition 5.5, which is preceded by some lemmas.

We introduce the function

$$x \mapsto \mathfrak{N}_{\mathfrak{U}}(x) = \sup_{\lambda \in l_x}\|\mathfrak{U}(x,\lambda);L_2(S(\mathfrak{M})_x)\to L_2(S(\mathfrak{M})_x)\|,$$

where $\mathfrak{U} \in \mathfrak{G}$, $x \in \mathfrak{M}$.

Lemma 5.2. *The function $\mathfrak{N}_{\mathfrak{U}}$ is upper semicontinuous.*

P r o o f. First suppose that $\mathfrak{U}$ can be represented as

$$\mathfrak{U}(x,\lambda) = \sum_j \prod_k \mathrm{E}(\lambda)^{-1}\Phi_{jk}(x,\cdot)\mathrm{E}(\lambda), \tag{5.2}$$

where $\Phi_{jk}(x,\cdot) \in C^\infty(S(\mathfrak{M})_x \times S^*(\mathfrak{M})_x)$, $\Phi_{jk} \in \mathfrak{M}$, while the subscripts j,k run over finite sets. Fix an arbitrary point x. By theorem 3.3.4, the operator $\mathfrak{U}(x,\lambda)$ can be rewritten as

$$\mathfrak{U}(x,\lambda) = \mathrm{E}(\lambda)^{-1}\sum_j \prod_k \Phi_{jk}(x,\cdot)\mathrm{E}(\lambda)+\mathrm{T}(\lambda), \tag{5.3}$$

moreover, $\|\mathrm{T}(\lambda);L_2(S(\mathfrak{M})_x)\to L_2(S(\mathfrak{M})_x)\| = O(|\lambda|^{-1})$ as $|\lambda|\to-\infty$. In order to estimate the first term on the righthand side in (5.3) we use (3.4). Note that in (3.4) for $\nabla g(\phi)$ we may take any vector orthogonal to ϕ, while the numbers μ and σ may be chosen arbitrarily. Hence we may assume that $|\mu|+|\sigma|\to\infty$ and that the vector $\mu\nabla g(\phi)+\sigma\phi$ remains on a given ray. This and (3.4) imply

$$\mathfrak{N}_{\mathfrak{U}}(x) \geqslant \left\| \sum_j \prod_k \Phi_{jk}(x,\cdot);C(S(\mathfrak{M})_x \times S^*(\mathfrak{M})_x) \right\|.$$

Recall that functions from $\mathfrak{M}$ depend everywhere, with the exception of a set of singular points $x \in \mathfrak{M}$ (which depends on the function under consideration) only on the cotangent vector ω. If x is not singular for any function Φ_{jk}, then

$$\sum_j \prod_k \mathrm{E}(\lambda)^{-1}\Phi_{jk}(x,\cdot)\mathrm{E}(\lambda) = \mathrm{E}(\lambda)^{-1}\sum_j \prod_k \Phi_{jk}(x,\cdot)\mathrm{E}(\lambda),$$

and, since $\mathrm{E}(\lambda)$ is unitary for any $\lambda \in l_x = \mathbf{R}$,

$$\mathfrak{N}_{\mathfrak{U}}(x) = \left\| \mathrm{E}(\lambda)^{-1}\sum_j \prod_k \Phi_{jk}(x,\cdot)\mathrm{E}(\lambda) \right\| =$$

$$= \left\| \sum_j \prod_k \Phi_{jk}(x,\cdot);C(S^*(\mathfrak{M})_x) \right\|.$$

If the nonsingular point x tends to a point x_0 that is singular for $\Sigma_j\Pi_k\Phi_{jk}$ along a direction $\phi \in S(\mathfrak{M})_{x_0}$, then, uniformly in ϕ,

$$\lim_{x\to x_0} \left\| \mathrm{E}(\lambda)^{-1}\sum_j \prod_k \Phi_{jk}(x,\cdot)\mathrm{E}(\lambda) \right\| =$$

$$= \left\| \left\| \sum_j \prod_k \Phi_{jk}(x_0, \phi, \cdot); C(S^*(\mathfrak{M})_{x_0}) \right\| \right\|.$$

So, the function $\mathfrak{N}_{\mathfrak{U}}$ is continuous at every point x that is nonsingular for the symbol $\Sigma_j \Pi_k \Phi_{jk}$, while at singular points the relation $\overline{\lim}_{x \to x_0} \mathfrak{N}_{\mathfrak{U}}(x) \leqslant \mathfrak{N}_{\mathfrak{U}}(x_0)$ holds. Thus, the lemma has been proved for the operators (5.2). In order to complete the proof it remains to use the denseness of the set of operators (5.2) in the algebra $\mathfrak{G}$. ■

Let I be a closed ideal in $\mathfrak{G}$. Denote by I_x, $x \in \mathfrak{M}$, the set of operator-functions on l_x defined by $I_x = \{\mathfrak{U}(x, \cdot): \mathfrak{U} \in I\}$.

L e m m a 5.3. *$\mathfrak{U} \in I$ if and only if $\mathfrak{U}(x, \cdot) \in I_x$ for all $x \in \mathfrak{M}$.*

P r o o f. It must be verified that $\mathfrak{U}(x, \cdot) \in I_x$ for all x implies $\mathfrak{U} \in I$.

For any point $x_0 \in \mathfrak{M}$ there is a function $\mathfrak{B} \in I$ such that $\mathfrak{U}(x_0, \cdot) = \mathfrak{B}(x_0, \cdot)$. Then, by lemma 5.2, the inequality

$$\sup_\lambda \| \mathfrak{U}(x, \lambda) - \mathfrak{B}(x, \lambda); L_2(S(\mathfrak{M})_x) \to L_2(S(\mathfrak{M})_x) \| < \epsilon$$

holds in a sufficiently small neighborhood of x_0. Hence, for an arbitrary $\epsilon > 0$ there are a finite cover $\{U_j\}$ of $\mathfrak{M}$ and a set of operator-functions $\{\mathfrak{B}_j\}$, $\mathfrak{B}_j \in I$, satisfying $\sup_{\lambda \in l_x} \| \mathfrak{U}(x, \lambda) - \mathfrak{B}_j(x, \lambda) \| < \epsilon$ for $x \in U_j$. Denote by $\{\eta_j\}$ a partition of unity subordinate to the cover $\{U_j\}$, and put $\mathfrak{B}(x, \cdot) = \Sigma_j \eta_j(x) \mathfrak{B}_j(x, \cdot)$. Clearly $\mathfrak{B} \in I$. We have

$$\sup_{\lambda \in l_x} \| \mathfrak{U}(x, \lambda) - \mathfrak{B}(x, \lambda); L_2(S(\mathfrak{M})_x) \to L_2(S(\mathfrak{M})_x) \| \leqslant$$

$$\leqslant \sum_j \eta_{ij}(x) \sup_{\lambda \in l_x} \| \mathfrak{U}(x, \lambda) - \mathfrak{B}_j(x, \lambda); L_2(S(\mathfrak{M})_x) \to L_2(S(\mathfrak{M})_x) \| < \epsilon.$$

Hence $\| \mathfrak{U} - \mathfrak{B}; \mathfrak{G} \| < \epsilon$. The assertion of the lemma now follows from the fact that I is a closed ideal. ■

Denote, as before, by $\mathfrak{G}(l_x)$ the algebra of restrictions $\{\mathfrak{U}(x, \cdot): \mathfrak{U} \in \mathfrak{G}\}$ to l_x of the elements of $\mathfrak{G}$, and by Y the set of points $x \in \mathfrak{M}$ for which $I_x \neq \mathfrak{G}(l_x)$.

L e m m a 5.4. *If I is a primitive ideal, then Y consists of a single point only.*

P r o o f. Assume the opposite holds. Let x_1, x_2 be two distinct points in Y and let U_1, U_2 be disjoint neighborhoods of them on $\mathfrak{M}$. Denote by $\mathcal{K}_i$ the closed two-sided ideal in $\mathfrak{G}$ formed by the functions which vanish outside U_i, $i = 1,2$. Clearly, $\mathcal{K}_1\mathcal{K}_2 = 0 \subset I$. Then, by proposition 1.3, at least one of $\mathcal{K}_i$ is contained in I. If $\mathcal{K}_1 \subset I$, then $(\mathcal{K}_1)_{x_1} \subset I_{x_1}$, which is impossible since, on the one hand, $(\mathcal{K}_1)_{x_1} = \mathfrak{G}(l_{x_1})$, and on the other hand $x_1 \in Y$. ∎

Let π be an irreducible representation of $\mathfrak{G}$ whose kernel is I, and let $x_0 \in \mathfrak{M}$ be a point for which $I_{x_0} \neq \mathfrak{G}(l_{x_0})$ (by lemma 5.4 there is at least one point with this property). If $\mathfrak{U}, \mathfrak{B}$ are elements of $\mathfrak{G}$ such that $\mathfrak{U}(x_0,\cdot) = \mathfrak{B}(x_0,\cdot)$, then $\pi(\mathfrak{U}) = \pi(\mathfrak{B})$, by lemmas 5.3 and 5.4. This allows us to consider π as a representation of the algebra $\mathfrak{G}(l_{x_0})$ of restrictions to l_{x_0} of the elements of $\mathfrak{G}$. This and theorem 4.6 imply

P r o p o s i t i o n 5.5. *Every irreducible representation of the algebra $\mathfrak{G}$ is equivalent to a representation listed in proposition* 5.1.

By theorem 4.4.1 the algebras $\mathcal{Q}/\mathcal{K}L_2(\mathfrak{M})$ and $\mathfrak{G}$ are isometrically isomorphic. Hence the first assertion of theorem 2.1 is that of proposition 5.1, and the second assertion that of proposition 5.5. Thus, it remains to prove the third assertion of the theorem only.

The description of the representations implies that the set of residue classes of equivalent irreducible representations can be parametrized by the points of $\hat{\mathfrak{G}}$. It is obvious that inequivalent representations have distinct kernels. Hence the spectrum of the algebra $\mathcal{Q}/\mathcal{K}L_2(\mathfrak{M})$ can be identified with the set of primitive ideals of this algebra. We verify that the topology on $\hat{\mathfrak{G}}$ coincides with the Jacobson topology. For this we need

P r o p o s i t i o n 5.6. *Let $\mathfrak{U}$ be an operator-function, defined by*

$$\mathfrak{U}(x,\lambda) = \begin{cases} 0, & x \neq x_0, \\ \mathrm{P}(\lambda), & x = x_0, \end{cases} \tag{5.4}$$

where $x \in \mathfrak{M}$ and P is a continuous function with compact range, given on l_{x_0} and tending to zero at infinity (in the operator norm in $L_2(S(\mathfrak{M})_{x_0})$). Then $\mathfrak{U} \in \mathfrak{G}$.

P r o o f. By lemma 4.3, $\mathcal{K}_0(l_{x_0}) \subset \mathfrak{G}(l_{x_0})$, hence there is a $\mathfrak{P} \in \mathfrak{G}$ coinciding on

l_{x_0} with P. Let $\{\zeta_n\}$ be a sequence of $C^\infty(\mathfrak{M})$ functions, $0 \leqslant \zeta_n \leqslant 1$, $\zeta_n(x_0) = 1$, with, moreover, the support of ζ_n contracting towards x_0. We will show that the sequence $\{\zeta_n \mathfrak{P}\}$ converges in $\mathfrak{G}$, and that $\lim \zeta_n \mathfrak{P} = \mathfrak{U}$. Let $\mathfrak{B}(x,\lambda) = \Sigma_j \Pi_k \mathrm{E}(\lambda)^{-1} \Phi_{jk}(x,\cdot)\mathrm{E}(\lambda)$, where $\Phi_{jk} \in \mathfrak{M}$, be a combination of generators of $\mathfrak{G}$ such that $\|\mathfrak{P}-\mathfrak{B};\mathfrak{G}\| < \epsilon$. The operator $\mathfrak{P}$, as does every element of $\mathfrak{G}$, has a symbol, $\Phi_\mathfrak{P} \in C(\hat{\mathfrak{M}})$. Clearly, $\Phi_\mathfrak{P}(x_0,\cdot) = 0$, since in the opposite case we cannot have $\mathrm{P} \in \mathcal{K}_0(l_{x_0})$. This and the inequality $\|\Phi_\mathfrak{P} - \Sigma_j \Pi_k \Phi_{jk}; C(\hat{\mathfrak{M}})\| \leqslant \|\mathfrak{P}-\mathfrak{B};\mathfrak{G}\| < \epsilon$ imply $|\Sigma_j \Pi_k \Phi_{jk}(x,\cdot)| < \epsilon$, which holds for all x in a neighborhood U of x_0. We have

$$\|\zeta_n \mathfrak{P} - \zeta_m \mathfrak{P};\mathfrak{G}\| \leqslant \tag{5.5}$$

$$\leqslant \|\zeta_m(\mathfrak{P}-\mathfrak{B});\mathfrak{G}\| + \|(\zeta_m - \zeta_n)\mathfrak{B};\mathfrak{G}\| + \|\zeta_n(\mathfrak{B}-\mathfrak{P});\mathfrak{G}\| <$$

$$< 2\epsilon + \|(\zeta_m - \zeta_n)\mathfrak{B};\mathfrak{G}\|.$$

We may assume that $\operatorname{supp}\zeta_n \cup \operatorname{supp}\zeta_m \subset U$ and that the Φ_{jk} have in U no points of discontinuity other than x_0. Then,

$$\|(\zeta_m - \zeta_n)\mathfrak{B};\mathfrak{G}\| = \sup_U \left|(\zeta_n - \zeta_m)\sum_j \prod_k \Phi_{jk}\right| < \epsilon.$$

Together with (5.5) this leads to $\|\zeta_n \mathfrak{P} - \zeta_m \mathfrak{P};\mathfrak{G}\| < 3\epsilon$, for sufficiently large m,n. Thus, $\{\zeta_n \mathfrak{P}\}$ is a Cauchy sequence in $\mathfrak{G}$. The relation $\lim \zeta_n \mathfrak{P} = \mathfrak{U}$ clearly holds. ■

We return to the description of the Jacobson topology on the spectrum of the algebra $\mathcal{Q}/\mathcal{KL}_2(\mathfrak{M})$. Recall that a base for this topology is formed by the sets $\{\pi: \|\pi(\mathfrak{B})\| > 1\}$, where $\mathfrak{B}$ runs over some dense subset of $\mathfrak{G}$. We must prove that the topology of $\hat{\mathfrak{G}}$, described in §2, coincides with the Jacobson topology.

Let λ_0 be an arbitrary point of l_{x_0}. In $\mathfrak{G}$, a fundamental system of neighborhoods of this points consists of the open intervals on l_{x_0} containing λ_0. The continuity of the functions $\lambda \mapsto \mathfrak{U}(x_0,\lambda)$, $\mathfrak{U} \in \mathfrak{G}$, implies that every neighborhood of λ_0 in the Jacobson topology is a neighborhood in the topology of $\hat{\mathfrak{G}}$ also. Proposition 5.6 means that, in particular, every open interval on l_{x_0} containing λ_0 is a neighborhood in the Jacobson topology also.

Now consider neighborhoods of a point (x_0,ϕ_0,θ_0), $x_0 \in \mathfrak{M}$, $\phi_0 \in S(\mathfrak{M})_{x_0}$, $\theta_0 \in S^*(\mathfrak{M})_{x_0}$. Let $\mathfrak{U} \in \mathfrak{G}$ and let

$$\mathfrak{U}(x,\lambda) = \sum_j \prod_k \mathrm{E}(\lambda)^{-1}\Phi_{jk}(x,\cdot)\mathrm{E}(\lambda),$$

where $\Phi_{jk} \in \mathfrak{M}$. Put $\Phi = \Sigma_j \Pi_k \Phi_{jk}$; we will assume that $|\Phi(x_0,\phi_0,\theta_0)| > 1$.

We verify that every neighborhood of (x_0,ϕ_0,θ_0) in the Jacobson topology is a neighborhood of this point in the space $\hat{\mathfrak{G}}$. As has been shown in the proof of 3) of theorem 4.6, a set $\{\lambda \in l_{x_0} : \|\mathfrak{U}(x_0,\lambda)\| > 1\}$ either coincides with l_{x_0} (if $\phi_0\theta_0 = 0$) or coincides with an interval $\{\lambda \in l_{x_0} : \lambda \gtreqless N\}$ (if $\phi_0\theta_0 \gtreqless 0$).

The set $\{(x_0,\phi,\theta) : |\Phi(x_0,\phi,\theta)| > 1\}$ contains a product $V(\phi_0) \times \mathfrak{W}$, where $V(\phi_0)$, $\mathfrak{W}$ are neighborhoods of ϕ_0, θ_0 on the spheres $S(\mathfrak{M})_{x_0}$, $S^*(\mathfrak{M})_{x_0}$, respectively.

The part of the bundle $S(\mathfrak{M}) \oplus S^*(\mathfrak{M})$ above a neighborhood U of a point x_0 can be represented as the product $U \times S^{n-1} \times S^{n-1}$. Since Φ is a continuous function in the topology $\mathfrak{T}(S,S^*)$, the inclusion $\{(x,\phi,\theta) : |\Phi(x,\phi,\theta)| > 1\} \supset \{(x,\phi,\theta) : x \in K(\phi_0),\ \phi \in S^{n-1},\ \theta \in \mathfrak{W}(\theta_0)\}$ holds. Here $K(\phi_0)$ is a subset of U whose image under some coordinate map $\kappa : U \to \mathbb{R}^n$ is the intersection of an open cone, with vertex at $\kappa(x_0)$ and containing all directions from $\kappa'(x_0)(V(x_0))$, and an open n-dimensional ball with center at $\kappa(x_0)$.

Finally, if $x \in K(\phi_0)$, $\theta \in \mathfrak{W}(\theta_0)$, we can choose $\phi \in S^{n-1}$ such that $\phi\theta = 0$. Since $|\Phi(x,\phi,\theta)| > 1$, the inequality $\|\mathfrak{U}(x,\lambda); L_2(S(\mathfrak{M})_x) \to L_2(S(\mathfrak{M})_x)\| > 1$ holds on all of l_x (cf. the proof of 3) the theorem 4.6).

All this implies that a neighborhood $\{\pi : \|\pi(\mathfrak{U})\| > 1\}$ of a point (x_0,ϕ_0,θ_0) in the Jacobson topology contains a neighborhood of this point in the topology of $\hat{\mathfrak{G}}$.

We will now show that every neighborhood Ω of a point (x_0,ϕ_0,θ_0) in $\hat{\mathfrak{G}}$ contains a neighborhood of this point in the Jacobson topology.

We will assume that Ω is a neighborhood from the fundamental system, and that in case $\phi_0\theta_0 \gtreqless 0$ the intersection $\Omega \cap (S(\mathfrak{M}) \oplus S^*(\mathfrak{M}))_{x_0}$ belongs to the set $\phi\theta \gtreqless 0$.

The intersection $\Omega \cap (S(\mathfrak{M}) \oplus S^*(\mathfrak{M}))$ is a neighborhood of $(x_0,\phi_0,\theta_0) \in S(\mathfrak{M}) \oplus S^*(\mathfrak{M})$ in the topology $\mathfrak{T}(S,S^*)$. There clearly exists a function $\Phi \in C(\hat{\mathfrak{M}})$ such that $(x_0,\phi_0,\theta_0) \in \{(x,\phi,\theta) : |\Phi(x,\phi,\theta)| > 1\} \subset \Omega \cap (S(\mathfrak{M}) \oplus S^*(\mathfrak{M}))$ and $\Phi = 0$ outside $\Omega \cap (S(\mathfrak{M}) \oplus S^*(\mathfrak{M}))$. If $\phi_0\theta_0 \neq 0$, we will assume that also $\Phi(x_0,\phi,\theta) = 0$ for $\phi\theta = 0$. Since $\mathfrak{M}$ is dense in $C(\hat{\mathfrak{M}})$, we may assume that a function Φ with such properties exists in $\mathfrak{M}$. Put $\mathfrak{U}(x,\lambda) = E(\lambda)^{-1}\Phi(x,\cdot)E(\lambda)$. In view of the inequality $\|\pi(x,\phi,\theta)\mathfrak{U}\| = |\Phi(x,\phi,\theta)|$ an element (x,ϕ,θ) which belongs to a Jacobson neighborhood of (x_0,ϕ_0,θ_0) belongs to Ω also.

If $\phi_0\theta_0 = 0$, then Ω contains the whole line l_{x_0} and the lines l_x, $x \in p(\Omega) \cap \mathcal{M}$ ($p: S(\mathcal{M}) \oplus S^*(\mathcal{M}) \to \mathcal{M}$ is projection). Clearly, in this case Ω contains also a neighborhood $\{\pi: \|\pi(\mathfrak{U})\| > 1\}$ of (x_0,ϕ_0,θ_0) in the Jacobson topology. If, however, $\phi_0\theta_0 \gtrless 0$, a part $\{\lambda \in l_{x_0}: \lambda \gtrless N\}$ of l_{x_0} belongs to Ω. By lemma 4.8, $\|\mathfrak{U}(x_0,\lambda); L_2(S(\mathcal{M})_x) \to L_2(S(\mathcal{M})_x)\| \to 0$ as $\lambda \to \mp\infty$. Therefore, by adding to $\mathfrak{U}$ the operator (5.4) from proposition 5.6 (with a suitable P) we obtain that the inequality $\|\mathfrak{U}(x_0,\lambda)\| < 1$ holds on the part of l_{x_0} not belonging to Ω. ■

Together with the description in theorem 2.1 of the irreducible representations, the following theorem gives a criterion in order that an arbitrary operator from $\mathcal{Q}$ be Fredholm.

T h e o r e m 5.7. *Let* $\mathcal{Q}$ *and* $\mathfrak{S}$ *be the same algebras as in theorem* 2.1. *An operator* $A \in \mathcal{Q}$ *is Fredholm if and only if for every irreducible representation* π *of* $\mathfrak{S}$ *the operator* $\pi(A)$ *has a bounded inverse on the representation space* H_π.

P r o o f. An operator A is Fredholm if and only if its residue class [A] is invertible in the Calkin algebra $\mathfrak{L}(L_2(\mathcal{M}))/\mathcal{K}L_2(\mathcal{M})$. (Here $\mathfrak{L}(L_2(\mathcal{M}))$ is the algebra of bounded operators on $L_2(\mathcal{M})$.) The element [A] is invertible simultaneously in the algebra $\mathfrak{L}(L_2(\mathcal{M}))/\mathcal{K}L_2(\mathcal{M})$ and in the subalgebra $\mathcal{Q}/\mathcal{K}L_2(\mathcal{M}) \approx \mathfrak{S}$ (proposition 1.11). It remains to use proposition 1.10. ■

§6. Ideals in the algebra of pseudodifferential operators with discontinuous symbols

Every closed two-sided ideal in a C^*-algebra is the intersection of the primitive ideals containing it. This and the theorems in §2, containing a description of the spectra of algebras of pseudodifferential operators, allow us to clarify the structure of all closed two-sided ideals in such algebras. In this paragraph we consider certain ideals of the algebra $\mathcal{Q}$ from theorem 2.1.

Let $\mathfrak{L}$ be the disjoint union of straight lines l_x, enumerated by the points of a manifold $\mathcal{M}$. On $\mathfrak{L}$ we introduce the topology induced by the topology of the space $\hat{\mathfrak{S}}$ (the spectrum of the algebra $\mathfrak{S}$ of operator symbols $\mathfrak{S} \approx \mathcal{Q}/\mathcal{K}L_2(\mathcal{M})$).

Clearly, $\mathfrak{L}$ becomes a locally compact space, and every line l_x (with the ordinary topology) is an open subset of $\mathfrak{L}$.

Let Q_x be a compact subset of l_x, and let $Q = \cup_x Q_x$, where x runs over a finite point set in $\mathfrak{M}$. Sets $\mathfrak{L} \setminus Q$ form a fundamental system of neighborhoods of the point at infinity, ∞, in $\mathfrak{L}$. The point ∞ clearly does not have a countable fundamental system of neighborhoods, hence $\mathfrak{L}$ is not metrizable.

We introduce the algebra $\mathcal{K}_0(\mathfrak{L})$ of operator-valued functions $\lambda \mapsto \mathfrak{U}(x,\lambda) \in \mathcal{K}L_2(S(\mathfrak{M})_x)$, $\lambda \in l_x$, given on $\mathfrak{L}$, continuous (in the operator norm), and tending to zero at infinity.

P r o p o s i t i o n 6.1. *Let $\mathfrak{U} \in \mathcal{K}_0(\mathfrak{L})$, and let ϵ be an arbitrary positive number. Then everywhere on $\mathfrak{L}$, except possibly on a finite set of lines, the following inequality holds:*

$$\|\mathfrak{U}(x,\lambda); L_2(S(\mathfrak{M})_x) \to L_2(S(\mathfrak{M})_x)\| < \epsilon, \quad \lambda \in l_x. \tag{6.1}$$

P r o o f. Since $\mathfrak{U}$ tends to zero at infinity, (6.1) holds in a neighborhood U of ∞. The complement $\mathfrak{L} \setminus U$ can intersect at most finitely many lines l_x. ■

P r o p o s i t i o n 6.2. *$\mathcal{K}_0(\mathfrak{L})$ is a closed two-sided ideal of the algebra $\mathfrak{S}$ of operator symbols.*

P r o o f. Let $\mathfrak{U}$ be an arbitrary element of $\mathcal{K}_0(\mathfrak{L})$. It suffices to prove that $\mathfrak{U}$ is the limit of a sequence of elements of $\mathfrak{S}$ (in the norm of $\mathfrak{S}$). Denote by $\{\epsilon_n\}$ a decreasing null sequence of positive numbers. Proposition 6.1 implies that there is an at most finite set $\mathfrak{L}_n$ of lines l_x on which (6.1) does not hold. Introduce the operator-valued function $\mathfrak{U}_n$ equal to $\mathfrak{U}$ on $\mathfrak{L}_n$ and equal to zero everywhere outside $\mathfrak{L}_n$. By proposition 5.6, $\mathfrak{U}_n \in \mathfrak{S}$. It is clear that $\mathfrak{U}$ is the limit of the sequence $\{\mathfrak{U}_n\}$. ■

P r o p o s i t i o n 6.3. *The following isomorphism holds:*

$$\mathfrak{S} / \mathcal{K}_0(\mathfrak{L}) \approx C(\hat{\mathfrak{M}}).$$

P r o o f. Let $\mathrm{j}_x : \mathfrak{U} \to \mathfrak{U}(x, \cdot)$ be restriction to l_x of an element $\mathfrak{U} \in \mathfrak{S}$. Define a map $\mathrm{h} : \mathfrak{S} / \mathcal{K}_0(\mathfrak{L}) \to C(\hat{\mathfrak{M}})$ by $\mathrm{h}([\mathfrak{U}])(m) = \pi(m) \circ \mathrm{j}_x(\mathfrak{U})$, where $[\mathfrak{U}]$ is the residue class of $\mathfrak{U} \in \mathfrak{S}$ in $\mathfrak{S} / \mathcal{K}_0(\mathfrak{L})$, $m \in (S(\mathfrak{M}) \oplus S^*(\mathfrak{M}))_x$, and $\pi(m)$ is a one-dimensional representation of $\mathfrak{G}(l_x)$ (the definition is correct, which follows from theorem 4.6; cf. also proposition 4.4). We will show that h is an isomorphism.

We verify that h is monomorphic. If $h([\mathfrak{U}]) = 0$, then, as follows from theorem 4.6, $\mathfrak{U}(x, \cdot) \in \mathcal{K}_0(l_x)$ for all $x \in \mathfrak{M}$. It remains to convince ourselves of the fact that $\mathfrak{U}$ tends to zero when passing to ∞ along $\mathfrak{L}$. Let $\Phi_{jk} \in \mathfrak{N}$, $\mathfrak{U}_{jk}(x, \lambda) = E(\lambda)^{-1}\Phi_{jk}(x, \phi, \theta)E(\lambda)$, $\phi \in S(\mathfrak{M})_x$, $\theta \in S^*(\mathfrak{M})_x$, and

$$\left\| \mathfrak{U} - \sum_j \prod_k \mathfrak{U}_{jk}; \mathfrak{G} \right\| < \epsilon/2, \tag{6.2}$$

where j,k run over finite sets. (The operator symbols of the form $\mathfrak{U}_{jk}$ generate the algebra $\mathfrak{G}$.) Then for any one-dimensional representation $\pi(m)$, $m \in \mathfrak{N}$, of $\mathfrak{G}$ the following inequality holds:

$$\left| \pi(m) \left(\sum_j \prod_k \mathfrak{U}_{jk} \right) \right| < \epsilon/2. \tag{6.3}$$

The symbols Φ_{jk} have at most finitely many points of discontinuity. At every point $x \in \mathfrak{M}$ at which all Φ_{jk} are continuous,

$$\sup_{\lambda \in l_x} \left\| \sum_j \prod_k \mathfrak{U}_{jk}(x, \lambda); L_2(S(\mathfrak{M})_x) \to L_2(S(\mathfrak{M})_x) \right\| =$$

$$= \max_{\phi, \theta} \left| \sum_j \prod_k \Phi_{jk}(x, \phi, \theta) \right|.$$

Taking into account (6.2) and (6.3), we thus obtain that $\sup_{\lambda \in l_x} \|\mathfrak{U}(x, \lambda)\| < \epsilon$ everywhere except possibly on a finite set of lines l_x. Thus, for any $\epsilon > 0$ there is in $\mathfrak{L}$ a neighborhood of ∞ in which (6.1) holds. Therefore $\mathfrak{U} \in \mathcal{K}_0(\mathfrak{L})$, and h is a monomorphism.

If $\Phi \in \mathfrak{N}$ and $\mathfrak{U}(x, \lambda) = E(\lambda)^{-1}\Phi(x, \phi, \theta)E(\lambda)$, then $h([\mathfrak{U}]) = \Phi$. The set $\mathfrak{N}$ generates $C(\hat{\mathfrak{N}})$. This and the fact that h is isometric gives that h is an epimorphism. ■

Denote by com $\mathfrak{B}$ the closed two-sided ideal of an algebra $\mathfrak{B}$ that is generated by the commutators of the elements of $\mathfrak{B}$.

C o r o l l a r y 6.4. com $\mathfrak{G} = \mathcal{K}_0(\mathfrak{L})$.

P r o o f. The ideal com $\mathfrak{G}$ is the intersection of the kernels of all irreducible one-dimensional representations of $\mathfrak{G}$. The algebra $\mathcal{K}_0(\mathfrak{L})$ does not have nonnull one-dimensional representations. Hence $\mathcal{K}_0(\mathfrak{L}) \subset \mathrm{com}\,\mathfrak{G}$. Since $\mathfrak{G}/\mathcal{K}_0(\mathfrak{L})$ is a commutative algebra, the opposite inclusion $\mathrm{com}\,\mathfrak{G} \subset \mathcal{K}_0(\mathfrak{L})$ holds. ∎

Let $p:\mathcal{A} \to \mathcal{A}/\mathcal{KL}_2(\mathcal{M})$ be projection and let $i:\mathcal{A}/\mathcal{KL}_2(\mathcal{M}) \approx \mathfrak{G}$ be the isomorphism mapping a pseudodifferential operator A to its operator symbol $\mathfrak{U}$.

P r o p o s i t i o n 6.5. $\mathrm{com}\,\mathcal{A} = p^{-1} \circ i^{-1}(\mathcal{K}_0(\mathfrak{L}))$.

P r o o f. Any one-dimensional representation of $\mathcal{A}$ can be regarded as a representation of the quotient algebra $\mathcal{A}/\mathcal{K} \approx \mathfrak{G}$. Since $\mathcal{K}_0(\mathfrak{L})$ belongs to the kernel of all one-dimensional representations, $p^{-1} \circ i^{-1}(\mathcal{K}_0(\mathfrak{L})) \subset \mathrm{com}\,\mathcal{A}$. We verify the opposite inclusion. If $A \in \mathrm{com}\,\mathcal{A}$, then $\pi(m)A = 0$ for all $m \in \hat{\mathfrak{M}}$. This implies (as has been shown in the proof of proposition 6.3) that $i \circ p(A) \in \mathcal{K}_0(\mathfrak{L})$. ∎

All results stated in this paragraph can be combined into the following commutative diagram

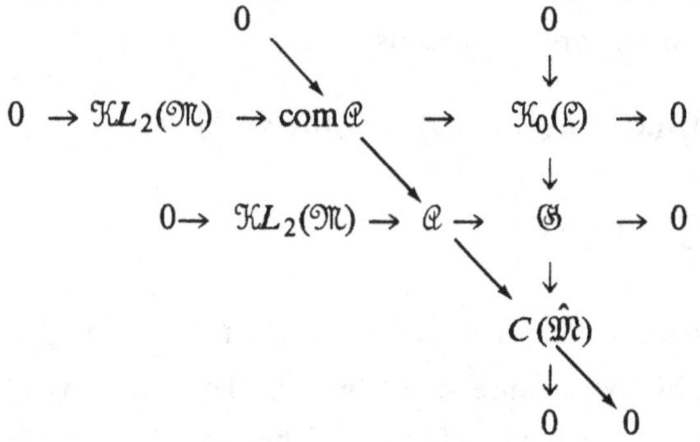

All straight paths ('without corners') in this diagram are made up by exact sequences.

Finally we note that all values of an operator symbol $\mathfrak{U} \in \mathfrak{G}$ are compact operators if and only if $\mathfrak{U}$ belongs to the kernel of every representation $\pi(m)$, where m is a point of $\hat{\mathfrak{M}}$ of the form (x, ϕ, θ), $\phi \in S(\mathcal{M})_x$, $\theta \in S^*(\mathcal{M})_x$, and $\phi\theta = 0$.

§7. Spectra of C^*-algebras of pseudodifferential operators on a manifold with conical points

In this paragraph we study the spectra of algebras generated by pseudodifferential operators with discontinuities of the first kind in the symbols, on manifolds with conical points. The operators act on the weighted normed spaces $H_\beta(\mathfrak{M})$ (cf. §4.4). In particular, we consider the case when at a nonsingular point $x^{(0)} \in \mathfrak{M}$ the weight takes the value 0 or ∞, while the symbols of the operators remain smooth. It turns out that also in this case infinite-dimensional representations appear - the functions space 'is to blame' - but they are 'twice as small' in comparison with the discontinuous situation (instead of the line $l_{x^{(0)}}$ the space $\hat{\mathfrak{G}}$ contains a half-line, cf. theorem 2.1).

The presence of conical points does not lead to new substantial problems. Therefore we consider in detail only the above-mentioned effect of the appearance of infinite-dimensional representations because of singularities of the weight. (In particular, we consider the local algebras $\mathfrak{G}_c(\lambda)$ for complex λ on the line $\{\lambda \in \mathbf{C}: \operatorname{Im}\lambda = \beta_0\}$, where β_0 is the exponent of the weight at $x^{(0)}$.)

We begin with studying the local algebras. Then we give a description of the spectrum of the algebra of pseudodifferential operators on a manifold with conical points. Finally we investigate the one-dimensional case: singular operators on a composite contour.

1. The algebra $\mathfrak{G}_c(\lambda)$. This algebra is generated by the operators $\mathfrak{U}_{\psi\to\phi}(\lambda) = \mathrm{E}_{\omega\to\phi}(\lambda)^{-1}\,\Phi(\omega)\mathrm{E}_{\psi\to\omega}(\lambda)$ on $L_2(S^{n-1})$; here, λ is a complex number, $\lambda \neq \pm i(k+n/2)$, $k = 0, 1, \cdots$, and $\Phi \in C^\infty(S^{n-1})$. The operator $\mathfrak{U}(\lambda)^*$ adjoint to $\mathfrak{U}(\lambda)$ has the form $\mathfrak{U}(\lambda)^* = \mathrm{E}(\bar{\lambda})^{-1}\overline{\Phi(\omega)}\mathrm{E}(\bar{\lambda})$. If λ is real, the algebra $\mathfrak{G}_c(\lambda)$ is, clearly, commutative. Moreover, for real λ the operator $\mathrm{E}(\lambda)$ is unitary, hence $\|\mathfrak{U}(\lambda); L_2(S^{n-1}) \to L_2(S^{n-1})\| = \|\Phi; C(S^{n-1})\|$. Thus, in this case $\mathfrak{G}_c(\lambda)$ is isomorphic to the ring $C(S^{n-1})$.

P r o p o s i t i o n 7.1. *If* $\operatorname{Im}\lambda \neq 0$, *then* $\mathfrak{G}_c(\lambda)$ *is an irreducible algebra.*

The p r o o f proceeds in several steps.

I. We show that every nontrivial invariant subspace $\mathcal{K} \subset L_2(S^{n-1})$ for $\mathfrak{G}_c(\lambda)$ contains a $C^\infty(S^{n-1})$ function f that is not identically equal to zero.

The operator $\mathrm{K} = i(\mathrm{E}(\lambda)^{-1}\Phi(\omega)\mathrm{E}(\lambda) - \mathrm{E}(\bar{\lambda})^{-1}\Phi(\omega)\mathrm{E}(\bar{\lambda}))$, where Φ is a real-valued function, is selfadjoint and belongs to $\mathfrak{G}_c(\lambda)$. By proposition 3.2.2 its order does not exceed -1. Denote by μ an arbitrary nonzero eigenvalue of K and by $\mathcal{H}_\mu$ the corresponding eigenspace. The projection operator $P: L_2(S^{n-1}) \to \mathcal{H}_\mu$ is an element of $\mathfrak{G}_c(\lambda)$. Since K is a smoothing operator, all functions from $\mathcal{H}_\mu$ belong to $C^\infty(S^{n-1})$. Since $P\mathcal{H} \subset \mathcal{H}_\mu \cap \mathcal{H}$, any nonzero element of $P\mathcal{H}$ can be taken as the function f. In case $P\mathcal{H} = 0$ the subspace $\mathcal{H}$ must be replaced by its orthogonal complement $\mathcal{H}^\perp$, which is also invariant under $\mathfrak{G}_c(\lambda)$.

II. In this step we show that there is a function $u \in C^\infty(S^{n-1})$ in $\mathcal{H}$ for which $(E(\lambda)u)(\phi_0) \neq 0$, where ϕ_0 is a given point on the sphere.

Let $f \in \mathcal{H} \cap C^\infty(S^{n-1})$, $f \neq 0$. Put $u = E(\bar{\lambda})^{-1}\Phi(\omega)E(\bar{\lambda})f$, where the smooth function Φ is chosen to satisfy: 1) $E(\bar{\lambda})f \neq 0$ on $\operatorname{supp}\Phi$; and 2) $\phi_0 \notin \operatorname{supp}\Phi$. (In view of the fact that all operators $E(\bar{\lambda})^{-1}\Phi E(\bar{\lambda})$ belong to $\mathfrak{G}_c(\lambda)$, the subspace $E(\bar{\lambda})\mathcal{H}$ is invariant under multiplication by smooth functions.)

We have $E(\lambda)u = E(\lambda)E(\bar{\lambda})^{-1}\Phi E(\bar{\lambda})f$. The operator $E(\lambda)E(\bar{\lambda})^{-1}$ can be written as

$$E(\lambda)E(\bar{\lambda})^{-1}v = \int_{S^{n-1}} G(\phi,\psi;\lambda)v(\psi)d\psi,$$

where

$$G(\phi,\psi;\lambda) = \int_0^\infty \frac{t^{-i\lambda+n/2-2\tau-1}}{|t\phi-\psi|^{n-2\tau}}dt,$$

$\tau = \operatorname{Im}\lambda$, $\phi,\psi \in S^{n-1}$, while the integral is understood in the sense of analytic extension in λ (cf. Chapter 3, §1.3).

So,

$$(E(\lambda)u)(\phi_0) = \int_{S^{n-1}} G(\phi_0,\omega;\lambda)\Phi(\omega)(E(\bar{\lambda})f)(\omega)d\omega. \tag{7.1}$$

The function $(\phi,\psi) \mapsto G(\phi,\psi;\lambda)$ is analytic on $(S^{n-1}\times S^{n-1})\setminus\Delta$ (Δ the diagonal). If for all Φ satisfying 1) and 2) the integral (7.1) would vanish, then the function $\omega \mapsto G(\phi_0,\omega;\lambda)$ would be identically zero on the set $(\operatorname{supp}(\Phi E(\bar{\lambda})f))\setminus\phi_0$, hence, by analyticity, would be equal to zero outside ϕ_0. On the other hand, the kernel G cannot have support concentrated on the diagonal, since the operator $E(\lambda)E(\bar{\lambda})^{-1}$ is smoothing ($E(\lambda)$ acts as a continuous operator $L_2(S^{n-1}) \to H^\tau(S^{n-1})$, and τ may be taken, without loss of generality, a positive number).

III. The previous step implies that the subspace $E(\lambda)\mathcal{H}$ contains a smooth function that is distinct from zero everywhere on the sphere. Thus, $C^\infty(S^{n-1}) \subset E(\lambda)\mathcal{H}$. This implies that $\mathcal{H} = L_2(S^{n-1})$. We have obtained a contradiction, which proves the proposition. ■

Corollary 7.2. *Every compact operator on $L_2(S^{n-1})$ belongs to $\mathfrak{G}_c(\lambda)$.*

Proof. The algebra $\mathfrak{G}_c(\lambda)$ contains the nonzero compact operator $E(\lambda)^{-1}\Phi(\omega)E(\lambda) - E(\bar{\lambda})^{-1}\Phi(\omega)E(\bar{\lambda})$, where Φ is a real-valued function. Since $\mathfrak{G}_c(\lambda)$ is irreducible, by proposition 1.5 every compact operator is an element of it. ■

Proposition 7.3. *Let* $\operatorname{Im}\lambda \neq 0$, $\lambda \neq \pm i(k+n/2)$, $k = 0, 1, \cdots$, *and* $\mathfrak{U}(\lambda) = E(\lambda)^{-1}\Phi(\omega)E(\lambda)$. *The map* $\mathfrak{U}(\lambda) \mapsto \Phi$ *induces an isomorphism between the algebras* $\mathfrak{G}_c(\lambda)/\mathcal{K}L_2(S^{n-1})$ *and* $C(S^{n-1})$. *In particular, the norm of a residue class* $[\mathfrak{U}]$ *is equal to* $\|\Phi; C(S^{n-1})\|$.

Proof. By proposition 3.2.2,

$$\mathfrak{U}(\lambda)^* = E(\bar{\lambda})^{-1}\overline{\Phi(\omega)}E(\bar{\lambda}) = E(\lambda)^{-1}\Phi(\omega)E(\lambda) + T, \tag{7.2}$$

$$T \in \mathcal{K}L_2(S^{n-1}).$$

Formulas (3.5) and (3.7) are also true in the case considered; their righthand sides must be replaced by $\|\Phi; C(S^{n-1})\|$ (in the proofs obvious changes must be introduced). Hence

$$\inf_{T \in \mathcal{K}L_2(S^{n-1})} \|\mathfrak{U}(\lambda) + T\| = \|\Phi; C(S^{n-1})\|. \tag{7.3}$$

It remains to compare (7.2) and (7.3) and to use the relation $\mathfrak{U}(\lambda)\mathfrak{B}(\lambda) = E(\lambda)^{-1}\Phi\Psi E(\lambda)$ for the operators $\mathfrak{U}(\lambda) = E(\lambda)^{-1}\Phi(\omega)E(\lambda)$ and $\mathfrak{B}(\lambda) = E(\lambda)^{-1}\Psi(\omega)E(\lambda)$. ■

Denote by $\hat{\mathfrak{G}}_c(\lambda)$ the disjoint union of the sphere S^{n-1} and the point λ. Introduce a topology on $\hat{\mathfrak{G}}_c(\lambda)$ in which the point λ forms an open set, while a fundamental system of neighborhoods of a point $\omega \in S^{n-1}$ is formed by sets $V(\omega) \cup \{\lambda\}$, where $V(\omega)$ is a neighborhood of ω on S^{n-1}.

Theorem 7.4. *Every irreducible representation of the algebra $\mathfrak{G}_c(\Lambda)$ is either one-dimensional or the identity representation $e(\Lambda)$. A one-dimensional representation can be realized as a map $\pi(\omega):\mathfrak{U}\mapsto\Phi(\omega)$, where $\Phi\in C^\infty(S^{n-1})$ is the function corresponding to the residue class $[\mathfrak{U}]$. The correspondence $\pi(\omega)\mapsto\omega$, $e(\Lambda)\mapsto\Lambda$ defines a bijection of the spectrum of $\mathfrak{G}_c(\Lambda)$ onto $\hat{\mathfrak{G}}_c(\Lambda)$. The topology on $\hat{\mathfrak{G}}_c(\Lambda)$ coincides with the Jacobson topology.*

The proof is obtained from proposition 7.3 by standard reasoning. ■

2. The algebra $\mathfrak{G}_c(m_\beta)$. Let $m_\beta=\{\lambda\in\mathbf{C}:\ \mathrm{Im}\lambda=\beta\}$ and $\mathfrak{G}_c(m_\beta)$ the algebra generated by the functions $\lambda\mapsto\mathfrak{U}(\lambda)=\mathrm{E}(\lambda)^{-1}\Phi(\omega)\mathrm{E}(\lambda)$ on m_β, $\beta\neq\pm(k+n/2)$, $k=0,1,\cdots$. The norm in $\mathfrak{G}_c(m_\beta)$ is defined by

$$\|\mathfrak{U};\mathfrak{G}_c(m_\beta)\|=\sup_{\lambda\in m_\beta}\|\mathfrak{U}(\lambda);L_2(S^{n-1})\to L_2(S^{n-1})\|.$$

Denote by $\pi(\lambda)$ the representation $\mathfrak{U}\mapsto\mathfrak{U}(\lambda)$ of $\mathfrak{G}_c(m_\beta)$ in $L_2(S^{n-1})$. For real λ all representations $\pi(\lambda)$ are equivalent. They are reducible, since for λ real $\mathfrak{G}_c(\lambda)$ is a commutative algebra. If, however, $\mathrm{Im}\lambda\neq0$, then $\pi(\lambda)$ is an irreducible representation in $L_2(S^{n-1})$ (cf. proposition 7.1).

Proposition 7.5. *Let $\lambda_+=\sigma+i\beta$, $\lambda_-=-\sigma+i\beta$, $\beta\neq0$. Then the representations $\pi(\lambda_+)$ and $\pi(\lambda_-)$ are equivalent.*

Proof. We show that the operator $\mathrm{U}=\mathrm{E}(\lambda_+)^{-1}\mathrm{E}(\lambda_-)$ is a unitary intertwining operator for $\pi(\lambda_+)$ and $\pi(\lambda_-)$, i.e. $\pi(\lambda_-)\mathfrak{U}=\mathrm{U}\pi(\lambda_+)\mathfrak{U}\mathrm{U}$ for all $\mathfrak{U}\in\mathfrak{G}_c(m_\beta)$. Clearly, for the operators $\mathfrak{U}(\lambda_\pm)=\mathrm{E}(\lambda_\pm)^{-1}\Phi(\omega)\mathrm{E}(\lambda_\pm)$ the equation $\mathfrak{U}(\lambda_-)=\mathrm{U}^{-1}\mathfrak{U}(\lambda_+)\mathrm{U}$ holds. Hence it suffices to show that U is unitary.

Formula (1.3.3) for an arbitrary harmonics Y_{mk} implies

$$\mathrm{U}Y_{mk}=2^{-2i\sigma}\frac{\Gamma\left(\dfrac{m+i\lambda_-+n/2}{2}\right)\Gamma\left(\dfrac{m-i\lambda_++n/2}{2}\right)}{\Gamma\left(\dfrac{m-i\lambda_-+n/2}{2}\right)\Gamma\left(\dfrac{m+i\lambda_++n/2}{2}\right)}Y_{mk}.$$

Note that

$$\left|\frac{\Gamma\left(\dfrac{m+i\lambda_-+n/2}{2}\right)}{\Gamma\left(\dfrac{m+i\lambda_++n/2}{2}\right)}\right| = \left|\frac{\Gamma\left(\dfrac{m-i\lambda_++n/2}{2}\right)}{\Gamma\left(\dfrac{m-i\lambda_-+n/2}{2}\right)}\right| = 1,$$

since $\Gamma(\overline{z}) = \overline{\Gamma(z)}$. It remains to take into account that the spherical harmonics $\{Y_{mk}\}$ form an orthonormal basis in $L_2(S^{n+1})$. ■

P r o p o s i t i o n 7.6. *Let μ,ν be two arbitrary, but distinct, points on m_β, $\beta \neq 0$, nonsymmetrically located about the imaginary axis. Then the representations $\pi(\mu)$ and $\pi(\nu)$ are inequivalent.*

P r o o f. Suppose the assertion is false, and that there exists a unitary operator $U: L_2(S^{n-1}) \to L_2(S^{n-1})$ such that $\pi(\mu)\mathfrak{U} = U^{-1}\pi(\nu)\mathfrak{U}U$ for all $\mathfrak{U} \in \mathfrak{G}_c(m_\beta)$.

We show that every intertwining operator must have the form $E(\nu)^{-1}\chi(\omega)E(\mu)$, where χ is a smooth function that does not vanish and is of constant modulus.

Put

$$K(\lambda) = (\mathfrak{U}(\lambda)\mathfrak{B}(\lambda)^* - \mathfrak{B}(\lambda)^*\mathfrak{U}(\lambda))\,(\mathfrak{U}(\lambda)\mathfrak{B}(\lambda)^* - \mathfrak{B}(\lambda)^*\mathfrak{U}(\lambda))^*,$$

where $\lambda \in m_\beta$, and $\mathfrak{U}(\lambda) = E(\lambda)^{-1}\Phi(\omega)E(\lambda)$, $\mathfrak{B}(\lambda) = E(\lambda)^{-1}\Psi(\omega)E(\lambda)$ are operator-functions from $\mathfrak{G}_c(m_\beta)$. Theorems 3.3.4 and 3.3.5 imply that $\|K(\lambda)u; H^2(\lambda, S^{n-1})\| \leqslant c_\beta \|u; L_2(S^{n-1})\|$ on m_β. Clearly, $K(\lambda)$ is selfadjoint for all $\lambda \in m_\beta$, and $U^{-1}K(\nu)U = K(\mu)$. If f is an eigenfunction of $K(\mu)$ corresponding to a nonzero eigenvalue, then Uf is an eigenfunction for $K(\nu)$ corresponding to the same eigenvalue; moreover, $f \in C^\infty(S^{n-1})$ and $Uf \in C^\infty(S^{n-1})$.

Formula (1.2.1) for $E(\lambda)$ implies that the operators $E(\lambda)^{\pm 1}$ commute with rotations of the sphere. Therefore, if g is an arbitrary rotation and $u(\phi) = E_{\omega\to\phi}(\lambda)^{-1}\Xi(\omega)E_{\psi\to\omega}(\lambda)v(\psi)$, then we have $u(g\phi) = E_{\omega\to\phi}(\lambda)^{-1}\Xi(g\omega)E_{\phi\to\omega}(\lambda)v(g\psi)$. Put

$$\mathfrak{U}_g(\lambda) = E(\lambda)^{-1}\Phi(g\omega)E(\lambda). \tag{7.4}$$

Then $\psi \mapsto f(g\psi)$ is an eigenfunction of the operator

$$K_g(\mu) = (\mathfrak{U}_g(\mu)\mathfrak{B}_g(\mu)^* - \mathfrak{B}_g(\mu)^*\mathfrak{U}_g(\mu)(\mathfrak{U}_g(\mu)\mathfrak{B}_g(\mu)^* - \mathfrak{B}_g(\mu)^*\mathfrak{U}_g(\mu))^*$$

where $\mathfrak{B}_g$ is defined as in (7.4).

The above-said implies that there is a nontrivial subspace $\mathcal{K}$ of $C^\infty(S^{n-1}) \cap UC^\infty(S^{n-1})$ which is invariant under rotations of the sphere.

Let f be a nonzero element in $\mathcal{K}$. Applying the operator $E(\nu)$ to both sides of the equation $\mathfrak{U}(\nu)Uf = U\mathfrak{U}(\mu)f$ and replacing f by $E(\mu)^{-1}h$, $h \in C^\infty(S^{n-1})$, leads to

$$\Phi E(\nu)UE(\mu)^{-1}h = E(\nu)UE(\mu)^{-1}\Phi h \tag{7.5}$$

(recall that $E(\mu)^{\pm 1}(C^\infty(S^{n-1})) \subset C^\infty(S^{n-1})$).

We show that we may set $h \equiv 1$. For this we must show that $E(\mu)^{-1}1 \in \mathcal{K}$. Let $\{g_i\}$ be a finite set of rotations of the sphere S^{n-1} such that the sets $\{\psi \in S^{n-1}: h(g_i\psi) \neq 0\}$ cover the sphere. Denote by $\{\zeta_i\}$ a partition of unity subordinate to this cover. Replacing in (7.5) Φ by $\zeta_i/(g_i^* h)$ and h by $g_i^* h$, where $(g_i^* h)(\psi) = h(g_i\psi)$, and adding the equations obtained, leads to

$$\sum_i \frac{\zeta_i}{g_i^* h} E(\nu)UE(\mu)^{-1}g_i^* h = E(\nu)UE(\mu)^{-1}1. \tag{7.6}$$

Since $g_i^* h \in g_i^* E(\mu)\mathcal{K} = E(\mu)g_i^*\mathcal{K} = E(\mu)\mathcal{K}$, we have $E(\mu)^{-1}g_i^* h \in \mathcal{K}$. Hence the lefthand side is an element of $C^\infty(S^{n-1})$, therefore the righthand side belongs to this space. Applying to both sides of (7.6) the operator $E(\nu)^{-1}$ we find that $UE(\mu)^{-1}1 \in C^\infty(S^{n-1})$. It remains to note that $E(\mu)^{-1}1 = \text{const}$ (in view of (1.3.3)). So, $E(\mu)1 \in \mathcal{K}$.

Substituting in (7.5) $h = 1$ gives that $E(\nu)UE(\mu)^{-1}$ coincides with the operator of multiplication by the smooth function $\chi = E(\nu)UE(\mu)^{-1}1$, whence $U = E(\nu)^{-1}\chi E(\mu)$.

We now verify that the function χ is nowhere vanishing. Interchanging the roles of μ and ν, we obtain, as before, that $U^{-1} = E(\mu)^{-1}\xi E(\nu)$, $\xi \in C^\infty(S^{n-1})$. Hence $I = UU^{-1} = E(\nu)^{-1}\chi\xi E(\nu)$, hence $\chi\zeta = 1$. So $|\chi| > 0$.

The fact that the operators $E(\lambda)^{\pm 1}$ commute with rotations implies that together with $U = E(\nu)^{-1}\chi E(\mu)$, the operator $U_g = E(\nu)^{-1}g^*\chi E(\mu)$ is intertwining. Consequently, the operator UU_g^{-1} commutes with an arbitrary operator from $\mathfrak{G}_c(\nu)$. Since this algebra is irreducible, the operator UU_g^{-1} is scalar (Chapter

5, §1.4), i.e. $UU_g^{-1} = c(g)^{-1}I$, $|c(g)| = 1$. Therefore $\chi(g\omega)/\chi(\omega) \equiv c(g)$. The map $SO(n) \ni g \mapsto c(g)$ defines a one-dimensional representation of the rotation group $SO(n)$. For $n \geqslant 3$, $SO(n)$ has only the trivial one-dimensional representation, i.e. $c(g) \equiv 1$, while for $n = 2$ we are led to $c(g) = e^{im\alpha}$, where m is an integer and g is identified with rotation over an angle α (cf. [10]). So $\chi(\omega) = \text{const}$ for $n \geqslant 3$, and $\chi(\omega) = c\, e^{im\omega}$ for $n = 2$.

Consider the case $n \geqslant 3$. We compute the value of the operator U on a spherical function Y_{mk}. By formula (1.3.3), $UY_{mk} = \mu_m(\mu)\mu_m(\nu)^{-1}Y_{mk}$, as

$$\mu_m(\lambda) = (-i)^m 2^{i\lambda}\Gamma\left[\frac{m+i\lambda+n/2}{2}\right]\Gamma\left[\frac{m-i\lambda+n}{2}\right]^{-1}.$$

If we would show that the modulus of the quotient $\mu_m(\mu)\mu_m(\nu)^{-1}$ depends on m, we find a contradiction with the fact that U is unitary. We have

$$\left|\frac{\mu_{m+2}(\mu)\mu_m(\nu)}{\mu_{m+2}(\nu)\mu_m(\mu)}\right| = \tag{7.7}$$

$$= \left|\frac{(m+n/2+i\sigma-\tau)(m+n/2-i\sigma_1+\tau)}{(m+n/2-i\sigma+\tau)(m+n/2+i\sigma_1-\tau)}\right|,$$

where $\mu = \sigma+i\tau$, $\nu = \sigma_1+i\tau$. The expression in (7.7) can be equal to one only if

$$(m+n/2)(\sigma-\sigma_1)+\tau(\sigma+\sigma_1) = \tag{7.8}$$

$$= \pm((m+n/2)(\sigma_1-\sigma)+\tau(\sigma+\sigma_1)),$$

which implies that $\sigma = \sigma_1$ or $\tau(\sigma+\sigma_1) = 0$. Since by assumption $\tau \neq 0$ and $\sigma \neq \pm\sigma_1$, (7.8) does not hold. Hence U cannot be unitary.

The case $n = 2$ is considered in a similar manner. ■

Corollary 7.7. *Let $\mathfrak{G}_c(m_\beta^+)$ be the algebra generated by the restrictions of the elements $\mathfrak{U} \in \mathfrak{G}_c(m_\beta)$ onto $m_\beta^+ = \{\lambda \in m_\beta \colon \operatorname{Re}\lambda \geqslant 0\}$, with norm*

$$\|\mathfrak{U};\mathfrak{G}_c(m_\beta^+)\| = \sup_{\lambda\in m_\beta^+}\|\mathfrak{U}(\lambda);L_2(S^{n-1})\to L_2(S^{n-1})\|.$$

Then the restriction morphism $\mathfrak{G}_c(m_\beta)\to\mathfrak{G}_c(m_\beta^+)$ is an isomorphism.

Proof. It suffices to note that, by proposition 7.5, $\|\mathfrak{U}(\lambda_+)\| = \|\mathfrak{U}(\lambda_-)\|$ for any element $\mathfrak{U} \in \mathfrak{G}_c(m_\beta)$. Therefore, $\|\mathfrak{U};\mathfrak{G}_c(m_\beta)\| = \|\mathfrak{U};\mathfrak{G}_c(m_\beta^+)\|$ (the element $\mathfrak{U} \in \mathfrak{G}_c(m_\beta)$ and its restriction onto m_β^+ are denoted by the same letter). ■

Denote by $\mathcal{K}_0(m_\beta^+)$ the algebra of continuous functions on the half-line $m_\beta^+ = \{\lambda \in m_\beta : \operatorname{Re}\lambda \geqslant 0\}$, with values in $\mathcal{KL}_2(S^{n-1})$, and tending to zero (in norm) at infinity. The norm in $\mathcal{K}_0(m_\beta^+)$ is introduced analogously to the norm in $\mathfrak{G}_c(m_\beta^+)$.

L e m m a 7.8. *The inclusion $\mathcal{K}_0(m_\beta^+) \subset \mathfrak{G}_c(m_\beta^+)$ holds, where $\beta \neq 0$ and $\mathfrak{G}_c(m_\beta^+)$ is the restriction of the algebra $\mathfrak{G}_c(m_\beta^+)$ onto m_β^+.*

The p r o o f is the same as the proof of lemma 4.3. We must only replace $\mathfrak{G}(l_x)$ by $\mathfrak{G}_c(m_\beta^+)$, and use propositions 7.1 and 7.6 instead of 3.1 and 4.2. ■

P r o p o s i t i o n 7.9. *Suppose the operator-function $\lambda \mapsto \mathfrak{U}(\lambda) = E(\lambda)^{-1}\Phi(\omega) E(\lambda)$ belongs to the algebra $\mathfrak{G}_c(m_\beta^+)$. Then the map $\mathfrak{U} \mapsto \Phi$ induces an isomorphism of the quotient algebra $\mathfrak{G}_c(m_\beta^+)/\mathcal{K}_0(m_\beta^+)$ onto $C(S^{n-1})$.*

P r o o f. Put $\mathfrak{B}(\lambda) = \Sigma_j \Pi_k \mathfrak{B}_{jk}(\lambda)$, where each of the operators $\mathfrak{B}_{jk}(\lambda)$ is either the operator $E(\lambda)^{-1}\Phi_{jk}(\omega)E(\lambda)$ or the operator $E(\bar\lambda)^{-1}\Phi_{jk}(\omega)E(\bar\lambda) = (E(\lambda)^{-1}\overline{\Phi_{jk}(\omega)}E(\lambda))^*$. Using the relation $E(\bar\lambda)^{-1}\Phi_{jk}(\omega)E(\bar\lambda) = E(\lambda)^{-1}\Phi_{jk}(\omega)E(\lambda) + T(\lambda)$ with $T(\lambda) \in \mathcal{K}_0(m_\beta^+)$ (proposition 3.2.2), we find

$$\mathfrak{B}(\lambda) = E(\lambda)^{-1}\sum_j\prod_k \Phi_{jk}(\omega)E(\lambda) + S(\lambda), \quad S \in \mathcal{K}_0(m_\beta^+).$$

This and (7.3) imply

$$\inf_{T \in \mathcal{K}_0(m_\beta^+)} \sup_{\lambda \in m_\beta^+} \|\mathfrak{B}(\lambda) + T(\lambda)\| \geqslant \left\| \sum_j\prod_k \Phi_{jk}; C(S^{n-1}) \right\|.$$

The opposite inequality is contained in theorem 3.6.5. ■

Put $\hat{\mathfrak{G}}_c(m_\beta) = S^{n-1} \cup m_\beta^+$. We introduce a topology on $\hat{\mathfrak{G}}_c(m_\beta)$. A fundamental system of neighborhoods of a point $\omega \in S^{n-1}$ is formed by sets $\mathcal{V}(\omega) \cup m_\beta^+$, where $\mathcal{V}(\omega)$ is a neighborhood of ω on S^{n-1}; a fundamental system of neighborhoods of a point $\lambda \in m_\beta^+$ is formed by open intervals on m_β^+ containing λ.

T h e o r e m 7.10. *Let* $\lambda \mapsto \mathfrak{U}(\lambda) = \mathrm{E}(\lambda)^{-1}\Phi(\omega)\mathrm{E}(\lambda)$ *be an operator-function from the algebra* $\mathfrak{G}_c(m_\beta)$. *The following assertions hold:*

1) *the maps* $\pi(\omega): \mathfrak{U} \mapsto \Phi(\omega)$, $\omega \in S^{n-1}$, *and* $\pi(\lambda): \mathfrak{U} \mapsto \mathfrak{U}(\lambda)$, $\lambda \in m_\beta^+$, *induce pairwise inequivalent irreducible representations of* $\mathfrak{G}_c(m_\beta)$;

2) *every irreducible representation of* $\mathfrak{G}_c(m_\beta)$ *is equivalent to a representation listed in* 1);

3) *the correspondence* $\pi(\omega) \mapsto \omega$, $\pi(\lambda) \mapsto \lambda$, $\lambda \in m_\beta^+$, *defines a bijection of the spectrum of* $\mathfrak{G}_c(m_\beta)$ *onto* $\hat{\mathfrak{G}}_c(m_\beta)$. *The topology on* $\hat{\mathfrak{G}}_c(m_\beta)$ *coincides with the Jacobson topology.*

The p r o o f of this assertion differs little from the proof of theorem 4.6 (we do not need lemmas 4.7 and 4.8 in this case), and is left to the reader. ■

3. The algebras $\mathfrak{C}(\lambda), \mathfrak{C}(m_\beta)$ related with a conical point. These algebras play the same role for a conical point as the algebras $\mathfrak{G}(\lambda)$, $\mathfrak{G}(l_x)$ in §3 and §4 for a nonsingular point $x \in \mathfrak{M}$. Irrespective of the fact that the real line l_x is replaced by a line $m_\beta = \{\lambda \in \mathbf{C}: \mathrm{Im}\lambda = \beta\}$ (the case $\beta = 0$ is not excluded), the dependence of symbols on two vectors - the tangent and the cotangent (and not just the cotangent) vector - makes the algebras $\mathfrak{C}(\lambda)$, $\mathfrak{C}(m_\beta)$ resemble more the algebras $\mathfrak{G}(\lambda)$, $\mathfrak{G}(l_x)$ mentioned than the algebras from the previous sections **1** or **2** of this paragraph. If $\mathfrak{C}(\lambda)$, $\mathfrak{C}(m_\beta)$ are considered for a nonsingular point $x \in \mathfrak{M}$ (they differ from $\mathfrak{G}(\lambda)$, $\mathfrak{G}(l_x)$ by the replacement of l_x by m_β only), the results of §3, §4 can be literally transferred to $\mathfrak{C}(\lambda)$, $\mathfrak{C}(m_\beta)$. The study of $\mathfrak{C}(\lambda)$, $\mathfrak{C}(m_\beta)$ for a conical point reduce also to this situation (in view of definition 4.2.1); it requires almost no additions to the assertions in §3, §4. Therefore we confine ourselves here to statements only (for a more detailed account see [57]).

Let X be a conical manifold, and Y a fixed, smooth, directing submanifold with measure ν. Denote by $\mathfrak{C}(\lambda)$ the algebra generated by all operators $\mathfrak{U}(\lambda)$ ($\mathfrak{U}$ a meromorphic pseudodifferential operator of order zero) on $L_2(Y, d\nu)$. For every λ, the identity representation of $\mathfrak{C}(\lambda)$ is irreducible. Every compact operator on $L_2(Y, d\nu)$ belongs to $\mathfrak{C}(\lambda)$.

Let ϕ be a unit vector in the directing cone, T_ϕ^* the cotangent space containing the dual vector ϕ^*, V the set of all pairs (ϕ, θ), where θ is a unit vector in T_ϕ^* for which, moreover, $\phi\theta = 0$. The following relation holds for the operators in $\mathfrak{C}(\lambda)$:

$$\inf_{T \in \mathcal{K}L_2(Y,dv)} \|\mathfrak{U}(\lambda)+T\| = \|\Phi;C(V)\|,$$

where Φ is the symbol of $\mathfrak{U}$. Therefore the map $\mathfrak{C}(\lambda)/\mathcal{K}L_2(Y,dv) \to C(V)$, assigning to a residue class the symbol of an element representing this class, is an isomorphism of the algebra $\mathfrak{C}(\lambda)/\mathcal{K}L_2(Y,dv)$ onto the algebra $C(V)$ of continuous functions.

Put $\hat{\mathfrak{C}}(\lambda) = V \cup \{\lambda\}$. We will assume that $\{\lambda\}$ is an open set, while a fundamental system of neighborhoods of a point $(\phi,\theta) \in V$ is constituted by sets of the form $\mathcal{W}(\phi,\theta) \cup \{\lambda\}$, where $\mathcal{W}(\phi,\theta)$ is a neighborhood of (ϕ,θ) on V.

Theorem 7.11. *Every irreducible representation of the algebra $\mathfrak{C}(\lambda)$ is either one-dimensional or the identity representation $e(\lambda)$. Any one-dimensional representation can be represented as a map $\pi(\phi,\theta):\mathfrak{U} \to \Phi(\phi,\theta)$, where Φ is the symbol of the class $[\mathfrak{U}] \in \mathfrak{C}(\lambda)/\mathcal{K}L_2(Y,dv)$, $(\phi,\theta) \in V$. The correspondence $\pi(\phi,\theta) \mapsto (\phi,\theta)$, $e(\lambda) \mapsto \lambda$ defines a bijection of the spectrum of $\mathfrak{C}(\lambda)$ onto $\hat{\mathfrak{C}}(\lambda)$. The topology on $\hat{\mathfrak{C}}(\lambda)$ coincides with the Jacobson topology.*

Introduce the algebra $\mathfrak{C}(m_\beta)$ of operator-functions on the line $m_\beta = \{\lambda \in \mathbf{C}: \operatorname{Im}\lambda = \beta\}$, generated by the meromorphic pseudodifferential operators of order zero, with norm $\|\mathfrak{U};\mathfrak{C}(m_\beta)\| = \sup_\lambda \|\mathfrak{U}(\lambda);L_2(Y,dv) \to L_2(Y,dv)\|$. Let $S^*(X)$ be the cotangent sphere bundle over $X \setminus 0$, and let $S^*|_Y$ be the part of $S^*(X)$ above the directrix Y of the cone X. Put $\hat{\mathfrak{C}}(m_\beta) = S^*|_Y \cup m_\beta$. We define a topology on $\hat{\mathfrak{C}}(m_\beta)$. A fundamental system of neighborhoods of a point $\lambda \in m_\beta$ is constituted by the open intervals containing λ. Let $\phi \in Y$ and ω a unit vector in T_ϕ^*, i.e. $(\phi,\omega) \in S^*|_Y$. It is obvious that ϕ may be regarded to be a tangent vector to X. If $\phi\omega > 0$ (respectively, $\phi\omega < 0$), a fundamental system of neighborhoods of (ϕ,ω) is constituted by sets $\mathcal{W} \cup \{\lambda \in m_\beta: \operatorname{Re}\lambda > N\}$ (respectively, $\mathcal{W} \cup \{\lambda \in m_\beta: \operatorname{Re}\lambda < N\}$), where $\mathcal{W}$ is a neighborhood of (ϕ,ω) on $S^*|_Y$ and N is a real number. If $\phi\omega = 0$, a fundamental system of neighborhoods is constituted by sets $\mathcal{W} \cup m_\beta$.

Theorem 7.12. *Let $\mathfrak{U}$ be an element of the algebra $\mathfrak{C}(m_\beta)$, and let Φ be its symbol. The following assertions hold.*

1) *The maps $\pi(\phi,\omega):\mathfrak{U} \to \Phi(\phi,\omega)$ and $\pi(\lambda):\mathfrak{U} \to \mathfrak{U}(\lambda)$ induce irreducible representations of $\mathfrak{C}(m_\beta)$. Any two of them are inequivalent.*

2) *Every irreducible representation of* $\mathfrak{C}(m_\beta)$ *is equivalent to a representation listed in* 1).

3) *The correspondence* $\pi(\phi,\cdot)\mapsto(\phi,\omega)$, $\pi(\lambda)\mapsto\lambda$ *defines a bijection of the spectrum of* $\mathfrak{C}(m_\beta)$ *onto* $\hat{\mathfrak{C}}(m_\beta)$. *The topology on* $\hat{\mathfrak{C}}(m_\beta)$ *coincides with the Jacobson topology.*

4. The main theorem. In this section we state the main theorem. Let $\mathfrak{M}$ be a $\check{C}^\infty$ manifold with finite set $Q = \{x^{(1)}, \cdots, x^{(l)}\}$ of conical points (cf. Chapter 4, §3.1). For symplicity of description we will assume that in a neighborhood of a point $x^{(j)} \in Q$ the manifold $\mathfrak{M}$ coincides with a cone X_j, $j = 1, \cdots, l$. Introduce the set $\mathfrak{R}$ of functions Φ that are given, and are smooth, on $S^*(\mathfrak{M}\setminus Q)$ except at a set $S^*(\mathfrak{M}\setminus Q)\,|\,\operatorname{sing}\Phi$; here $S^*(\mathfrak{M}\setminus Q)$ is the bundle of cotangent unit vectors over $\mathfrak{M}\setminus Q$, and $\operatorname{sing}\Phi$ is a finite subset of $\mathfrak{M}\setminus Q$ depending on Φ. At the points $x \in \operatorname{sing}\Phi$ the function Φ is allowed to have discontinuities of the first kind (cf. Chapter 4, §4.1). In a neighborhood of a conical point $x^{(j)} \in Q$ a function in $\mathfrak{R}$ is a smooth map $(\phi,\theta)\mapsto\Phi(\phi,\theta)$, where ϕ is a unit vector in the directrix of the cone X_j, θ is a unit vector in T^*_ϕ, and T^*_ϕ is the cotangent space containing the vector ϕ^* (dual for ϕ). We will also assume that $\operatorname{sing}\Phi \cap P = \varnothing$ for a fixed finite set $P = \{y^{(1)}, \cdots, y^{(k)}\} \subset \mathfrak{M}\setminus Q$ and any $\Phi \in \mathfrak{R}$. We may assume that functions in $\mathfrak{R}$ are given on the disjoint union $\hat{\mathfrak{R}}$ of the Whitney sum $S(\mathfrak{M}\setminus(P\cup Q))\oplus S^*(\mathfrak{M}\setminus(P\cup Q))$, the cotangent spheres $S^*(\mathfrak{M})_{y^{(j)}}$, $j = 1, \cdots, k$, and the bundles $S^*(X_j)\,|\,_{Y_j}$, $j = 1, \cdots, l$, where Y_j is a directing submanifold of X_j.

We introduce a topology on $\hat{\mathfrak{R}}$. Neighborhoods constituting a fundamental system of neighborhoods will be called *typical*. If $x_0 \notin P \cup Q$, a fundamental system of neighborhoods of (x_0,ϕ_0,θ_0), where $\phi_0 \in S(\mathfrak{M})_{x_0}$, $\theta_0 \in S^*(\mathfrak{M})_{x_0}$, is defined as in the topology $\mathfrak{T}(S,S^*)$ (§2). A typical neighborhood of a point $(y^{(j)},\theta_0)$ (where $y^{(j)} \in P$, $\theta_0 \in S^*(\mathfrak{M})_{y^{(j)}}$) is the union of sets $\{y^{(j)}\}\times\mathfrak{W}(\theta_0)$ and $(\mathfrak{U}\setminus\{y^{(j)}\})\times S^{n-1}\times\mathfrak{W}(\theta_0)$ (where $\mathfrak{U}$ is a neighborhood of $y^{(j)}$ on $\mathfrak{M}$; the part of $\hat{\mathfrak{R}}$ above $\mathfrak{U}\setminus\{y^{(j)}\}$ is regarded as the product $(U\setminus\{y^{(j)}\})\times S^{n-1}\times S^{n-1}$, and $\mathfrak{W}(\theta_0)$ is a neighborhood of θ_0 on the 'cotangent' sphere). Let now $x^{(j)} \in Q$, $\phi_0 \in Y_j$, and θ_0 be a point of the cotangent sphere $S^*_{x^{(j)},\phi_0}$ containing ϕ_0^*. Denote by $\mathfrak{V}(\phi_0)$ a neighborhood of ϕ_0 on Y_j, and by $K(\phi_0)$ the set of points $x \in X_j$ belonging to the directrix of X_j with directrices from $\mathfrak{V}(\phi_0)$ and such that $0<\operatorname{dist}(x,x^{(j)})<\epsilon$. By $\mathfrak{W}(\theta_0)$ we mean a neighborhood of θ_0 on $S^*_{x^{(j)},\phi_0}$. A *typical neighborhood* of a point $(x^{(j)},\phi_0,\theta_0)$ is the union of sets $\{x^{(j)}\}\times\mathcal{V}(\phi_0)\times\mathcal{W}(\theta_0)$ and $K(\phi_0)\times S^{n-1}\times\mathcal{W}(\theta_0)$. This topology on $\hat{\mathfrak{M}}$ is the weakest in which all functions in $\mathfrak{M}$ are continuous. Moreover, $\hat{\mathfrak{M}}$ is the maximal ideal space of the uniform algebra generated by the functions in $\mathfrak{M}$.

Let $\beta=(\beta_1,\cdots,\beta_l)$ and $\delta=(\delta_1,\cdots,\delta_k)$ be vectors with real coordinates, $\delta_j\neq\pm(q+n/2)$, $q=0,1,\cdots$. Denote by $H_{\beta,\delta}(\mathfrak{N})$ the space of functions on $\mathfrak{N}$, with norm defined by a partition of unity as follows. For a function u with support in a neighborhood of a conical point $x^{(j)}$ the norm in $H_{\beta,\delta}(\mathfrak{N})$ coincides with the norm in $H_{\beta_j}(X_j)$; for a u with support in a neighborhood of $y^{(j)}\in P$,

$$\|u;H_{\beta,\delta}(\mathfrak{N})\|=\left[\int_{\mathbf{R}^n}|x|^{2\delta_j}|(u\circ\kappa^{-1})(x)|^2dx\right]^{1/2};$$

finally, for a function u with support in a coordinate neighborhood, $\operatorname{supp}(u)\cap(P\cup Q)=\varnothing$, the norm in $H_{\beta,\delta}(\mathfrak{N})$ is equivalent to the norm of $u\circ\kappa^{-1}$ in $L_2(\mathbf{R}^n)$ (here, as before, κ is a coordinate homeomorphism). Let $\mathcal{Q}$ be the algebra generated on $H_{\beta,\delta}(\mathfrak{N})$ by the pseudodifferential operators of order zero whose symbols belong to $\mathfrak{M}$.

We introduce the space $\hat{\mathfrak{S}}$ (which is homeomorphic to the spectrum of the quotient algebra $\mathcal{Q}/\mathcal{K}H_{\beta,\delta}(\mathfrak{N})$). It is the union of the set $\hat{\mathfrak{M}}$ and sets l_x enumerated by the points $x\in\mathfrak{N}$. Here $l_x=\{\lambda\in\mathbf{C}\colon \operatorname{Im}\lambda=\beta_j\}$ for $x=x^{(j)}$, $l_x=\{\lambda\in\mathbf{C}\colon \operatorname{Im}\lambda=\delta_j,\operatorname{Re}\lambda\geqslant0\}$ for $x=y^{(j)}$, and, finally, l_x is identified with $\mathbf{R}$ if $x\in\mathfrak{N}\setminus(P\cup Q)$. We describe a topology on $\hat{\mathfrak{S}}$. Let $x_0\in\mathfrak{N}\setminus(P\cup Q)$, $(\phi_0,\theta_0)\in S(\mathfrak{N})_{x_0}\times S^*(\mathfrak{N})_{x_0}$, with, moreover, $\phi_0\theta_0=0$. A typical neighborhood of (x_0,ϕ_0,θ_0) in $\hat{\mathfrak{S}}$ is the union of a neighborhood U of this point in $\hat{\mathfrak{M}}$ and the set of lines l_x, $x\in p(U)$, where $p(U)$ is the projection of U on $\mathfrak{N}$. If $\phi_0\theta_0>0$ (respectively, $\phi_0\theta_0<0$), l_{x_0} is replaced by a set $\{\lambda\in l_{x_0}\colon \lambda>N\}$ (respectively, $\{\lambda\in l_{x_0}\colon \lambda>N\}$). Neighborhoods of a point $(x^{(j)},\phi_0,\theta_0)$ with $x^{(j)}\in Q$ are described similarly. A typical neighborhood of a point $(y^{(j)},\theta_0)$ (here $y^{(j)}\in P$, $\theta_0\in S^*(\mathfrak{N})_{y^{(j)}}$) is the union of a neighborhood U of $(y^{(j)},\theta_0)$ in $\hat{\mathfrak{M}}$, the set of lines l_x, $x\in p(U)\setminus\{y^{(j)}\}$, and the ray $l_{y^{(j)}}$. On the lines l_x and rays $l_{y^{(j)}}$ the ordinary topology is taken.

T h e o r e m 7.13. *Let* A *be a pseudodifferential operator with symbol* $\Phi\in\mathfrak{M}$, *and let* $\mathfrak{U}$ *be the operator symbol of* A *(cf. Chapter* 4, §4.4*). The following assertions* *hold.*

1) *The maps* $\pi(m): A \mapsto \Phi(m)$, $m \in \hat{\mathfrak{M}}$, $\pi(x,\lambda): A \mapsto \mathfrak{U}(x,\lambda)$, $\lambda \in l_x$, $x \in \mathfrak{M}$, *induce pairwise inequivalent irreducible representations of the algebra* $\mathfrak{A}/\mathcal{KH}_{\beta,\delta}(\mathfrak{M})$.

2) *Every irreducible representation of* $\mathfrak{A}/\mathcal{KH}_{\beta,\delta}(\mathfrak{M})$ *is equivalent to a representation listed in* 1).

3) *The correspondence* $\pi(m) \mapsto m$, $\pi(x,\lambda) \mapsto \lambda \in l_x$ *is a bijection of the spectrum of* $\mathfrak{A}/\mathcal{KH}_{\beta,\delta}(\mathfrak{M})$ *onto* $\hat{\mathfrak{S}}$. *The topology on* $\hat{\mathfrak{S}}$ *is the Jacobson topology.*

The considerations concerning local algebras, given in **1 - 3**, allow us to use in order to prove this theorem the reasoning given in the proof of theorem 2.1 (§5), with minor modifications. This is left to the reader.

5. The spectrum of an algebra of singular integral operators on a composite contour. We retain the notations from Chapter 4, §4.5, and that of theorem 2.5. Let γ_μ be an arc of a contour Γ. Two copies of it with initial point deleted are denoted by γ_μ^{--} and γ_μ^{-+}, and two other copies with end point deleted are denoted by γ^{+-} and γ_μ^{++}. We also introduce the set $\{l_z\}$ of lines, parametrized by the points of Γ. For every regular z, the line l_z is represented by the line $\mathrm{Im}\,\lambda = 0$ in the complex plane, and for $z = z_\nu$, l_z is understood to be the line $\mathrm{Im}\,\lambda = \beta_\nu$. The disjoint union $\cup_\mu\{\gamma_\mu^{--} \cup \gamma_\mu^{-+} \cup \gamma_\mu^{+-} \cup \gamma_\mu^{++}\} \cup \{\cup_{z\in\Gamma} l_z\}$ is denoted by $\hat{\mathfrak{S}}$. The topology on $\hat{\mathfrak{S}}$ is defined as in §2 (above theorem 2.5); moreover, if z is the initial (end) point of an arc, we can only speak of the points z^{++} and z^{+-} (respectively, z^{-+} and z^{--}).

T h e o r e m 7.14. *Let* A *be a singular integral operator on* $L_2(\Gamma,\rho)$ *with coefficients a and b, let* $\mathfrak{U}$ *be its (operator-) symbol, and let* $\gamma_{\nu,j}$ *be arcs intersecting at cusps* z_ν, $j = 1,\cdots,N_\nu$, $\nu = 1,\cdots,M$. *The following assertions hold.*

1) *The maps* $\pi(z,+,\pm): A \mapsto a_+(z) \pm b_+(z)$, $\pi(z,-,\pm): A \mapsto a_-(z) \pm b_-(z)$, $\pi(z,\lambda): A \mapsto \mathfrak{U}(z,\lambda)$, $\lambda \in l_z$, *where z is a regular point of the contour* Γ, *and the maps* $\pi(z_{\nu,j},\pm): A \mapsto a_j(z_\nu) \pm b_j(z_\nu)$, $j = 1,\cdots,N_\nu$, $\pi(z_\nu,\lambda): A \mapsto \mathfrak{U}(z_\nu,\lambda)$, $\lambda \in l_{z_\nu}$, *where* z_ν *is a cusp,* $\nu = 1,\cdots,M$, *induce pairwise inequivalent irreducible representations of the quotient algebra* $\mathfrak{A}/\mathcal{KL}_2(\Gamma,p)$. *(The representations* $\pi(z,\pm,\pm)$ *are one-dimensional, the* $\pi(z,\lambda)$ *are two-dimensional,* $\pi(z_\nu,\lambda)$ *is* N_ν*-dimensional.)*

2) *Every irreducible representation of $\mathfrak{A}/\mathcal{KL}_2(\Gamma,\rho)$ is equivalent to a representation listed in* 1).

3) *The correspondences $\pi(z,\pm,\pm)\mapsto z^{\pm\pm}$, $\pi(z,\lambda)\mapsto\lambda\in l_z$, where z is a regular point, and, in case of a cusp z_v, the correspondences $\pi(z_{v,j},+)\mapsto z_{v,j}^{++}$, $\pi(z_{v,j},-)\mapsto z_{v,j}^{+-}$, if $\gamma_{v,j}$ is an emanating arc (by $z_{v,j}^{+\pm}$ we understood the initial point of the arc $\gamma_{v,j}^{+\pm}$), or $\pi(z_{v,j},+)\mapsto z_{v,j}^{-+}$, $\pi(z_{v,j},-)\mapsto z_{v,j}^{--}$, if $\gamma_{v,j}$ is an incoming arc (by $z_{v,j}^{-\pm}$ we understand the endpoint of the arc $\gamma_{v,j}^{-\pm}$), $\pi(z_v,\lambda)\mapsto\lambda\in l_{z_v}$ define a bijection of the spectrum of $\mathfrak{A}/\mathcal{KL}_2(\Gamma,\rho)$ onto $\hat{\mathfrak{G}}$. The topology on $\hat{\mathfrak{G}}$ is the Jacobson topology.*

We will now assume that the coefficients of all singular integral operators considered are continuous at a certain regular point $z_0\in\Gamma$. Put $\rho_0(z) = |z-z_0|^{\delta_0}\rho(z)$, where $2\delta_0\in(-1,1)$. Introduce the algebra $\mathfrak{A}_0$ generated by such singular integral operators on $L_2(\Gamma,\rho_0)$. We identify the points z_0^{++} and z_0^{-+}, and denote the new point by z_0^+. The point z_0^- is obtained by identifying z_0^{+-} and z_0^{--}. If $\delta_0\neq 0$, we identify points on l_{z_0} that lie symmetric about the imaginary axis, while if $\delta_0=0$ we discard the line l_{z_0}. The set obtained from $\hat{\mathfrak{G}}$ by these operations will be denoted by $\hat{\mathfrak{G}}_0$; it is endowed with the quotient topology.

T h e o r e m 7.15. *All assertions of theorem 7.14 remain valid if $\mathfrak{A},\hat{\mathfrak{G}}$ are replaced by $\mathfrak{A}_0,\hat{\mathfrak{G}}_0$, the representations $\pi(z_0,+,\pm)$, $\pi(z_0,-,\pm)$ in 1) by the representations $\pi(z_0^+):\mathrm{A}\mapsto a(z_0)+b(z_0)$, $\pi(z_0^-):\mathrm{A}\mapsto a(z_0)-b(z_0)$, and the correspondences $\pi(z_0,\pm,\pm)\mapsto z_0^{\pm\pm}$ in 3) by the correspondences $\pi(z_0^{\pm})\mapsto z_0^{\pm}$.*

Some remarks concerning the p r o o f. The first assertions in theorems 7.14 and 7.15 must be proved without recourse to propositions 3.1, 4.2, etc. Moreover, the first assertion of theorem 7.14 is almost trivial, the inequivalence of representations $\pi(z_v,\lambda)$ for distinct λ can be established by comparing traces. This manner is not suitable for the representations $\pi(z_0,\lambda)$ in theorem 7.15 (for every singular integral operator A, the traces of the matrices $\pi(z_0,\lambda)\mathrm{A}$ for distinct λ coincide). We verify the part of the first assertion of theorem 7.15 that is related to the representations $\pi(z_0,\lambda)$.

We first prove that a representation $\pi(z_0,\lambda)$ is irreducible if and only if $\beta=\operatorname{Im}\lambda\neq 0$ (it is also assumed that $|\operatorname{Im}\lambda|<1$). The algebra $\pi(z_0,\lambda)\mathfrak{A}$ is generated by the matrices

$$\mathfrak{U}(\lambda) = \begin{bmatrix} -\tanh \pi\lambda & -i/\cosh \pi\lambda \\ i/\cosh \pi\lambda & \tanh \pi\lambda \end{bmatrix}.$$

If $\pi(z_0,\lambda)$ is irreducible, then there is a one-dimensional subspace that is invariant under the operators $\mathfrak{U}(\lambda)$ and $\mathfrak{U}(\lambda)^*$. Hence the representation is irreducible only if the operator has two orthogonal eigenvalues. The normalized vectors

$$f_1(\lambda) = (2\cosh \pi\sigma)^{-1/2} \begin{bmatrix} \exp\{i\pi(1/2-i\lambda)/2\} \\ \exp\{-i\pi(1/2-i\lambda)/2\} \end{bmatrix},$$

$$f_2(\lambda) = (2\cosh \pi\sigma)^{-1/2} \begin{bmatrix} \exp\{-i\pi(1/2-i\lambda)/2\} \\ \exp\{i\pi(1/2-i\lambda)/2\} \end{bmatrix},$$

where $\sigma = \mathrm{Re}\lambda$, are eigenvectors for $\mathfrak{U}(\lambda)$ and correspond to the eigenvalues $\mu_1 = 1$ and $\mu_2 = -1$. The scalar product of these vectors equals; for $\beta = 0$ these vectors are clearly orthogonal.

We now prove that for $\beta \neq 0$ two representations $\pi(z_0,\lambda_1)$ and $\pi(z_0,\lambda_2)$ are equivalent if and only if $\mathrm{Re}\lambda_1 = -\mathrm{Re}\lambda_2$. Suppose that the representations are equivalent, i.e. $\pi(\lambda_1)A = U^{-1}\pi(\lambda_2)(A)U$, where U is a unitary operator. Then the vectors $Uf_1(\lambda_1)$ and $Uf_2(\lambda_1)$ are eigenvectors for $\mathfrak{U}(\lambda_2)$ and correspond to the eigenvalues 1 and -1. Hence they can differ from the vectors $f_1(\lambda_2)$ and $f_2(\lambda_2)$ by factors whose moduli are 1 only. Thus, $|(f_1(\lambda_2), f_2(\lambda_2))| = |(f_1(\lambda_1), f_2(\lambda_1))|$; however, this is possible only if $\mathrm{Re}\lambda_1 = -\mathrm{Re}\lambda_2$. If the last equation is true, then the linear operator U mapping $f_j(\lambda_1)$ to $f_j(\lambda_2)$, $j = 1,2$, is unitary. Since the $f_j(\lambda)$ are eigenvectors for $\mathfrak{U}(\lambda)$, U is intertwining for the representations $\pi(\lambda_1)$ and $\pi(\lambda_2)$.

Note that similar considerations may be used also in the higher-dimensional situation in order to prove proposition 7.6. (The eigenvector of the operator $E_{\omega\to\phi}(\lambda)^{-1}\Phi(\omega)E_{\psi\to\omega}(\lambda)$ corresponding to an eigenvalue $\Phi(\omega_0)$ has the form $E_{\psi\to\omega}(\lambda)^{-1}\delta(\psi,\omega_0)$, where $\delta(\psi,\omega_0)$ is the Dirac-function on S^{n-1} concentrated at ω_0.)

It is left to the reader to become convinced of the fact that for $\beta = 0$ all representations $\pi(z_0,\lambda)$ are equivalent (and split in a sum of two one-dimensional representations).

The proof of assertions 2) and 3) in theorems 7.14 and 7.15 can be given along the general scheme (cf. §5).

§8. The spectrum of a C^*-algebra of pseudodifferential operators with oscillating symbols

1. The algebra $\mathfrak{S}(\lambda)$. Let $\Pi = d \times S^{n-1}$, d the interval $[1,e]$ with identified ends, and let $\Phi \in C^\infty(d \times S^{n-1} \times S^{n-1})$. Denote by $\mathfrak{S}(\lambda)$, $\lambda \in \mathbb{R}$, the C^*-algebra generated on $L_2(\Pi)$ by the operators

$$(\mathbf{A}(\lambda)u)(r,\phi) = \sum_{q=-\infty}^{+\infty} r^{2\pi i q}\mathfrak{U}_{\psi\to\phi}(r,\lambda+2\pi q)u_q(\psi), \tag{8.1}$$

where

$$u_q(\psi) = \int_1^l r^{-2\pi i q - 1}u(r,\psi)dr.$$

Note that the operator $\mathbf{A}(\lambda)^*$, adjoint to $\mathbf{A}(\lambda)$ with respect to the duality

$$<u,v> = \int_1^l \frac{dr}{r}\int_{S^{n-1}} u(r,\phi)\overline{v(r,\phi)}d\phi,$$

differs from the operator $u \mapsto \Sigma_q r^{2\pi i q}\mathrm{E}(\lambda+2\pi q)\overline{\Phi(r,\phi,\omega)}\mathrm{E}(\lambda+2\pi q)u_q$ by a compact term only (theorem 3.7.5).

P r o p o s i t i o n s 8.1. *For every $\lambda \in \mathbb{R}$ the algebra $\mathfrak{S}(\lambda)$ is irreducible.*

P r o o f. Irreducibility of the algebra is equivalent to the fact that every nonzero vector $u \in L_2(\Pi)$ is totalizing (§1.4). Let $u_{q_0} \in L_2(S^{n-1})$ be a nonzero Fourier coefficient of a vector u. Denote, as before, by $\mathcal{K}_0(\mathbb{R})$ the algebra of functions on $\mathbb{R}$ with as values compact operators in $L_2(S^{n-1})$ and tending to zero at infinity. Take for $\mathfrak{U}$ in (8.1) the element of $\mathcal{K}_0(\mathbb{R})$ such that $\mathfrak{U}(\lambda+2\pi q_0) = (\cdot, u_{q_0})u_{q_0}$ and $\mathfrak{U}(\lambda+2\pi q) = 0$ for $q \neq q_0$. By lemma 4.3, the corresponding operator $\mathbf{A}(\lambda)$ belongs to $\mathfrak{S}(\lambda)$. Clearly, $\mathbf{A}(\lambda)u = r^{2\pi i q_0}u_{q_0}$. The algebra generated on $L_2(S^{n-1})$ by the operators $\mathrm{E}(\lambda)^{-1}\Phi(\phi,\omega)\mathrm{E}(\lambda)$ is irreducible, for every $\lambda \in \mathbb{R}$ (proposition 3.1). The operator of multiplication by $r^{2\pi i q}$ belongs to $\mathfrak{S}(\lambda)$. This and all that has been said above imply that u is totalizing for $\mathfrak{S}(\lambda)$. ■

C o r o l l a r y 8.2. *The following inclusion holds: $\mathcal{K}L_2(\Pi) \subset \mathfrak{S}(\lambda)$.*

P r o o f. $\mathfrak{S}(\lambda)$ contains nonzero compact operators (commutators; cf. theorem

3.7.4). It remains to apply proposition 1.5. ■

Let $\mathbf{A}_{jk}(\lambda)$ be the operator (8.1) in which $\mathfrak{U}_{jk}(r,\lambda) = \mathrm{E}(\lambda)^{-1}\Phi_{jk}(r,\phi,\omega)\mathrm{E}(\lambda)$ plays the role of $\mathfrak{U}(r,\lambda)$. Put $\mathbf{A} = \Sigma_j\Pi_k\mathbf{A}_{jk}$, $\Phi = \Sigma_j\Pi_k\Phi_{jk}$ (j and k run over finite sets).

Proposition 8.3. *The following equation holds:*

$$\inf_{\mathrm{T}\in\mathcal{K}L_2(\Pi)} \|\mathbf{A}+\mathrm{T}\| = \|\Phi;C(d\times S^{n-1}\times S^{n-1})\|. \tag{8.2}$$

Hence the map $\mathbf{A}\mapsto\Phi$ establishes an isomorphism between the quotient algebra $\Theta(\lambda)/\mathcal{K}L_2(\Pi)$ and $C(d\times S^{n-1}\times S^{n-1})$.

Proof. We first show that the following inequality holds:

$$\inf_{\mathrm{T}}\|\mathbf{A}+\mathrm{T}\| \geqslant \|\Phi;C(d\times S^{n-1}\times S^{n-1})\|. \tag{8.3}$$

Let (r_0,ϕ_0,ω_0) be an arbitrary fixed point. Denote by $\{u^{(n)}\}$ a sequence of smooth functions on the sphere for which $\|u^{(n)};L_2(S^{n-1})\| = 1$, and for which the supports $\operatorname{supp} u^{(n)}$ contract towards the point ϕ_0. It is obvious that $\{u^{(n)}\}$ weakly converges to zero in $L_2(S^{n-1})$. Let also $\{v^{(n)}\}$ be a sequence of functions

$$v^{(n)}(r) = \sum_{k=-n}^{n} c_k^{(n)} r^{2\pi ik}, \qquad \int_1^l |v^{(n)}(r)|^2\frac{dr}{r} = 1,$$

with $|v^{(n)}(r)|^2/r\to\delta(r-r_0)$, where δ is the Dirac δ-function. (The Fejér kernel can be used to construct a sequence $\{v^{(n)}\}$.)

We will assume that $\phi_0\omega_0 \neq 0$. If $\phi_0\omega_0 > 0$ (respectively, $\phi_0\omega_0 < 0$), we choose a function g, smooth on $\mathbb{R}^n\setminus 0$ and homogeneous of degree zero, and a sequence $\{\mu_q\}$ of numbers $\mu_q\to+\infty$, such that the vector $\mu_q\nabla g(\phi_0)+(\lambda+2\pi q)\phi_0$ is parallel to ω_0 for all $q = 1,2,\cdots$ (respectively, $q = -1,-2,\cdots$). (Note that since g is homogeneous, vectors $\nabla g(\phi)$ and ϕ are orthogonal.) Theorem 3.5.1 implies that for $u \in C^\infty(S^{n-1})$,

$$(e^{i\mu_q g}\mathfrak{U}(r,\lambda+2\pi q)e^{i\mu_q g}u)(\phi)+$$
$$-\Phi(r,\phi,\mu_q\nabla g(\phi)+(\lambda+2\pi q)\phi)u(\phi) = o(1)$$

as $|q|\to\infty$, uniformly in $\phi\in S^{n-1}$, $r\in d$. Let $\{\epsilon_n\}$ be an arbitrary positive null sequence, and let $w^{(n)}(r,\psi) = u^{(n)}(\psi)v^{(n)}(r)e^{i\mu_q g(\psi)}r^{-2\pi iq}$, where $q = q(n)$ is chosen such that the inequality

$$\left| \sum_{k=q-n}^{q+n} r^{2\pi ik}\mathfrak{U}_{\psi\to\phi}(r,\lambda+2\pi k)w_k^{(n)}(\psi) - \Phi(r,\phi,\tau(\lambda,q,\phi))w^{(n)}(r,\phi) \right| < \epsilon_n$$

holds for all r and ϕ; here, $\tau(\lambda,q,\phi) = \mu_q \nabla g(\phi)+(\lambda+2\pi q)\phi$. This inequality can be rewritten as

$$|((\mathbf{A}(\lambda)-\mathrm{T})w^{(n)})(r,\phi) - \Phi(r,\phi,\tau(\lambda,q,\phi))w^{(n)}(r,\phi)| < \epsilon_n.$$

Since

$$\int |\Phi(r,\phi,\tau(\lambda,q,\phi))w^{(n)}(r,\phi)|^2 d\phi \frac{dr}{r} \to |\Phi(r_0,\phi_0,\omega_0)|^2,$$

we have $\|(\mathbf{A}-\mathrm{T})w^{(n)};L_2(\Pi)\| \to |\Phi(r_0,\phi_0,\omega_0)|$. The sequence $\{w^{(n)}\}$ is weakly convergent to zero in $L_2(\Pi)$, hence $\{\mathrm{T}w^{(n)}\}$ is norm convergent to zero, for every $\mathrm{T} \in \mathcal{K}L_2(\Pi)$. This implies inequality (8.3).

Now note that if $\inf|\Phi| > 0$, then the bounded operator $\mathbf{R}(\lambda):L_2(\Pi)\to L_2(\Pi)$, $\mathbf{R}(\lambda)u = \Sigma_q r^{2\pi iq}\mathfrak{R}(r,\lambda+2\pi q)u_q$, where $\mathfrak{R}(r,\lambda) = \mathrm{E}(\lambda)^{-1}\Phi(r,\phi,\omega)^{-1}\mathrm{E}(\lambda)$, is an inverse for $\mathbf{A}(\lambda)$, up to a compact term. This implies that $\operatorname{supp}\sigma_{\mathrm{ess}}(\mathbf{AA}^*) < \|\Phi;C(d\times S^{n-1}\times S^{n-1})\|$ (compare with lemma 3.6). Together with (8.3) this gives (8.2). ■

2. The algebra $\mathfrak{O}(S)$. Let S be a circle, which can be represented by the interval $0\leqslant\lambda\leqslant 2\pi$ with ends identified. Introduce the algebra $\mathfrak{O}(S)$ generated by the operator-functions $S\ni\lambda\mapsto\mathbf{A}(\lambda)$ of the form (8.1), with norm $\|\mathbf{A};\mathfrak{O}(S)\| = \sup_\lambda\|\mathbf{A}(\lambda);L_2(\Pi)\to L_2(\Pi)\|$. By $\pi(\lambda)$ we denote the representation $\mathbf{A}\mapsto\mathbf{A}(\lambda)$ of $\mathfrak{O}(S)$ in $L_2(\Pi)$.

P r o p o s i t i o n 8.4. *If λ,μ are distinct points on S, then $\pi(\lambda)$ and $\pi(\mu)$ are inequivalent.*

P r o o f. Suppose there is a unitary operator $\mathrm{V}:L_2(\Pi)\to L_2(\Pi)$ such that $\mathrm{V}^{-1}\mathbf{A}(\mu)\mathrm{V} = \mathbf{A}(\lambda)$ for all $\mathbf{A}\in\mathfrak{O}(S)$. Let also P_q be projection of $L_2(\Pi)$ onto the subspace H_q of functions $(r,\phi)\mapsto r^{2\pi iq}u(\phi)$, $u\in L_2(S^{n-1})$. Consider the subalgebra of $\mathfrak{O}(S)$ generated by the operator-functions (8.1) in which

$\mathfrak{U}(\lambda) = \mathrm{E}(\lambda)^{-1}\Phi(\phi,\omega)\mathrm{E}(\lambda)$ (Φ independent of r). For the elements of this subalgebra we have $\mathrm{V}^{-1}\mathbf{A}(\mu)\mathrm{V}\mathrm{P}_0 = \mathfrak{U}(\lambda)\mathrm{P}_0$, hence $\mathbf{A}(\mu)\mathrm{V}\mathrm{P}_0 = \mathrm{V}\mathrm{P}_0\mathfrak{U}(\lambda)\mathrm{P}_0$. Letting P_q act on both sides gives

$$\mathfrak{U}(\mu+2\pi q)\mathrm{P}_q\mathrm{V}\mathrm{P}_0 = \mathrm{P}_q\mathrm{V}\mathrm{P}_0\mathfrak{U}(\lambda)\mathrm{P}_0. \tag{8.4}$$

Replace $\mathbf{A}$ by the adjoint $\mathbf{A}^*$,

$$\mathbf{A}(\nu)^*v = \sum_q r^{2\pi i q}\mathfrak{U}(\nu+2\pi q)^*v_q.$$

Then (8.4) becomes

$$\mathfrak{U}(\mu+2\pi q)^*\mathrm{P}_q\mathrm{V}\mathrm{P}_0 = \mathrm{P}_q\mathrm{V}\mathrm{P}_0\mathfrak{U}(\lambda)^*\mathrm{P}_0. \tag{8.5}$$

Consider the representations $\mathfrak{U}\mapsto\mathfrak{U}(\lambda)$, $\mathfrak{U}\mapsto\mathfrak{U}(\mu+2\pi q)$ in $L_2(S^{n-1})$ of the algebra generated by the operator-functions $R\ni\nu\mapsto\mathfrak{U}(\nu)$. Equations (8.4), (8.5) imply that either these representations are equivalent, or $\mathrm{P}_q\mathrm{V}\mathrm{P}_0 = 0$. Equivalence of these representations means that $\mu+2\pi q = \lambda$ (proposition 4.2), and since both λ and μ belong to S, we find $q = 0$ and $\lambda = \mu$. But, by requirement, $\lambda\neq\mu$, hence $\mathrm{P}_q\mathrm{V}\mathrm{P}_0 = 0$ for all q. The last equation contradicts the unitarity of the operator V. ■

The verification of the following assertion is similar to the proof of lemma 4.3.

Proposition 8.5. *The following inclusion holds: $\mathfrak{K}(S)\subset\mathfrak{O}(S)$, where $\mathfrak{K}(S)$ is the algebra of continuous functions on S with compact values (in $\mathfrak{K}L_2(\Pi)$).*

Proposition 8.6. *Let $\mathbf{A}(\lambda)$ be an operator of the form* (8.1). *The map $\mathbf{A}\mapsto\Phi$ establishes an isomorphism between the algebras $\mathfrak{O}(S)/\mathfrak{K}(S)$ and $C(d\times S^{n-1}\times S^{n-1})$.*

Proof. Put $\mathbf{B}(\lambda) = \Sigma_j\Pi_k\mathbf{A}_{jk}(\lambda)$, where every operator $\mathbf{A}_{jk}$ is either an operator of the form (8.1) or the adjoint of such an operator. By theorems 3.7.4, 3.7.5, $\mathbf{B}(\lambda)v = \Sigma_q r^{2\pi i q}\mathfrak{B}(r,\lambda+2\pi q)v_q + \mathrm{T}(\lambda)v$, where $\mathfrak{B}(r,\lambda) = \mathrm{E}(\lambda)^{-1}\Sigma\Pi\Phi_{jk}(r,\phi,\omega)\mathrm{E}(\lambda)$ and $\mathrm{T}\in\mathfrak{K}(S)$. This and proposition 8.3 imply the inequality

$$\inf_{\mathrm{T}\in\mathfrak{K}(S)}\ \sup_{\lambda\in S}\|\mathbf{B}(\lambda)+\mathrm{T}(\lambda)\| \geqslant \sup\Big|\sum_j\prod_k\Phi_{jk}\Big|. \tag{8.6}$$

For a fixed $\lambda\in S$ we have, by (8.2),

$$\inf_{\mathbf{T}\in\mathcal{K}(S)} \|\mathbf{B}(\lambda)+\mathbf{T}(\lambda)\| = \sup\left|\sum_j\prod_k \Phi_{jk}\right|.$$

Hence, for any $\epsilon>0$ we can indicate a finite covering of S by open intervals Δ_α and a family of operator-functions $\mathbf{T}_\alpha \in \mathcal{K}(S)$, $\alpha = 1,\cdots,m$, such that $\|\mathbf{B}(\lambda)+\mathbf{T}_\alpha(\lambda)\| \leqslant \sup|\Sigma_j\Pi_k\Phi_{jk}|+\epsilon$ for all $\lambda \in \Delta_\alpha$. Let $\{\zeta_\alpha\}$ be a partition of unity subordinate to the cover $\{\Delta_\alpha\}$. Put $\mathbf{T}(\lambda) = \Sigma\zeta_\alpha(\lambda)\mathbf{T}_\alpha(\lambda)$. Then $\mathbf{T} \in \mathcal{K}(S)$ and $\|\mathbf{B}(\lambda)+\mathbf{T}(\lambda)\| \leqslant \sup|\Sigma_j\Pi_k\Phi_{jk}|+\epsilon$. This means that equality holds in (8.6). ■

3. The algebra of pseudodifferential operators with oscillating symbols. Let $\mathcal{Q}$ be the algebra of pseudodifferential operators introduced in Chapter 4, §4.6. (In the sequel we will use the notations from that section.) Let also $T_0(\mathfrak{M})$ be the bundle of nonzero cotangent vectors over $\mathfrak{M}$, and let G be the multiplicative group of the numbers e^n, n an integer, acting as an automorphism group of $T_0(\mathfrak{M})$ preserving fibers. Denote by $T_0(\mathfrak{M})/G$ the quotient space obtained from $T_0(\mathfrak{M})$ by factorization with respect to the action of G, and by ξ the Whitney sum $(T_0(\mathfrak{M})/G)\oplus S^*(\mathfrak{M})$. To every function $\Phi \in \mathfrak{N}$ a map $\Phi:\zeta\rightarrow\mathbf{C}$ is associated. Locally this map can be regarded as a map $(x,r,\phi,\theta)\mapsto\Phi(x,r,\phi,\theta)$, where $x \in \mathfrak{M}$, $\phi \in S(\mathfrak{M})_x$, $\theta \in S^*(\mathfrak{M})_x$, and r is a point of the interval d. If $x \notin \operatorname{sing}\Phi$ this map is independent of Φ and r.

Put $\hat{\Theta} = \{(T_0(\mathfrak{M})/G)\oplus S^*(\mathfrak{M})\} \cup \mathcal{C}$, where $\mathcal{C}$ is the disjoint union of a set of circles c_x enumerated by the points of $\mathfrak{M}$. Every circle can be regarded as the interval $0\leqslant\lambda\leqslant 2\pi$ with ends identified. We define a topology on $\hat{\Theta}$. The part of $(T_0(\mathfrak{M})/G)\oplus S^*(\mathfrak{M})$ lying above a neighborhood $\mathcal{U}$ of a point $x_0 \in \mathfrak{M}$ will be represented as a set of quadruples (x,r,ϕ,θ). Let $\mathcal{V}(\phi_0)$ and $\mathcal{W}(\theta_0)$ be neighborhoods of the points ϕ_0 and θ_0 on the unit spheres, let $K(\phi_0)$ be a subset of $\mathcal{U}$ whose image under some coordinate map $\kappa:\mathcal{U}\rightarrow\mathbf{R}^n$ is the intersection of an open cone with vertex at $\kappa(x_0)$ and containing all directions from $\kappa'(x_0)(\mathcal{V}(\phi_0))$, and the open n-dimensional ball with center at $\kappa(x_0)$. Let also $Q(r_0)$ be an ϵ-neighborhood of a point r_0 on d. We denote by $K(\phi_0,r_0,\epsilon)$ the subset of points of $K(\phi_0)$ satisfying one of the membership relations $e^m\operatorname{dist}(x_0,x) \in Q(r_0)$, $m = m_0$, $m_0+1,\cdots$. A typical neighborhood in $\hat{\Theta}$ of a point $(x_0,r_0,\phi_0,\theta_0)$ is the union of the sets $\{x_0\}\times Q(r_0)\times\mathcal{V}(\phi_0)\times\mathcal{W}(\theta_0)$, $K(\phi_0,r_0,\epsilon)\times d\times S^{n-1}\times\mathcal{W}(\theta_0)$, and the circles c_{x_0} and c_x, $x \in K(\phi_0,r_0,\epsilon)$. A neighborhood of a point $\lambda \in c_x$ is an open interval containing λ. All these neighborhoods form a base for a topology.

The space $(T_0(\mathfrak{M})/G) \oplus S^*(\mathfrak{M})$ with the topology induced by the topology of $\hat{\Theta}$ will be denoted by $\hat{\mathfrak{N}}$. This topology is the weakest topology in which all functions from $\mathfrak{N}$ are continuous. These functions generate a uniform algebra, for which $\hat{\mathfrak{N}}$ is the maximal ideal space.

Let A be a pseudodifferential operator of order zero, acting on $L_2(\mathfrak{M})$, let $\Phi \in \mathfrak{N}$ be its symbol, and **A** its operator symbol.

Theorem 8.7. *The maps* $\pi(v): \mathrm{A} \mapsto \Phi(v)$, $v \in \hat{\mathfrak{N}}$, *and* $\pi(x,\lambda): \mathrm{A} \mapsto \mathbf{A}(x,\lambda)$, $\lambda \in c_x$, *induce pairwise inequivalent irreducible representations of the quotient algebra* $\mathcal{A}/\mathcal{K}L_2(\mathfrak{M})$. *The correspondence* $\pi(v) \mapsto v$, $\pi(x,\lambda) \mapsto \lambda \in c_x$ *defines a bijection of the spectrum* $(\mathcal{A}/\mathcal{K}L_2(\mathfrak{M}))\hat{}$ *onto* $\hat{\Theta}$. *The topology on* $\hat{\Theta}$ *coincides with the Jacobson topology.* $\mathcal{A}/\mathcal{K}L_2(\mathfrak{M})$ *is a type I algebra.*

Using the results of sections **1** and **2**, the proof of this theorem can be given along the lines of the proof of theorem 2.1. ∎

Chapter 6

The spectrum of a C^*-algebra of pseudodifferential operators on a manifold with boundary

This chapter is devoted to the study of the spectra of algebras generated by pseudodifferential operators on a smooth manifold with boundary. Discontinuities of the first kind are allowed in the symbols of these pseudodifferential operators. Dependence of the spectrum on the discontinuities of the symbols, as well as on the choice of a (weighted) function space, is elucidated.

In §1 - §4 local algebras are studied. In the last §5, theorems concerning the spectrum of an algebra of pseudodifferential operators on $\mathfrak{M}$ are given.

§1. The algebras $\mathfrak{S}_c^+(\lambda)$

1. Integral transforms. We introduce the operators

$$(\Pi^{\mp} f)(s) = \mp\frac{i}{2\pi}\int_{-\infty}^{+\infty}\frac{f(t)dt}{s \mp i0 - t} = \tag{1.1}$$

$$= \frac{f(s)}{2} \mp \frac{i}{2\pi}\text{p.v.}\int_{-\infty}^{+\infty}\frac{f(s)ds}{s-t}.$$

Denote by $\chi_\pm$ the characteristic function of the semi-axis $\mathbf{R}_\pm$, and put $(\mathrm{P}^{\pm}u)(x) = \chi_\pm(x_n)u(x',x_n)$, $x = (x',x_n) \in \mathbf{R}^n$. The following formulas hold:

$$\mathrm{F}_{x\to\xi}(\mathrm{P}^{\pm}u)(x) = \Pi^{\mp}_{\eta_n\to\xi_n}(\mathrm{F}u)(\xi',\eta_n), \tag{1.2}$$

$$\mathrm{F}^{-1}_{\xi\to x}(\mathrm{P}^{\pm}u(x) = \Pi^{\pm}_{y_n\to x_n}(\mathrm{F}^{-1}v)(x',y_n).$$

If w is a function on the sphere S^{n-1}, then, by definition, $(\mathrm{P}^{\pm}w)(\theta) = \chi_\pm(\theta_n)w(\theta',\theta_n)$, $|\theta| = 1$.

P r o p o s i t i o n 1.1. *For* $u \in C_0^\infty(\mathbf{R}^n\setminus\{x{:}x'=0\})$ *we have*

$$\mathrm{P}^{\pm}\mathrm{E}(\lambda)\tilde{u}(in/2+\lambda,\cdot) = \mathrm{E}(\lambda)\mathrm{M}_{|x|\to in/2+\lambda}\Pi^{\pm}_{y_n\to x_n}u(x',y_n), \tag{1.3}$$

$$\mathrm{P}^{\pm}\mathrm{E}(\lambda)^{-1}\tilde{u}(in/2-\lambda,\cdot) = \mathrm{E}(\lambda)^{-1}\mathrm{M}_{|x|\to in/2-\lambda}\Pi^{\mp}_{y_n\to x_n}u(x',y_n).$$

Moreover, in the first equation $\lambda \neq i(k+n/2)$, *and in the second* $\lambda \neq -i(k+n/2)$, *where* $k = 0, 1, \cdots$.

P r o o f. We verify, e.g., the first formula in (1.3). It suffices to do this for λ real (the formula is generalized to other λ by analytic extension). By (1.2.8),

$$\mathrm{P}^{\pm}\mathrm{E}(\lambda)\tilde{u}(in/2+\lambda,\cdot) =$$

$$= \mathrm{P}^{\pm}\mathrm{M}_{|\xi|\to in/2-\lambda}\mathrm{F}_{x\to\xi}(\mathrm{M}^{-1})_{in/2+\lambda\to|x|}\tilde{u}(in/2+\lambda,\cdot) =$$

$$= M_{|\xi|\to in/2-\lambda}\mathrm{P}^{\pm}\mathrm{F}_{x\to\xi}u(x).$$

Since by (1.2), $\mathrm{P}^{\pm}\mathrm{F} = \mathrm{F}\Pi^{\pm}$, we have

$$\mathrm{P}^{\pm}\mathrm{E}(\lambda)\tilde{u}(in/2+\lambda,\cdot) = \tag{1.4}$$

$$= \mathrm{M}_{|\xi|\to in/2-\lambda}\mathrm{F}_{x\to\xi}\Pi^{\pm}_{y_n\to x_n}u(x',y_n).$$

Note that the support of the function $x \mapsto v(x',x_n) = \Pi^{\pm}_{y_n\to x_n}u(x',y_n)$ lies in a cylinder containing $\operatorname{supp} u$; the directrix of this cylinder is parallel to the x_n-axis. It is obvious that every ray emanating from the coordinate origin has compact intersection with the support of v. Hence the Mellin transform $\mathrm{M}_{|x|\to\lambda+in/2}v$ is defined for a ray. Insert the identity operator $\mathrm{M}^{-1}_{\lambda+in/2\to|x|}\mathrm{M}_{|x|\to\lambda+in/2}$ in between F and $\Pi^{\pm}$ in (1.4). It remains to use (1.2.8). ■

P r o p o s i t i o n 1.2. *Let* $u \in C_0^{\infty}(\mathbf{R}^n \setminus \{x:x'=0\})$ *and* $v(x',x_n) = \Pi^{\pm}_{y_n\to x_n}u(x',y_n)$. *Then, for* $\lambda \in \mathbf{R}$,

$$\tilde{v}(in/2+\lambda,\kappa(\psi,\eta)) = (1+\eta^2)^{(n/2-i\lambda)/2} \times \tag{1.5}$$

$$\times \Pi^{\pm}_{\zeta\to\eta}(1+\zeta^2)^{(i\lambda-n/2)/2}\tilde{u}(in/2+\lambda,\kappa(\psi,\zeta)).$$

Here, as in Chapter 1, §6 *the map* $\kappa: S^{n-2}\times\mathbf{R} \to S^{n-1}\setminus\{(0,\pm)\}$ *is defined by* $S^{n-2}\times\mathbf{R} \ni (\mathring{\omega},t) \mapsto \kappa(\mathring{\omega},t) = (\mathring{\omega}(1+t^2)^{-1/2},\, t(1+t^2)^{-1/2}) = \omega \in S^{n-1}$.

P r o o f. We have

$$\tilde{v}(in/2+\lambda,\psi) = \frac{1}{\sqrt{2\pi}} \int_0^\infty r^{-i(in/2-\lambda)-1} dr \times \tag{1.6}$$

$$\times \left[\pm\frac{i}{2\pi}\right] \int_{-\infty}^{+\infty} \frac{dy_n}{x_n \pm i0 - y_n} \times$$

$$\times \frac{1}{\sqrt{2\pi}} \int_{-\infty}^{+\infty} |z|^{i\mu-n/2} \tilde{u}(in/2+\mu, z/|z|) d\mu,$$

where $z = (x', y_n)$, $r = |x|$. Put $\zeta = y_n/r|\psi'|$, $\eta = \psi_n/|\psi'|$. Now (1.6) can be rewritten as

$$\tilde{v}(in/2+\lambda,\kappa(\psi,\eta)) = \frac{1}{\sqrt{2\pi}} \int_0^\infty r^{-i\lambda-1} dr \left[\pm\frac{i}{2\pi}\right] \int_{-\infty}^{+\infty} \frac{d\zeta}{\eta \pm i0 - \zeta} \times$$

$$\times \frac{1}{\sqrt{2\pi}} \int_{-\infty}^{+\infty} \left[\frac{1+\zeta^2}{1+\eta^2}\right]^{(i\mu-n/2)/2} r^{i\mu} \tilde{u}(in/2+\mu, \kappa(\psi,\zeta)) d\mu.$$

By changing the order of integration with respect to the variables ζ and μ we obtain (1.5). ■

For smooth bounded functions on $\mathbb{R}$ we introduce the operators

$$(Q^{\pm}(\lambda)f)(\eta) = (1+\eta^2)^{(n/2-i\lambda)/2} \Pi^{\pm}_{\zeta\to\eta} (1+\zeta^2)^{(i\lambda-n/2)/2} f(\zeta),$$

and denote by $\kappa(Q^{\pm}(\lambda))$ the following operators, acting on functions $u \in C^\infty(S^{n-1})$:

$$\kappa(Q^{\pm}(\lambda))u = [Q^{\pm}(\lambda)(u \circ \kappa)] \circ \kappa^{-1}.$$

The following assertion can be verified by combining propositions 1.1 and 1.2.

Proposition 1.3. *For all $u \in C^\infty(S^{n-1})$ we have*

$$P^{\pm}E(\lambda)u = E(\lambda)\kappa(Q^{\pm}(\lambda))u, \quad P^{\pm}E(\lambda)^{-1}u = E(\lambda)^{-1}\kappa(Q^{\mp}(-\lambda))u.$$

Let $\mathcal{H}^{\pm}(\mathbb{R})$ be a subspace of $L_2(\mathbb{R})$, $\mathcal{H}^{\pm}(\mathbb{R}) = \{h \in L_2(\mathbb{R}): h = Ff, \operatorname{supp} f \subset \overline{\mathbb{R}_{\mp}}\}$, i.e. $\mathcal{H}^{\pm}(\mathbb{R})$ consists of the functions that are holomorphic in the upper (lower) halfplane. Let also $L_2(S^{n-2}, \mathcal{H}^{\pm}(\mathbb{R}))$ be the space of square-integrable functions on S^{n-2} with values in $\mathcal{H}^{\pm}(\mathbb{R})$, and let $\mathcal{H}_{\pm}(-i\lambda-n/2)$ be the space of functions homogeneous of degree $-i\lambda-n/2$ and analytic in the halfplane $\operatorname{Im}\xi_n \gtrless 0$, which was introduced above theorem 1.6.5.

P r o p o s i t i o n 1.4. *For every $\lambda \in \mathbb{R}$ the map $J(\lambda):\mathcal{H}_\pm(-i\lambda-n/2)\to L_2(S^{n-1},\mathcal{H}^\pm(\mathbb{R}))$ defined by*

$$(J(\lambda)u)(\psi,\eta) = (1+\eta^2)^{-(i\lambda+n/2)/2}(u\circ\kappa)(\psi,\eta)$$

is unitary.

The p r o o f follows from the definitions of the spaces and the relation $d\psi = (1+\eta^2)^{-n/2}d\eta d\psi$, where $\psi = \kappa(\psi,\eta)$ and $d\psi$, $d\psi$ are the volume elements on S^{n-1} and S^{n-2}, respectively. ■

2. The algebra $\mathfrak{G}_c^+(\lambda)$ for $\operatorname{Im}\lambda = 0$. Let $\Phi \in C^\infty(S^{n-1})$ and let $\Pi^-_{\zeta\to\eta}\Phi(\kappa(\psi,\zeta)):\mathcal{H}^-(\mathbb{R})\to\mathcal{H}^-(\mathbb{R})$ be the Toeplitz operator composed of the operator of multiplication by the function $\zeta\mapsto\Phi(\kappa(\psi,\zeta))$ and the projection operator Π^-. Denote by $\mathfrak{F}$ the algebra of operator-functions $S^{n-2}\ni\psi\mapsto\Pi^-_{\zeta\to\eta}\Phi(\kappa(\psi,\zeta))$ with norm

$$\|\Pi^-\Phi(\kappa(\cdot));\mathfrak{F}\| = \max_{\psi\in S^{n-2}}\|\Pi^-\Phi(\kappa(\psi,\cdot));\ \mathcal{H}^-(\mathbb{R})\to\mathcal{H}^-(\mathbb{R})\|.$$

Let $L_2^+(S^{n-1})$ be the subspace of $L_2(S^{n-1})$ consisting of the functions that vanish on the lower hemisphere of S^{n-1}; the elements of $L_2^-(S^{n-1})$ are assumed to vanish on the upper hemisphere of S^{n-1}. Let also $\mathfrak{G}_c^+(\lambda)$ be the algebra generated by the operators $P^+E(\lambda)^{-1}\Phi(\omega)E(\lambda)$ on $L_2^+(S^{n-1})$.

P r o p o s i t i o n 1.5. *The algebras $\mathfrak{G}_c^+(\lambda)$ and $\mathfrak{F}$ are isomorphic for any $\lambda \in \mathbb{R}$. Thus, $\mathfrak{G}_c^+(\lambda)$ and $\mathfrak{G}_c^+(\mu)$ are isomorphic for all real λ and μ.*

P r o o f. By proposition 1.3,

$$P^+E(\lambda)^{-1}\Phi(\omega)E(\lambda)u = E(\lambda)^{-1}\kappa(Q^-(-\lambda))\Phi E(\lambda)u = \tag{1.7}$$

$$= E(\lambda)^{-1}\{[J(\lambda)^{-1}\Pi^-_{\zeta\to\eta}\Phi(\kappa(\mathring{\omega},\zeta))\,J(\lambda)(E(\lambda)u)(\kappa(\mathring{\omega},\zeta))]\circ\kappa^{-1}(\omega)\},$$

where $\omega = \kappa(\mathring{\omega},\zeta)$. Theorem 1.6.5 and proposition 1.4 imply that the operator $U(\lambda):L_2^+(S^{n-1})\to L_2(S^{n-2},\mathcal{H}_-(\mathbb{R}))$ defined by $U(\lambda)u = J(\lambda)[E(\lambda)u)\circ\kappa]$ is unitary. It is obvious that $U(\lambda)^{-1}v = E(\lambda)^{-1}[(J(\lambda)^{-1}v)\circ\kappa^{-1}]$. This and (1.7) imply that $U(\lambda)$ is the required intertwining operator for the identity representations of

$\mathfrak{G}_c^+(\lambda)$ and $\mathfrak{J}$. ∎

Introduce the algebra $\mathfrak{J}(\psi)$ generated by the operators $\Pi^- \Phi(\kappa(\psi,\cdot)):\mathcal{H}^-(\mathbf{R})\to\mathcal{H}^-(\mathbf{R})$ (with $\psi\in S^{n-1}$ fixed). The following theorem describes the spectrum of $\mathfrak{J}(\psi)$. The proof of this theorem can be found in [19] (see also [25]).

The meridian $m(\psi) = \{\psi=(\psi,\psi_n) \in S^{n-1}: \psi'/|\psi'|=\psi\}$ of S^{n-1} passing through the point ψ on the equator is suplemented by the diameter $d = \{x\in\mathbf{R}^n: x=(0,-t), -1\leqslant t\leqslant 1\}$ joining the north and south points. On the closed curve $\gamma(\mathring{\omega}) = m(\mathring{\omega}) \cup d$ thus obtained we define the function $\sigma_\Phi(\cdot,\mathring{\omega})$ by $\sigma_\Phi(z,\mathring{\omega}) = \Phi(\omega)$ if $z = \psi \in m(\mathring{\omega})$, and $\sigma_\Phi(z,\mathring{\omega}) = (1-t)\Phi(0,-1)/2+(1+t)\Phi(0,1)/2$ if $z = (0,t) \in d$.

T h e o r e m 1.6. *The algebra $\mathfrak{J}(\mathring{\omega})$ is irreducible and contains the ideal $\mathcal{K}\mathcal{H}^-(\mathbf{R})$ of compact operators on $\mathcal{H}^-(\mathbf{R})$. The quotient algebra $\mathfrak{J}(\mathring{\omega})/\mathcal{K}\mathcal{H}^-(\mathbf{R})$ and the algebra $C(\gamma(\mathring{\omega}))$ of continuous functions on $\gamma(\mathring{\omega})$ are isomorphic. The isomorphism is realized by mapping a function $\sigma_\Phi(\cdot,\mathring{\omega}) \in C(\gamma(\mathring{\omega}))$ to the residue class containing the operator $\Pi^{-1}\Phi(\kappa(\mathring{\omega},\cdot)) \in \mathfrak{J}(\mathring{\omega})$.*

So, the maps $\pi(z):\Pi^{-1}\Phi(\kappa(\mathring{\omega},\cdot))\mapsto\sigma_\Phi(z,\mathring{\omega})$, $z \in \gamma(\mathring{\omega})$, define pairwise inequivalent one-dimensional representations of $\mathfrak{J}(\mathring{\omega})$. Together with the identity representation of $\mathfrak{J}(\mathring{\omega})$ these exhaust, up to equivalence, all representations of $\mathfrak{J}(\mathring{\omega})$. Hence the spectrum $\hat{\mathfrak{J}}(\mathring{\omega})$ can be bijectively mapped onto the disjoint union of $\gamma(\mathring{\omega})$ and a point $e(\mathring{\omega})$ corresponding to the identity representation. In the Jacobson topology, $\{e(\mathring{\omega})\}$ is an open set, while a fundamental system of neighborhoods of a point $z \in \gamma(\mathring{\omega})$ is formed by sets $\mathfrak{W}(z) \cup \{e(\mathring{\omega})\}$, where $\mathfrak{W}(z)$ is an ordinary neighborhood of z on $\gamma(\mathring{\omega})$.

We now turn to a description of the spectrum of the algebra $\mathfrak{G}_c^+(\lambda)$ for $\operatorname{Im}\lambda = 0$. Let $\hat{\mathfrak{G}}_c^+$ be the disjoint union of the sphere S^{n-1}, a diameter d joining the north and south poles, and a second, individual, copy of the equatorial sphere, denoted by ∂S_+^{n-1}. The ends of the diameter are identified with the respective poles. Endow $S^{n-1} \cup d$ with the topology inherited from $\mathbf{R}^n$. Denote by p the projection of the sphere punctured at the poles onto the equator ∂S_+^{n-1}; the points ψ and $p(\psi)$ belong to the same meridian.

We introduce a topology on $\hat{\mathfrak{G}}_c^+$. A subset on ∂S_+^{n-1} is regarded as being open in $\hat{\mathfrak{G}}_c^+$ if and only if it is open in the ordinary topology of the sphere

∂S^{n-1}_{+}. For a point $\psi \in S^{n-1}$ distinct from a pole, sets $\mathfrak{W}(\psi) \cup p(\mathfrak{W}(\psi))$, where $\mathfrak{W}(\psi)$ is a neighborhood of ψ on S^{n-1}, constitute a fundamental system of neighborhoods of ψ (we may assume that $\mathfrak{W}(\psi)$ does not contain poles). A neighborhood of a point $z \in d$ (including poles) is a set $\mathfrak{W}(z) \cup \partial S^{n-1}$ (where $\mathfrak{W}(z)$ is a neighborhood of z on $S^{n-1} \cup d$). The topology on $\hat{\mathfrak{G}}_c^+$ is not separable. It is clear that $\hat{\mathfrak{G}}_c^+$ is a T_0-space.

T h e o r e m 1.7. *Let* $\mathfrak{U}(\lambda) = \mathrm{E}_{\psi\to\phi}(\lambda)^{-1}\Phi(\psi)\mathrm{E}_{\theta\to\psi}(\lambda)$. *Then* $\mathrm{P}^+\mathfrak{U}(\lambda) \in \mathfrak{G}_c^+(\lambda)$, *and the following assertions hold.*

1) *the maps*

$$\pi(\psi): \mathrm{P}^+\mathfrak{U}(\lambda) \mapsto \Phi(\psi),\ \psi \in S^{n-1},\ \psi \neq (0, \pm 1);$$

$$\pi(t): \mathrm{P}^+\mathfrak{U}(\lambda) \mapsto \frac{(1-t)}{2}\Phi(0,-1) + \frac{(1+t)}{2}\Phi(0,1),\ -1 \leqslant t \leqslant 1;$$

$$\tau(\hat{\omega}): \mathrm{P}^+\mathfrak{U}(\lambda) \mapsto \Pi^-\Phi(\kappa(\hat{\omega},\cdot)),\ \hat{\omega} \in \partial S^{n-1}_{+},$$

induce pairwise inequivalent irreducible representations of $\mathfrak{G}_c^+(\lambda)$. *(Both* $\pi(t)$ *and* $\pi(\psi)$, $\psi \in S^{n-1}$, *are one-dimensional,* $\tau(\hat{\omega})$, $\hat{\omega} \in S^{n-1}_{+}$, *is a representation in* $\mathcal{H}^-(\mathbf{R})$.*)*

2) *Every irreducible representation of* $\mathfrak{G}_c^+(\lambda)$ *is equivalent to a representation listed in* 1).

3) *The correspondence* $\pi(\psi) \mapsto \psi \in S^{n-1}$, $\pi(t) \mapsto (0,t)$, $\tau(\hat{\omega}) \mapsto \hat{\omega} \in \partial S^{n-1}_{+}$ *defines a bijection of the spectrum of* $\mathfrak{G}_c^+(\lambda)$ *onto the space* $\hat{\mathfrak{G}}_c^+$. *The topology on* $\hat{\mathfrak{G}}_c^+$ *coincides with the Jacobson topology.*

P r o o f. The first assertion follows by combining proposition 1.5 and theorem 1.6. We verify the second assertion.

Let Ψ be a function given, and smooth, on the cylinder $S^{n-2} \times m$, where S^{n-2} is the equator and m is a meridian of the sphere S^{n-1}. In other words, Ψ is smooth on S^{n-1} outside the poles, and has discontinuities 'of the first kind' at the poles. Introduce the algebra $\mathfrak{F}'$ generated by the operator-functions $S^{n-2} \ni \hat{\omega} \mapsto \Pi^-_{\zeta\to\eta}\Psi(\kappa(\hat{\omega},\zeta))$. It is obvious that $\mathfrak{F}$ is a subalgebra of $\mathfrak{F}'$.

Denote (for a fixed $\hat{\omega} \in S^{n-2}$) by $\mathfrak{F}'(\hat{\omega})$ the algebra generated by the operators $\Pi^-\Psi(\kappa(\hat{\omega},\cdot)): \mathcal{H}^-(\mathbf{R}) \to \mathcal{H}^-(\mathbf{R})$. For each $\hat{\omega} \in S^{n-1}$ and any irreducible representations π of $\mathfrak{F}'(\dot{\omega})$ we define a representation ρ_π of $\mathfrak{F}'$ by $\rho_\pi(x) = \pi(x(\dot{\omega}))$, $x \in \mathfrak{F}'$. Our nearest aim is to convince ourselves of the fact that the map $\pi \mapsto \rho_\pi$ is a bijection of the union of spectra $\cup \hat{\mathfrak{F}}'(\dot{\omega})$, $\dot{\omega} \in S^{n-1}$, onto the spectrum $\hat{\mathfrak{F}}'$.

Let $\mu \in \hat{\mathfrak{F}}'(\dot{\omega}_1)$, $\nu \in \hat{\mathfrak{F}}'(\dot{\omega}_2)$, with $\dot{\omega}_1 \neq \dot{\omega}_2$. Let also $f \in C^\infty(S^{n-1})$, $f(\dot{\omega}_1) = 1$, $f(\dot{\omega}_2) = 0$. Choose an element $x \in \mathfrak{F}'$ for which $\mu(x(\dot{\omega}_2)) \neq 0$. It is clear that $fx \in \mathfrak{F}'$ and that $\rho_\mu(fx) = \mu(x(\dot{\omega}_1)) \neq 0$, $\rho_0(fx) = \nu(0) = 0$, i.e. ρ_μ and ρ_ν are inequivalent representations. Hence the map $\pi \mapsto \rho_\pi$ is injective.

Now assume that $\rho \in \mathfrak{F}'$ and that I is the kernel of ρ. Put $I(\dot{\omega}) = \{x(\dot{\omega}): x \in I\}$. Just as in the proof of lemma 5.5.3 it can be established that $x \in I$ if and only if $x(\dot{\omega}) \in I(\dot{\omega})$ for all $\dot{\omega} \in S^{n-2}$. The set $Y = \{\dot{\omega}: I(\dot{\omega}) \neq \mathfrak{F}'(\dot{\omega})\}$ consists of a single point, θ (cf. lemma 5.5.4). This implies that if two elements x,y of $\mathfrak{F}'$ coincide on θ, then $\rho(x) = \rho(y)$. This allows us to regard ρ as a representation of $\mathfrak{F}'(\theta)$. In other words, $\rho(x) = \pi(x(\theta))$ for a certain $\pi \in \hat{\mathfrak{F}}'(\theta)$, i.e. the map $\pi \mapsto \rho_\pi$ is surjective.

So, the map indicated is bijective. Together with theorem 1.6 this implies that every irreducible representation of $\mathfrak{F}'$ is equivalent to one of the following representations: $\tau(\dot{\omega}):\Pi^-\Psi(\kappa(\cdot)) \mapsto \Pi^-\Psi(\kappa(\dot{\omega},\cdot)), \dot{\omega} \in S^{n-2}$; $\pi(\psi):\Pi^-\Psi(\kappa(\cdot)) \mapsto \Psi(\psi)$, $\psi \in S^{n-1}, \psi \neq (0, \pm 1)$; $\pi(t,\dot{\omega}):\Pi^-\Psi(\kappa(\cdot)) \mapsto (1-t)\Psi(m_-(\dot{\omega}))/2 + (1+t)\Psi(m_+(\dot{\omega}))/2$, $-1 \leq t \leq 1$, where $m_+(\dot{\omega})$ (respectively, $m_-(\dot{\omega})$) is the north (respectively, south) end of the meridian $m(\dot{\omega})$. The truth of assertion 2) now follows from proposition 5.1.6.

The third assertion can be verified by directly comparing the topology on $\mathfrak{G}_c^+$ with the Jacobson topology. ■

3. The algebra $\mathfrak{G}_c^+(\lambda)$ for $\operatorname{Im}\lambda \neq 0$. In this section it is assumed that λ is an arbitrary complex number subject to the requirements $\lambda \neq i(k + n/2)$, $k = 0, 1, \cdots, \operatorname{Im}\lambda \neq 0$.

P r o p o s i t i o n 1.8. *The algebra $\mathfrak{G}_c^+(\lambda)$ generated by the operators* $P^+E(\lambda)^{-1}\Phi(\omega)E(\lambda)$ *on $L_2(S^{n-1})$ is irreducible.*

P r o o f. Let $\mathfrak{K} \subset L_2^+(S^{n-1})$ be an arbitrary invariant subspace of $\mathfrak{G}_c^+(\lambda)$ (hence its orthogonal complement $\mathfrak{K}^\perp$ in $L_2^+(S^{n-1})$ is also an invariant subspace for $\mathfrak{G}_c^+(\lambda)$). We must convince ourselves that either $\mathfrak{K} = 0$ or $\mathfrak{K} = L_2^+(S^{n-1})$.

I. In this step we prove that there is in at least one of the spaces $\mathcal{K}$ and $\mathcal{K}^{\perp}$ a function $u \not\equiv 0$ that is sufficiently smooth on S^{n-1}.

Put $\Psi(x) = |x'|^{\nu}(x_n - i|x'|)^{-\nu} f(x'/|x'|)$, where $f \in C^{\infty}(S^{n-2})$ and $\nu > 0$ is large. For every $x' \neq 0$, the function Ψ has an analytic extension to the halfplane $\operatorname{Im} x_n < 0$, which remains to be homogeneous of degree zero. By increasing ν we may assume that the restriction of Ψ onto S^{n-1} has any given (finite) smoothness. By proposition 3.1.1, for ν large the operator $\mathrm{P}^+ \mathrm{E}(\lambda)^{-1}\Psi\mathrm{E}(\lambda)$ belongs to $\mathfrak{G}_c^+(\lambda)$ (as does the operator $\mathrm{P}^+ \mathrm{E}(\bar\lambda)^{-1}\Psi\mathrm{E}(\bar\lambda) = (\mathrm{P}^+ \mathrm{E}(\lambda)^{-1}\bar\Psi\mathrm{E}(\lambda))^*$).

For $v \in C_0^{\infty}(S_+^{n-1})$, the function $x \mapsto \Psi(x)(\mathrm{E}(\lambda)v)(x)$, given and homogeneous of degree $-i\lambda - n/2$ on $\mathbb{R}^n \setminus 0$, can be analytically extended to the halfplane $\operatorname{Im} x_n < 0$, the support of $\mathrm{E}(\lambda)^{-1}\Psi\mathrm{E}(\lambda)v$ lies on $\overline{S_+^{n-1}}$ (theorem 1.6.5). Thus, the operator $\mathrm{E}(\lambda)^{-1}\Psi\mathrm{E}(\lambda)$ coincides on $L_2^+(S^{n-1})$ with the operator $\mathrm{P}^+ \mathrm{E}(\lambda)^{-1}\Psi\mathrm{E}(\lambda)$, and belongs to $\mathfrak{G}_c^+(\lambda)$. Similarly, $\mathrm{E}(\bar\lambda)^{-1}\Psi\mathrm{E}(\bar\lambda) \in \mathfrak{G}_c^+(\lambda)$. This implies that the selfadjoint operator

$$\mathrm{K} = (\mathrm{E}(\lambda)^{-1}\Psi\mathrm{E}(\lambda) - \mathrm{E}(\bar\lambda)^{-1}\Psi\mathrm{E}(\bar\lambda))^l \times \tag{1.8}$$

$$\times \{\mathrm{P}^+(\mathrm{E}(\bar\lambda)^{-1}\bar\Psi\mathrm{E}(\bar\lambda) - \mathrm{E}(\lambda)^{-1}\bar\Psi\mathrm{E}(\lambda))\}^l,$$

where l is an arbitrary natural number, also belongs to $\mathfrak{G}_c^+(\lambda)$. If the number ν participating in the definition of Ψ is large (i.e. Ψ is sufficiently smooth), then the operator $\mathrm{E}(\lambda)^{-1}\Psi\mathrm{E}(\lambda) - \mathrm{E}(\bar\lambda)^{-1}\Psi\mathrm{E}(\bar\lambda)$ acts continuously from $H^s(S^{n-1})$ into $H^{s+1}(S^{n-1})$ for $0 \leqslant s \leqslant N$, where N can be taken arbitrarily large by increasing ν (proposition 3.2.2). Hence the map $\mathrm{K}: H^0(S^{n-1}) \to H^l(S^{n-1})$ is continuous for ν large. Consequently, every eigenfunction of $\mathrm{K}: L_2^+(S^{n-1}) \to L_2^+(S^{n-1})$ corresponding to a nonzero eigenvalue is sufficiently smooth on S^{n-1}.

Let μ be a nonzero eigenvalue of K, let $\mathcal{K}_\mu$ be the subspace spanned on all eigenvectors corresponding to this eigenvalue, and let $\mathrm{P}_\mu: L_2^+(S^{n-1}) \to \mathcal{K}_\mu$ be orthogonal projection. Since $\mathrm{P}_\mu \in \mathfrak{G}_c^+(\lambda)$, we have $\mathrm{P}_\mu\mathcal{K} \subset \mathcal{K}_\mu \cap \mathcal{K}$. If $\mathrm{P}_\mu\mathcal{K} \neq 0$, then $\mathcal{K}$ contains a nonzero smooth function. If, however, $\mathrm{P}_\mu\mathcal{K} = 0$, then $\mathrm{P}_\mu\mathcal{K}^{\perp} \neq 0$ and there is a smooth function in $\mathrm{P}_\mu\mathcal{K}^{\perp} \subset \mathcal{K}_\mu \cap \mathcal{K}^{\perp}$.

II. We prove the existence of a set $\Omega \subset S^{n-1}$ of positive measure and satisfying the requirements: a) Ω contains, for every point $\omega \neq (0, \pm 1)$ of it, the meridian passing through ω and the poles $(0, \pm 1)$; b) $\mathrm{P}^+ \mathrm{E}(\bar\lambda)^{-1}h \in \mathcal{K}$ for all $h \in L_2(S^{n-1})$ with support in Ω.

By I above, we may assume that the invariant subspace $\mathcal{K}$ contains a sufficiently smooth function u, $u \not\equiv 0$, on S^{n-1}. Put $v = \mathrm{E}(\bar{\lambda})u$. By theorem 1.6.5, the function v (which is smooth and homogeneous of degree $-i\bar{\lambda}-n/2$) has, for $x' \neq 0$, an analytic extension to the halfplane $\operatorname{Im} x_n < 0$. Hence, if for a certain x' the function $x_n \mapsto v(x', x_n)$ is not identically equal to zero, then the zero set of this function on $\operatorname{Im} x_n = 0$ has measure zero.

Denote by $Z(v)$ the zero set of v on S^{n-1}, and by Ω the subset of the sphere equal to the union of the meridians passing through the poles $(0, \pm 1)$ and not completely contained in $Z(v)$. It is obvious that $Z(v)$ is a closed subset of S^{n-1}, and that the set obtained from Ω by delecting the poles $(0, \pm 1)$ is open; moreover, $\operatorname{mes}_{n-1}(\Omega \cap Z(v)) = 0$. This implies that the set $\{\Phi v \colon \Phi \in C^\infty(S^{n-1})\}$ is dense in the space $\chi_\Omega L_2(S^{n-1})$, where χ_Ω is the characteristic function of Ω.

Since $u \in \mathcal{K}$, we have $\mathrm{P}^+\mathrm{E}(\bar{\lambda})^{-1}\Phi\mathrm{E}(\bar{\lambda})u \in \mathcal{K}$, i.e. $\mathrm{P}^+\mathrm{E}(\bar{\lambda})^{-1}\Phi v \in \mathcal{K}$ for all $\Phi \in C^\infty(S^{n-1})$. Without loss of generality we may assume that $\operatorname{Im}\lambda > 0$ (in the opposite case we interchange the roles of λ and $\bar{\lambda}$ from the very beginning). The continuity of the map $\mathrm{E}(\bar{\lambda})^{-1} \colon L_2(S^{n-1}) \to H^{\operatorname{Im}\lambda}(S^{n-1}) \subset L_2(S^{n-1})$ (proposition 1.5.5) and the denseness of the set $\{\Phi v\}$ given above imply that $\mathrm{P}^+\mathrm{E}(\bar{\lambda})^{-1}h \in \mathcal{K}$ for all $h \in \chi_\Omega L_2(S^{n-1})$.

III. We prove that if $\mathrm{P}_\mu\mathcal{K} = \mathcal{K}_\mu$, or, equivalently, $\mathcal{K}_\mu \subset \mathcal{K}$, then $\mathcal{K} = L_2^+(S^{n-1})$ (cf. the notation in the first part of the proof). Introduce the operator K_q obtained from K by replacing Ψ by $g^*\Psi = \Psi \circ g$, where g is an element of the orthogonal group $\mathrm{SO}(n-1)$, regarded as a subgroup of $\mathrm{SO}(n)$: $g(x) = (g(x'), x_n)$. The operators $\mathrm{E}(\lambda)$, $\mathrm{E}(\bar{\lambda})$, and P^+ commute with g^*, $g \in \mathrm{SO}(n-1)$, hence $g^*\mathcal{K}_\mu = \{f \circ g \colon f \in \mathcal{K}_\mu\}$ is an eigenspace of K_g for the same eigenvalue μ. Let $\mathrm{P}_{\mu,g} \colon L_2^+(S^{n-1}) \to g^+\mathcal{K}_\mu$ be orthogonal projection, and let U be the set of elements $g \in \mathrm{SO}(n-1)$ for which $\mathrm{P}_{\mu,g}\mathcal{K} = g^*\mathcal{K}_\mu$. By assumption, the identity element $e \in \mathrm{SO}(n-1)$ belongs to U, hence $U \neq \varnothing$. The inclusion $g^*\mathcal{K}_\mu \subset \mathcal{K}$ (true for $g \in U$) implies that U is closed. Now assume that $g_0 \in U$ and let $h_1, \cdots, h_m$ be elements of $\mathcal{K}$ such that $\{\mathrm{P}_{\mu,g_0}h_j\}_{j=1}^m$ is a basis in $g_0^*\mathcal{K}_\mu$. It is obvious that for any $g \in \mathrm{SO}(n-1)$ near g_0 the system $\{\mathrm{P}_{\mu,g}h_j\}_{j=1}^m$ is linearly independent. In other words, a neighborhood of g_0 belongs to U, i.e. U is open. Consequently, $U = \mathrm{SO}(n-1)$ (under the assumption that $\mathrm{P}_\mu\mathcal{K} = \mathcal{K}_\mu$, which is assumed to hold in this step of the proof).

Let u, v, and Ω be as in the first step, and let $g_1, \cdots, g_l$ be a set of elements of

SO$(n-1)$ such that the sets $g_j\Omega$, $j = 1, \cdots, l$, cover the sphere. The set $g_j\Omega$ corresponds to the function $g_j^* u$ in the same sense in which Ω corresponds to u. In particular, $g_j^* v = \mathrm{E}(\bar{\lambda}) g_j^* u$, and $\mathrm{P}^+ \mathrm{E}(\bar{\lambda})^{-1} h \in \mathcal{K}$ for all $h \in \chi_{g_j\Omega} L_2(S^{n-1})$.

Denote by $\{\eta_j\}$ a partition of unity on the equator ∂S_+^{n-1} subordinate to the cover $\{g_j\Omega \cap \partial S_+^{n-1}\}$. Extend the η_j onto S^{n-1} by taking them independent of x_n. For any $f \in L_2(S^{n-1})$ we have $\mathrm{P}^+ \mathrm{E}(\bar{\lambda})^{-1} f = \Sigma_j \mathrm{P}^+ \mathrm{E}(\bar{\lambda})^{-1} \eta_j f \in \mathcal{K}$, since every term belongs to $\mathcal{K}$. Since the set $\{h\colon h = \mathrm{P}^+ \mathrm{E}(\bar{\lambda})^{-1} f,\ f \in L_2(S^{n-1})\}$ is dense in $L_2^+(S^{n-1})$, we have $\mathcal{K} = L_2^+(S^{n-1})$.

IV. Replacing, if necessary, $\mathcal{K}$ by $\mathcal{K}^\perp$, we will assume that $\mathrm{P}_\mu \mathcal{K} \neq 0$. The equality $\dim \mathcal{K} = 1$ implies that the condition $\mathrm{P}_\mu \mathcal{K} = \mathcal{K}_\mu$ holds. If K has an eigenvalue μ such that $\dim \mathcal{K}_\mu = 1$, the proof of the proposition would be finished. Suppose there is no such eigenvalue; we will show how to modify K such that an eigenvalue as required appears.

Let, as before, $\mathcal{K}_\mu$ be an eigenspace of the operator (1.8), $\mu \neq 0$, and $\dim \mathcal{K}_\mu \geqslant 2$. Put $\mathrm{Q} = \mathrm{P}_\mu \mathfrak{B} \mathrm{P}_\mu$, where $\mathfrak{B}$ is a selfadjoint operator from $\mathfrak{S}_c^+(\lambda)$. Clearly, Q is finite-dimensional and selfadjoint, and $\mathrm{Q} \in \mathfrak{S}_c^*(\lambda)$. All eigenvectors of Q corresponding to nonzero eigenvalues lie in $\mathcal{K}_\mu$. We show that by choosing $\mathfrak{B}$ suitably, we may also assume Q to have at least two distinct eigenvalues. Assume the opposite holds. Then for any $\mathfrak{B} = \mathfrak{B}^* \in \mathfrak{S}_c^+(\lambda)$ we have $\mathrm{Q}u = \nu u$, where u is any vector from $\mathcal{K}_\mu$ and ν is a number. The real and imaginary parts of every operator in $\mathfrak{S}_c^+(\lambda)$ belong to $\mathfrak{S}_c^+(\lambda)$. Hence an equality $\mathrm{Q}u = \nu u$ holds if and only if $\mathfrak{B}$ is an arbitrary operator in $\mathfrak{S}_c^+(\lambda)$; in particular, $\mathfrak{B} = \mathrm{P}^+ \mathrm{E}(\lambda)^{-1} \Phi \mathrm{E}(\lambda)$. This implies that the quantity $c = (\mathrm{E}(\lambda)^{-1} \Phi \mathrm{E}(\lambda) u, u) = (\Phi \mathrm{E}(\lambda) u, \mathrm{E}(\bar{\lambda}) u)$ is independent of $u \in \mathcal{K}_\mu$, $\|u\| = 1$. Take orthonormal vectors u, v in $\mathcal{K}_\mu$ and complex numbers α, β, $|\alpha|^2 + |\beta|^2 = 1$. We have

$$c = (\Phi \mathrm{E}(\lambda)(\alpha u + \beta v),\ \mathrm{E}(\bar{\lambda})(\alpha u + \beta v)) =$$

$$= c + \alpha\bar{\beta}(\Phi \mathrm{E}(\lambda) u,\ \mathrm{E}(\bar{\lambda}) v) + \bar{\alpha}\beta(\Phi \mathrm{E}(\lambda) v,\ \mathrm{E}(\bar{\lambda}) u).$$

Suppose that Φ is real-valued. Then the last formula implies that $\mathrm{Re}\, \alpha\bar{\beta}(\Phi \mathrm{E}(\lambda) u,\ \mathrm{E}(\bar{\lambda}) v) = 0$ for all $\alpha, \beta \in \mathbf{C}$, $|\alpha|^2 + |\beta|^2 = 1$. This is possible only if $(\Phi \mathrm{E}(\lambda) u,\ \mathrm{E}(\bar{\lambda}) v) = 0$. Since Φ is an arbitrary smooth function,

$$\mathrm{E}(\lambda) u \overline{\mathrm{E}(\bar{\lambda}) v} = 0. \tag{1.9}$$

Moreover, the equations $c = (\Phi E(\lambda)u,\ E(\bar\lambda)u) = (\Phi E(\lambda)v,\ E(\bar\lambda)v)$ imply

$$\overline{E(\lambda)u}E(\bar\lambda)u = \overline{E(\lambda)v}E(\bar\lambda)v. \tag{1.10}$$

Combining (1.9) and (1.10) gives

$$\left[E(\lambda)u\overline{E(\bar\lambda)u}\right]^2 = E(\lambda)u\overline{E(\bar\lambda)u}E(\lambda)v\overline{E(\bar\lambda)v} = 0.$$

Thus, $0 = (E(\lambda)u,\ \overline{E(\bar\lambda)u}) = (u,\ u) = 1$. This contradiction shows that there is an operator Q with the properties listed above.

Let ν_1, ν_2 be distinct eigenvalues of Q. Then the operator $K+\epsilon Q$ has at least two distinct eigenvalues $\mu_j = \mu+\epsilon\nu_j$, $j = 1,2$. For ϵ small these numbers are distinct from zero, and the sum of their multiplicities does not exceed $\dim \mathcal{H}_\mu$ ([16], theorem 1.3.1 about stability of root multiplicities). Denote by $\mathcal{H}_{\mu_1}$ the eigenspace of $K+\epsilon Q$ corresponding to μ_1. Clearly, $\dim \mathcal{H}_{\mu_1} < \dim \mathcal{H}_\mu$, $\mathcal{H}_{\mu_1} \subset \mathcal{H}_\mu$. If $\dim \mathcal{H}_{\mu_1} > 1$, by repeating this reasoning we find an operator $K+\epsilon Q+\epsilon_1 Q_1$ which will have an eigenspace of dimension less than $\dim \mathcal{H}_{\mu_1}$ and contained in $\mathcal{H}_{\mu_1}$. In a finite number of steps we have constructed an operator with a one-dimensional eigenspace. ■

C o r o l l a r y 1.9. *The algebra* $\mathfrak{S}_c^+(\lambda)$ *contains the ideal* $\mathcal{K}L_2^+(S^{n-1})$ *of compact operators.*

P r o o f. $\mathfrak{S}_c^+(\lambda)$ contains the nonzero compact operator $P^+E(\lambda)^{-1}\Phi(\omega)E(\lambda) - P^+E(\bar\lambda)^{-1}\Phi(\omega)E(\bar\lambda)$, where Φ is a real-valued function. The irreducibility of $\mathfrak{S}_c^+(\lambda)$ and proposition 5.1.5 guarantee that $\mathcal{K}L_2(S^{n-1}) \subset \mathfrak{S}_c^+(\lambda)$. ■

P r o p o s i t i o n 1.10. *The map* $P^+E(\lambda)^{-1}\Phi(\omega)E(\lambda) \mapsto \Pi^-\Phi(\kappa(\cdot))$ *induces an isomorphism between the algebras* $\mathfrak{S}_c^+(\lambda)/\mathcal{K}L_2^+(S^{n-1})$ *and* $\mathfrak{J}$.

P r o o f. Let $\mathfrak{U}_{jk}(\lambda)$ be an operator of the form $E(\lambda)^{-1}\Phi_{jk}(\omega)E(\lambda)$ or $E(\bar\lambda)^{-1}\Phi_{jk}(\omega)E(\bar\lambda)$, and let $\sigma = \mathrm{Re}\,\lambda$. By proposition 3.2.2, $P^+\mathfrak{U}_{jk}(\lambda) = P^+\mathfrak{U}_{jk}(\sigma)+T_j$, where $T_{jk} \in \mathcal{K}L_2^+(S^{n-1})$. This and proposition 1.5 imply

$$\inf_{T\in\mathcal{K}L_2^+(S^{n-1})}\left\|\sum_j\prod_k P^+\mathfrak{U}_{jk}(\lambda)+T;L_2^+(S^{n-1})\to L_2^+(S^{n-1})\right\|\leqslant$$

$$\leqslant\left\|\sum_j\prod_k\Pi^{-1}\Phi_{jk}(\kappa(\cdot));\mathfrak{S}\right\|,$$

where the indices j,k run over finite sets.

We now prove the opposite inequality. Let $\mathcal{F}(\mathring{\omega})=\Sigma_j\Pi_k\Pi^-\Phi_{jk}(\kappa(\mathring{\omega},\cdot)):\mathcal{K}^-(\mathbb{R})\to\mathcal{K}^-(\mathbb{R})$, $\mathring{\omega}\in\partial S_+^{n-1}$. Let also $u\in\mathcal{K}^-(\mathbb{R})$, $\|u\|=1$, and $\|\mathcal{F}(\mathring{\omega}_0)u;\ \mathcal{K}^-(\mathbb{R})\|\geqslant\|\mathcal{F}(\mathring{\omega}_0);\ \mathcal{K}^-(\mathbb{R})\to\mathcal{K}^-(\mathbb{R})\|-\epsilon$, for given $\mathring{\omega}_0\in\partial S_+^{n-1}$ and $\epsilon>0$. Put $v(\omega',\omega_n)=(1+(\omega_n/|\omega'|)^2)^{(i\sigma+n/2)/2}\chi(\omega'/|\omega'|)u(\omega_n/|\omega'|)$; here $\chi\in C^\infty(\partial S_+^{n-1})$, the support of χ is concentrated in a neighborhood of $\mathring{\omega}_0$, and $\|\chi;L_2(\partial S_+^{n-1})\|=1$. By proposition 1.4, $v\in\mathcal{K}_-(-i\sigma-n/2)$, and $\|v;L_2(S^{n-1})\|=1$. By theorem 1.6.5, $w=\mathrm{E}(\sigma)^{-1}v$ belongs to $L_2^+(S^{n-1})$, and $\|w;L_2(S^{n-1})\|=1$ (since $\mathrm{E}(\sigma)$ is a unitary operator). By formula (1.7),

$$\left\|\sum_j\prod_k P^+\mathfrak{U}_{jk}(\sigma)w;L_2(S^{n-1})\right\|=\|\mathcal{F}(\cdot)\chi u;L_2(\partial S_+^{n-1},\mathcal{K}^-(\mathbb{R}))\|.$$

Taking into account that the support of χ is small, as well as the choice of u, we find

$$-\epsilon<\left\|\sum_j\prod_k P^+\mathfrak{U}_{jk}(\sigma)w;L_2(S^{n-1})\right\|=\|\mathcal{F}(\mathring{\omega}_0);\mathcal{K}^-(\mathbb{R})\to\mathcal{K}^-(\mathbb{R}))\|<\epsilon.$$

Let $\{\epsilon_n\}$ be a null sequence of positive numbers, and let $\{w_n\}\subset L_2^+(S^{n-1})$ be a sequence of functions corresponding to it (i.e. satisfying the inequalities above); we may moreover assume that the supports of the functions χ_n contract towards the point $\mathring{\omega}_0\in S^{n-2}$. It is obvious that $\{w_n\}$ has the following properties: $\|w_n;L_2(S^{n-1})\|=1$; $\{w_n\}$ is weakly convergent to zero in $L_2^+(S^{n-1})$; and, as $n\to\infty$,

$$\left\|\sum_j\prod_k P^+\mathfrak{U}_{jk}(\sigma)w_n;L_2(S^{n-1})\right\|\to\left\|\mathcal{F}(\mathring{\omega}_0);\mathcal{K}^-(\mathbb{R})\to\mathcal{K}^-(\mathbb{R})\right\|.$$

Recall that $\Sigma_j\Pi_kP^+\mathfrak{U}_{jk}(\lambda)=\Sigma_j\Pi_kP^+\mathfrak{U}_{jk}(\sigma)+T$, where $T\in\mathcal{K}L_2^+(S^{n-1})$. This leads to the inequality opposite to (1.11). ■

C o r o l l a r y 1.11. *Let* $\lambda \in \mathbb{C}$, $\lambda \neq \pm i(k+n/2)$, $k = 0, 1, \cdots$, *and* $\mathrm{Im}\lambda \neq 0$. *There is an isomorphism* $\mathfrak{G}_c^+(\lambda)/\mathcal{K}L_2^+(S^{n-1}) \approx \mathfrak{G}_c^+(\mu)$, *where* μ *is an arbitrary real number.*

T h e o r e m 1.12. *Theorem 1.7 remains valid if* λ *is an arbitrary complex number,* $\lambda \neq \pm i(k+n/2)$, $k = 0, 1, \cdots, \mathrm{Im}\lambda \neq 0$, *if the representations listed in* 1) *are supplemented by the identity representation of the algebra* $\mathfrak{G}_c^+(\lambda)$, *and if the space* $\hat{\mathfrak{G}}_c^+$ *is replaced by the union* $\hat{\mathfrak{G}}_c^+ \cup \{e(\lambda)\}$, *where* $e(\lambda)$ *is the point corresponding to the identity representation. This point constitutes an open set, whose closure coincides with* $\hat{\mathfrak{G}}_c^+ \cup \{e(\lambda)\}$.

In order to verify this theorem it suffices to combine proposition 1.8, corollary 1.11, and proposition 5.1.4.

§2. The algebras $\mathfrak{G}^+(\lambda)$

1. The algebras $\mathfrak{G}^*(\lambda)$ and $\mathfrak{P}$. Let $\mathfrak{G}^+(\lambda)$ be the algebra generated on $L_2^+(S^{n-1})$ by the operators $\mathrm{P}^+\mathfrak{U}(\lambda)$, where $\mathfrak{U}(\lambda) = \mathrm{E}(\lambda)^{-1}\Phi(\phi,\omega)\mathrm{E}(\lambda)$, $\Phi \in C^\infty(S^{n-1}\times S^{n-1})$, $\lambda \in \mathbb{C}$, $\lambda \neq \pm i(k+n/2)$, $k = 0, 1, \cdots$. The following assertion can be proved similarly to proposition 5.3.1.

P r o p o s i t i o n 2.1. *The algebra* $\mathfrak{G}^+(\lambda)$ *is irreducible.*

It is obvious that operators of the form $\zeta \mathrm{P}^+\mathfrak{U}(\lambda)\eta$, with $\zeta,\eta \in C_0^\infty(S_+^{n-1})$, belong to $\mathfrak{G}^+(\lambda)$. Such operators with $\mathrm{supp}\,\zeta \cap \mathrm{supp}\,\eta = \varnothing$ are compact on $L_2^+(S^{n-1})$. This and propositions 2.1 and 5.1.5 imply

C o r o l l a r y 2.2. *The following inclusion holds:* $\mathcal{K}L_2^+(S^{n-1}) \subset \mathfrak{G}^+(\lambda)$.

P r o p o s i t i o n 2.3. *Let* $\mathfrak{U}(\lambda) = \mathrm{E}(\lambda)^{-1}\Phi(\phi,\omega)\mathrm{E}(\lambda)$, $\mathfrak{B}(\lambda) = \mathrm{E}(\lambda)^{-1}\Psi(\phi,\omega)\mathrm{E}(\lambda)$. *Then*

$$\mathrm{P}^+\mathfrak{U}(\lambda)\mathrm{P}^+\mathfrak{B}(\lambda)u - \mathrm{T}(\lambda)u =$$

$$= \mathrm{E}_{\omega\to\phi}(\sigma)^{-1}\{[\mathrm{J}(\sigma)^{-1}\Pi^-_{\chi\to\zeta}\Phi(\phi,\kappa(\hat{\omega},\chi)) \times$$

$$\times \Pi^-_{\eta\to\chi}\Psi(\phi,\kappa(\hat{\omega},\eta))\mathrm{J}(\sigma)(\mathrm{E}(\lambda)u)(\kappa(\hat{\omega},\eta))]\circ\kappa^{-1}(\omega)\},$$

where $\sigma = \mathrm{Re}\lambda$, $\omega = \kappa(\hat{\omega},\zeta)$, $u \in C_0^\infty(S^{n-1})$, $\mathrm{T}(\lambda) \in \mathcal{KL}_2^+(S^{n-1})$. *If* $\mathrm{Im}\lambda = \mathrm{const}$ *and* $\lambda \to \infty$, *then* $\|\mathrm{T}(\lambda)\| \to 0$. *Thus, up to a compact term, to the composite operator* $\mathrm{P}^+\mathfrak{U}(\lambda)\mathrm{P}^+\mathfrak{B}(\lambda)$ *corresponds the composite function* $\Pi^-\Phi(\kappa(\cdot))\Pi^-\Psi(\kappa(\cdot))$.

P r o o f. Put $\mathrm{T}_1(\lambda) = \mathrm{P}^+\mathfrak{U}(\lambda) - \mathrm{P}^+\mathfrak{U}(\sigma)$, $\mathrm{T}_2(\lambda) = \mathrm{P}^+\mathfrak{B}(\lambda) - \mathrm{P}^+\mathfrak{B}(\sigma)$. Let, initially, $\Psi(\theta,\omega) = a(\theta)Y(\omega)$. We have

$$\mathrm{P}^+\mathfrak{U}_{\theta\to\phi}(\sigma)\mathrm{P}^+\mathfrak{B}_{\psi\to\theta}(\sigma) =$$

$$= \mathrm{P}^+\mathrm{E}_{\omega\to\phi}(\sigma)^{-1}\Phi(\phi,\omega)\mathrm{E}_{\theta\to\omega}(\sigma) \times$$

$$\times \mathrm{P}^+\mathrm{E}_{\omega\to\theta}(\sigma)^{-1}\Psi(\phi,\omega)\mathrm{E}_{\psi\to\omega}(\sigma) + \mathrm{T}_3(\sigma),$$

where $\mathrm{T}_3(\sigma) = \mathrm{P}^+(\mathfrak{U}(\sigma)a - a\mathfrak{U}(\sigma))\mathrm{P}^+\mathrm{E}(\sigma)^{-1}Y\mathrm{E}(\sigma)$. Note that each of the operators T_j, $j = 1,2,3$, has the properties stipulated for T. (For $\mathrm{T}_1, \mathrm{T}_2$ this follows from proposition 3.2.2, and for T_3 from proposition 3.2.1.) Writing each of the operators $\mathrm{P}^+\mathrm{E}_{\omega\to\phi}(\sigma)^{-1}\Phi(\phi,\omega)\mathrm{E}_{\theta\to\omega}(\sigma)$ and $\mathrm{P}^+\mathrm{E}_{\omega\to\theta}(\sigma)^{-1}\Psi(\phi,\omega)\mathrm{E}_{\psi\to\omega}(\sigma)$ using formula (1.7), our assertion is proved in this particular case. The general case reduces to this one by expansion of the function in a series in spherical harmonics: $\Psi(\theta,\omega) = \Sigma a_{mk}(\theta)Y_{mk}(\omega)$. ■

P r o p o s i t i o n 2.4. *Let* $\mathfrak{U}(\lambda) = \mathrm{E}_{\omega\to\phi}(\lambda)^{-1}\Phi(\phi,\omega)\mathrm{E}_{\psi\to\omega}(\lambda)$. *Then the following formula holds for the operator* $(\mathrm{P}^+\mathfrak{U}(\lambda))^*$ *adjoint to* $\mathrm{P}^+\mathfrak{U}(\lambda)$ *with respect to the scalar product in* $L_2^+(S^{n-1})$:

$$(\mathrm{P}^+\mathfrak{U}(\lambda))^*u - \mathrm{T}(\lambda)u = \mathrm{E}_{\omega\to\phi}(\sigma)^{-1} \times$$

$$\times \{[\mathrm{J}(\sigma)^{-1}\Pi^-_{\zeta\to\eta}\overline{\Phi(\phi,\kappa(\hat{\omega},\zeta))}\mathrm{J}(\sigma)(\mathrm{E}(\sigma)u)\circ\kappa(\hat{\omega},\zeta)]\circ\kappa^{-1}(\omega)\};$$

here $\sigma = \mathrm{Re}\lambda$, $\mathrm{T}(\lambda) \in \mathcal{KL}_2^+(S^{n-1})$, *and* $\|\mathrm{T}(\lambda)\| \to 0$ *as* $\lambda \to \infty$, $\mathrm{Im}\lambda = \mathrm{const}$.

P r o o f. Put $\mathrm{T}_1(\lambda) = \mathrm{P}^+\mathfrak{U}(\lambda) - \mathrm{P}^+\mathfrak{U}(\sigma)$, $\mathrm{T}_2(\sigma) = (\mathrm{P}^+\mathfrak{U}(\sigma))^* - \mathrm{E}(\sigma)^{-1}\overline{\Phi(\phi,\omega)}\mathrm{E}(\sigma)$. The operators $\mathrm{T}_1, \mathrm{T}_2$ have the properties stipulated for T (propositions 3.2.2 and 3.2.5). It remains to use (1.7). ■

R e m a r k 2.5. The operator P^+ (of multiplication by the characteristic function of the sphere S_+^{n-1}) is continuous on $H^s(S^{n-1})$ for $|s| < 1/2$ (cf., e.g.,

[74]). This and propositions 3.2.1 and 3.2.2 imply that on a line $\operatorname{Im}\lambda = \text{const}$ the following estimate holds for the operators T participating in the statements of propositions 2.3 and 2.4:

$$\|\mathrm{T}(\lambda); H^{-\delta}(\lambda, S^{n-1}) \to L_2^+(S^{n-1})\| \leqslant c(\delta);$$

here $\delta < 1/2$ is arbitrary positive and $c(\delta)$ is a constant depending on δ but not on λ.

Our nearest aim is to give a description of the algebra $\mathfrak{B}$, which, as will be shown in the sequel, is isomorphic to the quotient algebra $\mathfrak{G}^+(\lambda)/\mathcal{K}L_2^+(S^{n-1})$. Let $\phi \in S^{n-1}$ and $S^{n-2}(\phi) = \{\omega \in S^{n-1}: \phi\omega = 0\}$. For $\phi \in S_+^{n-1}$ we denote by $\mathcal{P}(\phi)$ the algebra $C(S^{n-2}(\phi))$ of continuous functions. For $\phi \in \partial S_+^{n-1}$, by $\mathcal{P}(\phi)$ we mean the algebra of continuous operator-functions $\omega \mapsto \mathcal{T}(\omega)$ on the $(n-3)$-dimensional sphere $\{\omega: \omega \in S^{n-2}(\phi) \cap \partial S_+^{n-1}\}$; here $\mathcal{T}(\omega)$ is a Toeplitz operator on $\mathcal{H}^-(\mathbf{R})$ of the form $(\mathcal{T}(\omega)u)(\eta) = \Pi^-_{\zeta\to\eta}\, \Xi(\omega,\zeta)u(\zeta)$, where Ξ is a continuous complex-valued function on $\{(\omega,\zeta): (\omega,\zeta) \in (S^{n-2}(\phi) \cap \partial S_+^{n-1}) \times [-\infty, +\infty]\}$. By definition, $\|\mathcal{T};\mathcal{P}(\phi)\| = \sup_\omega \|\mathcal{T}(\omega): \mathcal{H}^-(\mathbf{R}) \to \mathcal{H}^-(\mathbf{R})\|$.

Let $V^+(n,2)$ be the manifold of pairs (ϕ,ω) of orthogonal unit vectors in $\mathbf{R}^n$ with $\phi \in S_+^{n-1}$. Every function $\Phi \in C(V^+(n,2))$ induces a vector field, i.e. a map $\overline{S_+^{n-1}} \ni \phi \mapsto B_\Phi(\phi) \in \mathcal{P}(\phi)$. For $\phi \in S_+^{n-1}$ this field is defined by $B_\Phi(\phi) = \Phi(\phi,\cdot)$, while for $\phi \in \partial S_+^{n-1}$ by $B_\Phi(\phi)$ we mean the operator-function $S^{n-2}(\phi) \cap \partial S_+^{n-1} \ni \mathring{\omega} \mapsto \Pi^- \Phi(\phi, \kappa(\mathring{\omega}, \cdot))$. Finally, we introduce the algebra $\mathfrak{B}$ generated by such vector fields; it is an algebra under 'pointwise' multiplication and involution, endowed with the norm

$$\|B_\Phi(\cdot);\mathfrak{B}\| = \sup_{\phi \in \overline{S_+^{n-1}}} \|B_\Phi(\phi);\mathcal{P}(\phi)\|.$$

2. The isomorphism between $\mathfrak{G}^+(\lambda)/\mathcal{K}L_2^+(S^{n-1})$ and $\mathfrak{B}$. Let $\mathfrak{U}(\lambda) = \mathrm{E}(\lambda)^{-1}\Phi(\phi,\omega)\mathrm{E}(\lambda)$, and let B_Φ be the element of $\mathfrak{B}$ determined by the restriction of Φ to $V^+(n,2)$. Denote by $[\mathrm{P}^+\mathfrak{U}(\lambda)]$ the residue class in the quotient algebra $\mathfrak{G}^+(\lambda)/\mathcal{K}L_2^+(S^{n-1})$ containing $P^+\mathfrak{U}(\lambda)$.

P r o p o s i t i o n 2.6. *The map $[P^+\mathfrak{U}(\lambda)] \mapsto B_\Phi$ induces an isomorphism between $\mathfrak{G}^+(\lambda)/\mathcal{K}L_2^+(S^{n-1})$ and $\mathfrak{B}$.*

P r o o f. Preservation by this map of the algebraic operations is implied, in particular, by propositions 2.3 and 2.4. It remains to verify that it is an isometry. This we do in several steps.

I. Put $\mathfrak{U}_{jk}(\lambda) = \mathrm{E}(\lambda)^{-1}\Phi_{jk}(\phi,\omega)\mathrm{E}(\lambda)$; we prove the inequality

$$\inf_{\mathrm{T}}\left\|\sum_{j=1}^{M}\prod_{k=1}^{N}\mathrm{P}^{+}\mathfrak{U}_{jk}(\lambda)+\mathrm{T};L_2^{+}(S^{n-1})\to L_2^{+}(S^{n-1})\right\| \geqslant \tag{2.1}$$

$$\geqslant \left\|\sum_{j=1}^{M}\prod_{k=1}^{N}B_{\Phi_{jk}};\mathfrak{B}\right\|,$$

where Φ_{jk} are arbitrary $C^{\infty}(S^{n-1}\times S^{n-1})$ functions, and T runs through $\mathcal{K}L_2^{+}(S^{n-1})$.

Let $(\phi_0,\omega_0) \in V^{+}(n,2)$ with $\phi_0 \in S_{+}^{n-1}$. We show the existence of a sequence $\{v_q\} \subset C_0^{\infty}(S_{+}^{n-1})$ such that $\|v_q;L_2(S^{n-1})\| = 1$, $\{v_q\}$ is weakly convergent to zero in $L_2^{+}(S^{n-1})$, and $\|\Sigma_j\Pi_k\mathrm{P}^{+}\mathfrak{U}_{jk}(\lambda)v_q\| \to |\Sigma_j\Pi_k\Phi_{jk}(\phi_0,\omega_0)|$. Denote by ζ_k, $k = 1,\cdots,N+1$, $C_0^{\infty}(S_{+}^{n-1})$ functions satisfying $\zeta_k\zeta_{k+1} = \zeta_{k+1}$, $k = 1,\cdots,N$, and $\zeta_{N+1} = 1$ in a neighborhood of ϕ_0. Let also $v \in C_0^{\infty}(S_{+}^{n-1})$ with $v\zeta_{N+1} = v$. Recall that $\zeta\mathfrak{U}_{jk}\eta \in \mathcal{K}L_2^{+}(S^{n-1})$ for any smooth functions ζ and η with disjoint supports (proposition 3.1.6). We have

$$\left[\prod_{k=1}^{N}\mathrm{P}^{+}\mathfrak{U}_{jk}\right]v = \left[\prod_{k=1}^{N}\zeta_k\mathfrak{U}_{jk}\right]\zeta_{N+1}v+\mathrm{T}v,$$

where $T \in \mathcal{K}L_2^{+}(S^{n-1})$. By applying theorem 3.3.4 we find that

$$\left[\sum_j\prod_k\mathrm{P}^{+}\mathfrak{U}_{jk}\right]v = \zeta_1\mathrm{E}(\lambda)^{-1}\sum_j\prod_k\Phi_{jk}(\phi,\omega)\mathrm{E}(\lambda)\zeta_{N+1}v+\mathrm{T}_1v;$$

where also $\mathrm{T}_1 \in \mathcal{K}L_2^{+}(S^{n-1})$. Now the existence of a sequence $\{v_q\}$ is proved literally as in lemma 5.3.3. An obvious consequence is the inequality

$$\inf_{\mathrm{T}}\left\|\sum_j\prod_k\mathrm{P}^{+}\mathfrak{U}_{jk}(\lambda)+\mathrm{T};L_2^{+}(S^{n-1})\to L_2^{+}(S^{n-1})\right\| \geqslant \tag{2.2}$$

$$\geqslant \sup_{(\phi,\omega)\in V^{+}(n,2)}\left|\sum_j\prod_k\Phi_{jk}(\phi,\omega)\right|.$$

This inequality gives the necessary estimate for $\|\Sigma_j\Pi_kB_{\Phi_{jk}}(\phi);\ \mathcal{P}(\phi)\|$ for $\phi \in S_{+}^{n-1}$ (cf. (2.1)). It remains to estimate this quantity for $\phi \in \partial S_{+}^{n-1}$.

Introduce the operator $\mathcal{F}(\phi,\mathring{\omega}) = \Sigma_j \Pi_k \Pi^- \Phi_{jk}(\phi,\kappa(\mathring{\omega},\cdot))\colon \mathcal{H}^-(\mathbb{R}) \to \mathcal{H}^-(\mathbb{R})$. Let $(\phi_0,\omega_0) \in V^+(n,2)$ and $\phi_0,\omega_0 \in \partial S_+^{n-1}$ (so that $\omega_0 = \mathring{\omega}_0$). Let also $u \in \mathcal{H}^-(\mathbb{R})$, $\|u\| = 1$, and $\|\mathcal{F}(\phi_0,\omega_0)u; \mathcal{H}^-(\mathbb{R})\| > \|\mathcal{F}(\phi_0,\omega_0); \mathcal{H}^-(\mathbb{R}) \to \mathcal{H}^-(\mathbb{R})\| - \epsilon$ for a given $\epsilon > 0$. Put, as in the proof of proposition 1.10, $v(\omega',\omega_n) = (1+(\omega_n/|\omega'|)^2)^{(i\sigma+n/2)/2}\chi(\omega'/|\omega'|)u(\omega_n/|\omega'|)$; here $\chi \in C^\infty(S^{n-2})$, the support of χ is concentrated near ω_0, and $\|\chi;L_2(S^{n-1})\| = 1$. Then $v \in \mathcal{H}_-(-i\sigma-n/2)$, $\|v;L_2(S^{n-1})\| = 1$. Denote by g a smooth real-valued function of a point $(\omega',\omega_n) \in S^{n-1}$ which is independent of ω_n outside small neighborhoods of the poles $(0,\pm 1)$. We will assume that g has been extended onto $\mathbb{R}^n \setminus 0$ as a homogeneous function of degree zero, and that $\nabla g(\omega_0) = \phi_0$. We have

$$e^{i\mu g}(\phi)\mathrm{E}_{\psi\to\phi}(\sigma)\Psi(\phi,\psi)\mathrm{E}_{\omega\to\psi}(\sigma)^{-1}e^{-i\mu g(\omega)}v(\omega) + \tag{2.3}$$
$$-\Psi(\phi,\mu\nabla g(\phi)+\sigma\phi)v(\phi) = o(1)$$

as $|\mu| + |\sigma| \to \infty$, $\mu \in \mathbb{R}$ (cf. (5.3.4)); in the first term of this formula we may replace $\mathrm{E}(\sigma)^{\pm 1}$ by $\mathrm{E}(\sigma)^{\mp 1}$ and, simultaneously, μ by $-\mu$. Relation (2.3) means that as $\mu \to \infty$ and σ is fixed, the support of the function $w(\psi) = \mathrm{E}_{\omega\to\psi}(\sigma)^{-1}e^{-i\mu g(\omega)}v(\omega)$ 'contracts' towards the point $\phi_0 = \nabla g(\omega_0)$; more precisely, for $\epsilon > 0$ and a small neighborhood $\mathcal{U}$ of ϕ_0 given, and for μ large, the inequality $\int_{S^{n-1}\setminus\mathcal{U}} |w(\psi)|^2 d\psi < \epsilon$ holds.

Since $v \in \mathcal{H}_-(-\sigma-n/2)$, up to a small term belonging to $L_2(S^{n-1})$ the function w belongs to $L_2^+(S^{n-1})$. Therefore we may assume that for $\mu >> 1$,

$$\int_{S^{n-1}\setminus(\mathcal{U}\cap S_+^{n-1})} |w(\psi)|^2 d\psi < \delta,$$

where δ is an arbitrarily given positive number.

Now take μ sufficiently large and modify the function w constructed to be equal to zero outside a small neighborhood $\mathcal{V}$ of ϕ_0 on $\overline{S_+^{n-1}}$. After normalization in $L_2(S^{n-1})$, the function thus obtained will also be denoted by w.

Let $\zeta,\eta \in C^\infty(S^{n-1})$, with both functions vanishing outside a small neighborhood of ϕ_0, with $\zeta = 1$ on $\mathcal{V}$, and with $\zeta\eta = \zeta$. Check that

$$\left[\sum_j \prod_k \mathrm{P}^+ \mathfrak{U}_{jk}(\lambda)\right]\zeta = \tag{2.4}$$

$$= \left[\sum_j \prod_k \mathrm{P}^+ \mathrm{E}(\lambda)^{-1}\Phi_{jk}(\phi,\cdot)\mathrm{E}(\lambda)\right]\zeta + \mathrm{T}_1 =$$

$$= \eta\left[\sum_j \prod_k \mathrm{P}^+ \mathrm{E}(\lambda)^{-1}\Phi_{jk}(\phi_0,\cdot)\mathrm{E}(\lambda)\right]\zeta + \mathrm{T} + \mathrm{S},$$

where $\mathrm{T},\mathrm{T}_1 \in \mathcal{K}L_2^+(S^{n-1})$ and $\mathrm{S}: L_2^+(S^{n-1}) \to L_2^+(S^{n-1})$ is an operator with small norm. By applying (2.4) and (1.7) we find that

$$\left[\sum_j \prod_k \mathrm{P}^+ \mathfrak{U}_{jk}(\lambda)\right] w - \mathrm{T}w - \mathrm{S}w =$$

$$= \mathrm{E}(\sigma)^{-1}\{[\mathrm{J}(\sigma)^{-1}\mathcal{F}(\phi_0,\mathring{\omega})\mathrm{J}(\sigma)(\mathrm{E}(\sigma)w)(\kappa(\mathring{\omega},\cdot))]\circ\kappa^{-1}(\omega)\}.$$

It follows from the way by which w was constructed that the norm in $L_2^+(S^{n-1})$ of the righthand side of this equation can be estimated from below by $\|\mathcal{F}(\sigma_0,\omega_0);\mathcal{H}^-(\mathbb{R})\to\mathcal{H}^-(\mathbb{R})\| - \delta$, for $\delta > 0$ small (cf. the proof of proposition 1.10).

These considerations allow us to become convinced of the existence of a sequence $\{w_n\}$ of normalized functions converging weakly to zero in $L_2^+(S^{n-1})$ and such that $\|(\Sigma_j\Pi_k\Pi^-\mathfrak{U}_{jk}(\lambda))w_n;\ L_2(S^{n-1})\| \to \|\mathcal{F}(\phi_0,\omega_0);\ \mathcal{H}^-(\mathbb{R})\to\mathcal{H}^-(\mathbb{R})\|$. This implies the inequality

$$\inf_{\mathrm{T}} \left\|\sum_j \prod_k \mathrm{P}^+ \mathfrak{U}_{jk}(\lambda) + \mathrm{T}; L_2^+(S^{n-1}) \to L_2^+(S^{n-1})\right\| \geqslant$$

$$\geqslant \sup_{\substack{(\phi,\omega)\in V^*(n,2)\\ \phi\in\partial S_+^{n-1}}} \left\|\sum_j \prod_k \Pi^- \Phi_{jk}(\phi,\kappa(\mathring{\omega},\cdot));\ \mathcal{H}^-(\mathbb{R})\to\mathcal{H}^-(\mathbb{R})\right\|.$$

Again using (2.2), we are led to (2.1).

II. We now show that if

$$\alpha > \left\|\sum_j \prod_k B_{\Phi_{jk}}; \mathfrak{B}\right\|, \tag{2.5}$$

then $\mathrm{A} = \Sigma_j\Pi_k\mathrm{P}^+\mathfrak{U}_{jk}(\lambda) - \alpha\mathrm{I}$ is a Fredholm operator. It suffices to construct a left and a right regularizer, i.e. operators that invert A up to compact terms.

Let $\{U_q\}_{q=1}^{Q}$ be a finite open cover of the hemisphere $\overline{S_+^{n-1}}$, with $\partial S_+^{n-1} \cap U_1 = \varnothing$, while the sets $U_2, \cdots, U_Q$ have small diameters and

$U_q \cap \partial S_+^{n-1} \neq \emptyset$, $q = 2, \cdots, Q$. Let also $\{\zeta_q\}$ be a partition of unity subordinate to this cover.

Inequality (2.5) implies that $|\Sigma_j \Pi_k \Phi_{jk}(\phi,\omega)| - \alpha < 0$ in a neighborhood of $V^+(n,2)$. Denote by Ψ a $C^\infty(S^{n-1} \times S^{n-1})$ function which equals $(\Sigma_j \Pi_k \Phi_{jk} - \alpha)^{-1}$ in a neighborhood of $V^+(n,2)$. Put $R_1 = P^+ E(\lambda)^{-1} \Psi(\phi,\omega) E(\lambda)$. We have

$$R_1 \zeta_1 A = P^+ E(\lambda)^{-1} \Psi(\phi,\omega) E(\lambda) \zeta_1 \left[\sum_j \prod_k \mathfrak{U}_{jk}(\lambda) - \alpha I\right] + T,$$

where $T \in \mathcal{KL}_2^+(S^{n-1})$ (we have used proposition 3.1.6). In the sequel, distinct compact operators are denoted by the same letter T. Applying theorem 3.3.4 we find

$$R_1 \zeta_1 A = \zeta_1 E(\lambda)^{-1} \Psi(\phi,\omega) \left[\sum_j \prod_k \Phi_{jk}(\phi,\omega) - \alpha\right] E(\lambda) + T.$$

The function $\Psi(\Sigma_j \Pi_k \Phi_{jk} - \alpha)$ equals 1 in a neighborhood of $V^+(n,2)$. This and lemma 5.3.5 imply that

$$R_1 \zeta_1 A = \zeta_1 I + T. \tag{2.6}$$

Now consider the operator $\zeta_q A$ for $q \geqslant 2$; it can be written as

$$(\zeta_q A u)(\phi) = \zeta_q(\phi) \left[\sum_j \prod_k P^+ E(\sigma)^{-1} \Phi_{jk}(\phi,\cdot) E(\sigma) - \alpha I\right] u + Tu \tag{2.7}$$

(we have used proposition 3.2.2 in order to replace λ by $\sigma = \mathrm{Re}\lambda$, and proposition 3.2.1 in order to give all functions Φ_{jk} the same argument ϕ). Choose an arbitrary point $\phi_q \in \operatorname{supp} \zeta_q \cap \partial S_+^{n-1}$, and introduce the set $\Omega(q,\epsilon) = \{w \in S^{n-1}: |\phi_q \hat{\omega}| < \epsilon\} \cup \{w \in S^{n-1}: \omega = (\omega', \omega_n), |\omega_n| > 1 - \epsilon\}$, where ϵ is given and small. Let $p: S^{n-1} \times S^{n-1} \to S^{n-1}$ be projection on the second factor. We may assume that the diameter of U_q is small such that $p(\{(\phi,\omega) \in S^{n-1} \times S^{n-1}: \phi \in U_q, \phi\omega = 0\}) \subset \Omega(q,\epsilon)$. Let $\chi_{q,\epsilon}$ be the characteristic function of $\Omega(q,\epsilon)$, and let $\eta_q \in C_0^\infty(U_q)$, $\zeta_q \eta_q = \zeta_q$. Represent each operator $\eta_q E(\sigma)^{-1} \Phi_{jk}(\phi,\omega) E(\sigma)$ as a sum

$$\eta_q(\phi) E(\sigma)^{-1} \Phi_{jk}(\phi_q,\omega) \chi_{q,\epsilon}(\omega) E(\sigma) + \tag{2.8}$$

$$+ \eta_q(\phi) E(\sigma)^{-1} \Phi_{jk}(\phi_q,\omega)(1 - \chi_{q,\epsilon}(\omega)) E(\sigma) +$$

$$+\eta_q(\phi)\mathrm{E}(\sigma)^{-1}(\Phi_{jk}(\phi,\omega)-\Phi_{jk}(\phi_q,\omega))\mathrm{E}(\sigma).$$

By lemma 5.3.5, the second term in (2.8) is a compact operator, while the third term has small norm.

Consider the operator $\mathfrak{F}(\phi_q,\mathring{\omega}) = \Sigma_j \Pi_k \Pi^- \Phi_{jk}(\phi_q,\kappa(\mathring{\omega},\cdot))$. If $\phi_q\mathring{\omega} = 0$, then $\|\mathfrak{F}(\phi_q,\mathring{\omega});\mathcal{H}^-(\mathbb{R})\rightarrow\mathcal{H}^-(\mathbb{R})\| < \alpha$ by requirement. Consequently, this inequality holds for $|\phi_q\mathring{\omega}| < \epsilon$, since ϵ is small. Modify (if necessary) the Φ_{jk} on $S^{n-1}\setminus\Omega(q,\epsilon)$ such that the inequality $\|\mathfrak{F}(\phi_q,\mathring{\omega})\| < \alpha$ holds for all $\mathring{\omega} \in \partial S_+^{n-1}$; the new functions will be denoted, as before, by Φ_{jk}. It is clear that we have to add a compact operator to (2.8) as a result of this modification. This and (1.7) implies

$$(\zeta_q\mathrm{A}u)(\phi) = \zeta_q(\phi)\mathrm{E}(\sigma)^{-1}\{[\mathrm{J}(\sigma)^{-1}(\mathfrak{F}(\phi_q,\mathring{\omega})-\alpha\mathrm{I})]\mathrm{J}(\sigma)\times \tag{2.9}$$

$$\times(\mathrm{E}(\sigma)u)(\kappa(\mathring{\omega},\cdot))]\circ\kappa^{-1}(\omega)\}+\mathrm{T}u+\mathrm{S}u,$$

where $\mathrm{T} \in \mathcal{K}L_2^+(S^{n-1})$, S is an operator with small norm, and $\|\mathfrak{F}(\phi_q,\mathring{\omega});\mathcal{H}^-(\mathbb{R})\rightarrow\mathcal{H}^-(\mathbb{R})\| < \alpha$ for all $\mathring{\omega} \in \partial S_+^{n-1}$.

Put $\Xi(\phi_q,\mathring{\omega}) = (\mathfrak{F}(\phi_q,\mathring{\omega})-\alpha\mathrm{I})^{-1}$,

$$(\mathrm{R}_q v)(\phi) = \mathrm{E}(\sigma)^{-1}\{[\mathrm{J}(\sigma)^{-1}\Xi(\phi_q,\mathring{\omega})\mathrm{J}(\sigma)(\mathrm{E}(\sigma)v)(\kappa(\mathring{\omega},\cdot))]\circ\kappa^{-1}(\omega)\}$$

and introduce the operator $\mathrm{R} = \Sigma_{q=1}^{Q}\mathrm{R}_q\zeta_q\colon L_2^+(S^{n-1})\rightarrow L_2^+(S^{n-1})$. Note that by using proposition 3.2.1 we may replace in the first righthand term in (2.9) the function ζ_q by u. (A compact operator should be added to the righthand side of (2.9) in this case.)

Therefore, for $q = 2,\cdots,Q$,

$$\mathrm{R}_q\zeta_q\mathrm{A} = \zeta_q\mathrm{I}+\mathrm{T}+\mathrm{S}. \tag{2.10}$$

Combining (2.6) and (2.10) we find that $\mathrm{RA} = \Sigma_{q=1}^{Q}\mathrm{R}_q\zeta_q\mathrm{A} = \mathrm{I}+\mathrm{T}+\mathrm{S}$, where $\mathrm{T} \in \mathcal{K}L_2^+(S^{n-1})$, and S is an operator with small norm. Thus, $(\mathrm{I}+\mathrm{S})^{-1}\mathrm{R}$ inverts A up to a compact term. The existence of a right regularizer is proved similarly.

III. Finally, we will prove the equation

$$\inf_{\mathrm{T}}\left\|\sum_j\prod_k \mathrm{P}^+\mathfrak{U}_{jk}(\lambda)+\mathrm{T};L_2^+(S^{n-1})\rightarrow L_2^+(S^{n-1})\right\| = \tag{2.11}$$

$$= \left\|\sum_j\prod_k \mathrm{B}_{\Phi_{jk}};\mathfrak{P}\right\|.$$

The preceding part of the proof implies that the inequality

$$\sup \sigma_{\text{ess}}(D^*D) \leqslant$$

holds for the operator $D = \Sigma_j \Pi_k P^+ \mathfrak{U}_{jk}(\lambda)$ ($\sigma_{\text{ess}}(G)$ is the essential spectrum of an operator G). This means that

$$\inf_{T \in \mathcal{K}L_2^+(S^{n-1})} \|D+T; L_2^+(S^{n-1}) \to L_2^+(S^{n-1})\| \leqslant \left\|\sum_j \prod_k B_{\Phi_{jk}}; \mathfrak{P}\right\|$$

(compare with the proof of lemma 5.3.6). Using (2.1) we are led to (2.11). ■

3. The spectrum of the algebra $\mathfrak{G}^+(\lambda)$. By proposition 2.6 the description of the spectrum of $\mathfrak{G}^+(\lambda)$ reduces to the description of the spectrum of $\mathfrak{P}$. Denote by $\hat{\mathfrak{P}}$ the disjoint union of the manifolds $V^+(n,2) = \{(\phi,\omega): \phi \in \overline{S^{n-1}_+},\ w \in S^{n-1},\ \phi\omega = 0\}$, $V(n-1,2) = \{(\phi,\omega) \in \partial S^{n-1}_+ \times \partial S^{n-1}_+ :\ \phi\omega = 0\}$ and the set $\partial S^{n-1}_+ \times [-1,1]$. Identify the sets $\partial S^{n-1}_+ \times \{\pm 1\} \subset \partial S^{n-1}_+ \times [-1,1]$ with the sets $\partial S^{n-1}_+ \times \{(0,\pm 1)\} \subset \overline{S^{n-1}_+} \times S^{n-1}$, respectively. We will regard $[-1,1]$ as the diameter of S^{n-1} with end points at the poles $(0,\pm 1)$. We introduce a topology on $\hat{\mathfrak{P}}$. A fundamental system of neighborhoods on $\hat{\mathfrak{P}}$ of a point $(\phi_0,\omega_0) \in V^+(n,2)$, $\phi_0 \notin \partial S^{n-1}_+$, is formed by ordinary neighborhoods of this point on $V^+(n,2)$. Let $\phi_0 \in \partial S^{n-1}_+$, $(\phi_0,\omega_0) \in V^+(n,2)$, with, moreover, ω_0 not one of the poles $(0,\pm 1)$. For such a point (ϕ_0,ω_0) a fundamental system of neighborhoods is formed by sets $\mathcal{U}(\phi_0,\omega_0) \cup \mathcal{V}(\phi_0,\omega_0)$, where $\mathcal{U}(\phi_0,\omega_0)$ is a neighborhood of (ϕ_0,ω_0) on $V^+(n,2)$ and $\mathcal{V}(\phi_0,\omega_0)$ is a neighborhood of $(\phi_0,\mathring{\omega}_0)$ on $V(n-1,2)$. (Recall that $\mathring{\omega}$ denotes the point of intersection of the meridian passing through $\omega = (\omega',\omega_n)$, $\omega' \neq 0$, with the equator ∂S^{n-1}_+, i.e. $\mathring{\omega} = \omega'/|\omega'|$.) If $(\phi_0,\omega_0) \in V(n-1,2)$, a neighborhood $\mathcal{V}(\phi_0,\omega_0)$ of it on $V(n-1,2)$ will also be a neighborhood on $\mathfrak{P}$. Further, let $p: V(n-1,2) \to \partial S^{n-1}_+$ be the composite of the inclusion $V(n-1,2) \subset \partial S^{n-1}_+ \times \partial S^{n-1}_+$ and projection $\partial S^{n-1}_+ \times \partial S^{n-1}_+ \to \partial S^{n-1}_+$ on the first factor. Neighborhoods of $(\phi_0,t_0) \in \partial S^{n-1}_+ \times (-1,1)$, forming a fundamental system of neighborhoods, are unions $\mathcal{W}(\phi_0) \times \mathcal{T}(t_0) \cup p^{-1}(\mathcal{W}(\phi_0))$, where $\mathcal{T}(t_0)$ is a neighborhood of t_0 on the open interval $(-1,1)$, and $\mathcal{W}(\phi_0)$ is a neighborhood of ϕ_0 on ∂S^{n-1}_+. Finally, for a point (ϕ_0,ω_0) where $\phi_0 \in \partial S^{n-1}_+$ and ω_0 is one of the poles $(0,\pm 1)$, sets $\mathcal{W}(\phi_0) \times \mathcal{T}(\omega_0) \cup \mathcal{U}(\phi_0,\omega_0) \cup p^{-1}(\mathcal{W}(\phi_0))$ form a fundamental system of neighborhoods. Here $\mathfrak{T}(\omega_0)$ is a neighborhood of ω_0 on the closed interval $[-1,1]$ (we require in the sequel that the end points of this interval are identified with the poles).

T h e o r e m 2.7. *Let* $\mathfrak{U}(\lambda) = \mathrm{E}_{\omega\to\phi}(\lambda)^{-1}\Phi(\phi,\omega)\mathrm{E}_{\psi\to\omega}(\lambda)$. *Then* $\mathrm{P}^+\mathfrak{U}(\lambda) \in \mathfrak{G}^+(\lambda)$, *and the following assertions hold:*

1) *the maps*

$$\pi(\phi,\omega):\mathrm{P}^+\mathfrak{U}(\lambda)\mapsto\Phi(\phi,\omega),\ (\phi,\omega)\in V^+(n,2);$$

$$\pi(\phi,t):\mathrm{P}^+\mathfrak{U}(\lambda)\mapsto\frac{1-t}{2}\Phi(\phi,(0,-1))+\frac{1+t}{2}\Phi(\phi,(0,1)),$$

$$\phi\in\partial S_+^{n-1},\ t\in[-1,1];$$

$$\tau(\phi,\omega):\mathrm{P}^+\mathfrak{U}(\lambda)\mapsto\Pi^-\Phi(\phi,\kappa(\omega,\cdot)),\ (\phi,\omega)\in V(n-1,2),$$

induce pairwise inequivalent irreducible representations of the quotient algebra $\mathfrak{G}^+(\lambda)/\mathcal{K}L_2^+(S^{n-1})$ *(the representations* $\pi(\phi,\omega)$, $\pi(\phi,t)$ *are one-dimensional, and the* $\tau(\phi,\omega)$ *are representations in* $\mathcal{H}^-(\mathbb{R})$*).*

2) *Every irreducible representation of* $\mathfrak{G}^+(\lambda)$ *is equivalent to either a representation listed in* 1) *or to the identity representation.*

3) *The correspondence* $\pi(\phi,\omega)\mapsto(\phi,\omega)$, $\pi(\phi,t)\mapsto(\phi,t)$, $\tau(\phi,\omega)\mapsto(\phi,\omega)$ *defines a bijection of the spectrum of* $\mathfrak{G}^+(\lambda)/\mathcal{K}L_2^+(S^{n-1})$ *onto* $\hat{\mathfrak{P}}$. *The topology on* $\hat{\mathfrak{P}}$ *is the Jacobson topology.*

P r o o f. The validity of the first assertion follows from proposition 2.6 and theorem 1.6. The second assertion can be verified by a reasoning similar to the corresponding part of the proof of theorem 1.7. Direct comparison of the topology on $\hat{\mathfrak{P}}$ and the Jacobson topology establishes the third assertion. ■

Note that, for any irreducible representation π, the algebra $\pi(\mathfrak{G}^+(\lambda))$ contains all compact operators on the representation space. The same is true for the algebra $\mathfrak{G}_0^+(\lambda)$. Hence $\mathfrak{G}^+(\lambda)$ and $\mathfrak{G}_0^+(\lambda)$ are type I algebras ([24], theorem 9.1).

§3. The algebras $\mathfrak{G}_c^+(l_\beta)$

We introduce the algebra $\mathfrak{G}_c^+(l_\beta)$ of operator-functions on the line $l_\beta = \{\lambda \in \mathbf{C}: \operatorname{Im}\lambda = \beta\}$, $\beta \neq \pm(k+n/2)$, $k = 0, 1, \cdots$, generated by the functions $\lambda \mapsto \mathrm{P}^+\mathfrak{U}(\lambda)$ where $\mathfrak{U}(\lambda) = \mathrm{E}(\lambda)^{-1}\Phi(\omega)\mathrm{E}(\lambda)$, $\Phi \in C^\infty(S^{n-1})$. The operations in this algebra are pointwise, and the norm is defined by

$$\|\mathfrak{B};\mathfrak{G}_c^+(l_\beta)\| = \sup_{\lambda \in l_\beta} \|\mathfrak{B}(\lambda); L_2^+(S^{n-1}) \to L_2^+(S^{n-1})\|.$$

For $\beta = 0$ this algebra is isomorphic to any of the algebras $\mathfrak{G}_c^+(\lambda)$, $\operatorname{Im}\lambda = 0$. Therefore we will assume that also $\beta \neq 0$. By proposition 1.8 a representation $\pi(\lambda): \mathrm{P}^+\mathfrak{U} \mapsto \mathrm{P}^+\mathfrak{U}(\lambda)$ of $\mathfrak{G}_c^+(l_\beta)$ is irreducible. It turns out that representations corresponding to distinct λ's are inequivalent. The proof of this assertion is preceded by a lemma.

Define in the halfplane $\operatorname{Im} t \leq 0$ with slit along the interval $\{t: \operatorname{Re} t = 0, -1 \leq \operatorname{Im} t \leq 0\}$ the function $t \mapsto f_1(t) = \exp\{[(-i\lambda - n/2)/2]\ln(1+t^2)\}$, where $\ln(1+t^2)$ is understood to mean the principal branch on the positive real line.

L e m m a 3.1. *There exists for any sufficiently large number k a function $u \in C^\infty(S^{n-1}) \cap E(\lambda)L_2^+(S^{n-1})$ satisfying: 1) the analytic extension of $t \mapsto (1+t^2)^{-(i\lambda+n/2)/2}(u \circ \kappa)(\hat\omega, t)$ is zero free in the halfplane $\operatorname{Im} t \leq 0$ (for all $\hat\omega \in S^{n-2}$); and 2) as $t \to \infty$ in the halfplane $\operatorname{Im} t \leq 0$, then*

$$(1+t^2)^{-(i\lambda+n/2)/2}(u \circ \kappa)(\hat\omega, t) = f_1(t) + O(|t|^{\operatorname{Im}\lambda - k - 1 - n/2}).$$

P r o o f. Introduce, in the halfplane $\operatorname{Im} t < 1$, the function $f_2(t) = \exp\{(-i\lambda - n/2)\ln(t-i)\}$ (where the branch of the logarithm is chosen as before). The quotient $t \mapsto f_3(t) = f_1(t)/f_2(t)$ can be extended as a holomorphic function to the plane with slit along the interval $[-i, i]$ of the imaginary axis. Hence, in the domain $|t-i| > 2$ the function f_3 has a Laurent expansion:

$$f_3(t) = \sum_{j=0}^{\infty} a_j(t-i)^{-j}. \tag{3.1}$$

This implies that outside the slit in the lower halfplane we have $f_1(t) = f(t) + R_k(t)$, where

$$f(t) = f_2(t)\sum_{j=0}^{k} a_j(t-i)^{-j},$$

$|R_k(t)| = O(|t|^{\mathrm{Im}\lambda-k-1-n/2})$ as $t\to\infty$, $\mathrm{Im}\,t\leqslant 0$. The function f_2 is zero free in $\mathrm{Im}\,t\leqslant 0$, while a partial sum of the series (3.1) can have at most finitely many zeros in this halfplane (being a polynomial in $(t-i)^{-1}$). This means that f is holomorphic for $\mathrm{Im}\,t\leqslant 0$ and has at most finitely many zeros.

If $f(t)\neq 0$ for $\mathrm{Im}\,t\leqslant 0$, we put $u(x) = |x'|^{-i\lambda-n/2}f(x_n/|x'|)$. Outside the poles $(0,\pm 1)$ the function u is infinitely differentiable on S^{n-1}. The estimate for R_k and the definitions of f and u imply that in a neighborhood of the north pole $u(x) = 1+O(|x'|^{k+1})$; this ensures that u has continuous derivatives up to order k inclusive at the north pole. Near the south pole we have $u(x) = \exp\{-\pi i(i\lambda+n/2)\}+O(|x'|^{k+1})$, and the same conclusion holds.

The function u is clearly homogeneous of degree $-i\lambda-n/2$, and can be analytically extended to the halfplane $\mathrm{Im}\,x_n\leqslant 0$. The formula $f = f_1-R_k$ and the estimate for R_k give the inequality necessary in order to conclude $u\in\mathcal{H}_-(-i\lambda-n/2)$; thus, $u = \mathrm{E}(\lambda)v$, where $v\in H_+^{-\mathrm{Im}\lambda}(S^{n-1})$ (theorem 1.6.5). Since k may be chosen arbitrarily large, by proposition 1.5.5 we have $v = \mathrm{E}(\lambda)^{-1}u\in L_2(S^{n-1})$. So, $v\in H_+^{-\mathrm{Im}\lambda}(S^{n-1})\cap L_2(S^{n-1})\subset L_2^+(S^{n-1})$, i.e. u is the function looked for.

Now assume that f has, in $\mathrm{Im}\,t\leqslant 0$, only the zeros $t_1,\cdots,t_m$ with multiplicities $\kappa_1,\cdots,\kappa_m$. Let q be a fixed natural number. If a_j is a negative number that is large in absolute value, and if $t_j^* = t_j-ia_j$, then the function

$$p_j(t) = ((t-t_j^*)^q+(a_ji)^q)((t-t_j^*)^q-(a_ji)^q)^{-1}$$

has the unique pole $t = t_j^*$ in $\mathrm{Im}\,t\leqslant 0$, and is zero free there. The function $g(t) = f(t)p_1(t)^{\kappa_1}\cdots p_m(t)^{\kappa_m}$ is holomorphic and zero free in $\mathrm{Im}\,t\leqslant 0$. Moreover, $p_j(t) = 1+O(|t|^{-q})$ as $t\to\infty$. Therefore, by choosing q large we may define u by $u(x) = |x'|^{-i\lambda-n/2}g(x_n/|x'|)$. ■

P r o p o s i t i o n 3.2. *Let $\pi(\lambda)$, $\lambda\in l_\beta$, be the representation of $\mathfrak{G}_c^+(l_\beta)$ defined by $\pi(\lambda)\mathfrak{B} = \mathfrak{B}(\lambda)$, $\mathfrak{B}\in\mathfrak{G}_c^+(l_\beta)$. For distinct numbers μ and ν on l_β the representations $\pi(\mu)$ and $\pi(\nu)$ are inequivalent.*

P r o o f. $\mathfrak{G}_c^+(l_\beta)$ contains both functions $\lambda\mapsto \mathrm{P}^+\mathfrak{U}(\lambda) = \mathrm{P}^+\mathrm{E}(\lambda)^{-1}\Phi(\omega)\mathrm{E}(\lambda)$ and $\lambda\mapsto\mathrm{P}^+\mathfrak{U}(\lambda)^* = \mathrm{P}^+\mathrm{E}(\bar\lambda)^{-1}\overline{\Phi(\omega)}\mathrm{E}(\bar\lambda)$, hence we may assume, without loss of generality, that $\beta = \mathrm{Im}\,\lambda$ is positive.

Suppose $\pi(\mu)$ and $\pi(\nu)$ are equivalent. Then there is a unitary operator $U: L_2^+(S^{n-1}) \to L_2^+(S^{n-1})$ such that $U^* P^+ E(\nu)^{-1} \Phi E(\nu) U = P^+ E(\mu)^{-1} \Phi E(\mu)$. Assume that the function $\mathbb{R}^n \setminus 0 \ni \xi \to \Phi(\xi)$, homogeneous of degree zero, is sufficiently smooth on S^{n-1} in order that $\mathfrak{U}(\lambda)$ be continuous (cf. proposition 3.1.1), and that the function $\xi_n \mapsto \Phi(\xi', \xi_n)$ can be analytically extended, for every $\xi' \neq 0$, to the halfplane $\operatorname{Im} \xi_n < 0$, while remaining bounded. Then, by (1.6.5), P^+ may be cancelled. Thus, $U^* E(\nu)^{-1} \Phi E(\nu) U = E(\mu)^{-1} \Phi E(\mu)$, i.e. $\Phi E(\nu) U E(\mu)^{-1} = E(\nu) U E(\mu^{-1}) \Phi$ (here Φ denotes the operator of multiplication by the function Φ). This implies

$$E(\nu) U E(\mu)^{-1} \Phi u = h \Phi u, \tag{3.2}$$

where $u \in E(\mu) L_2^+(S^{n-1})$ is the function of lemma 3.1, $h = (E(\nu) U E(\mu)^{-1} u)/u \in H^\beta(S^{n-1})$; the membership is valid, for $E(\nu): L_2(S^{n-1}) \to H^\beta(S^{n-1})$ is continuous (proposition 1.5.5), while u may be assumed smooth such that u and u^{-1} are multipliers in $H^\beta(S^{n-1})$.

Consider the set $\mathfrak{L} = \{v : v = \Phi u\}$, where u is the, fixed, function of lemma 3.1 (as $\lambda = \mu$), while Φ runs through the set of functions homogeneous of degree zero and having only the properties listed. The set $\mathfrak{L}$ is dense in the space $E(\mu) L_2^+(S^{n-1})$ (with respect to the norm of $H^\beta(S^{n-1})$). Indeed, $E(\mu) C_0^\infty(S_+^{n+1}) \subset C^\infty(S^{n-1})$ and for a $v \in E(\mu) C_0^\infty(S_+^{n-1})$, the function $t \mapsto (1+t^2)^{-(i\mu+n/2)/2} (v \circ \kappa)(\mathring{\omega}, t)$ can be analytically extended to the halfplane $\operatorname{Im} t \leq 0$ (theorem 1.6.5). It follows immediately from the form of the kernel of the operator $E(\mu)$ (cf. (1.2.1)) that for $|t|$ large, $\operatorname{Im} t \leq 0$, we have $(1+t^2)^{-(i\mu+n/2)/2} (v \circ \kappa)(\mathring{\omega}, t) = t^{-i\mu - n/2} b(\mathring{\omega}, t)$, with $|b| \leq \text{const.}$ Thus, $t \mapsto (\Phi \circ \kappa)(\mathring{\omega}, t) = (v \circ \kappa)(\mathring{\omega}, t)/(u \circ \kappa)(\mathring{\omega}, t)$ is a bounded analytic function in the lower halfplane. In other words, the function $\Phi = v/u$, which is homogeneous of degree zero, has the properties listed at the beginning of the proof. Thus, any element $v \in E(\mu) C_0^\infty(S_+^{n-1})$ can be written as $v = \Phi u$, and hence $\mathfrak{L}$ is dense in $E(\mu) L_2^+(S^{n-1})$. Together with (3.2) this leads to: $E(\nu) U E(\mu)^{-1} w = h w$ for every $w \in E(\mu) L_2^+(S^{n-1})$. Whence,

$$U = E(\nu)^{-1} h E(\mu). \tag{3.3}$$

Note that the proposition concerning the existence of an intertwining operator U implies that $(U^{-1})^* P^+ E(\bar{\mu})^{-1} \Phi E(\bar{\mu}) U^{-1} = P^+ E(\bar{\nu})^{-1} \Phi E(\bar{\nu})$. Now, by a reasoning similar as in the proof of (3.3) we obtain an analogous expression for U^{-1}:

$$U^{-1} = E(\bar{\mu})^{-1} h_1 E(\bar{\nu}), \tag{3.4}$$

where $h_1 = (E(\bar{\mu})U^{-1}E(\bar{\nu})^{-1}u_1)/u_1$ and $u_1 \in E(\bar{\nu})L_2^+(S^{n-1})$ is a function whose existence is ensured by lemma 3.1.

Using the relation $U^* = U^{-1}$ and (3.3), (3.4), we are led to the formula $(E(\nu)^{-1}hE(\mu))^* = E(\bar{\mu})^{-1}h_1E(\bar{\nu})$, which can be rewritten as $E(\bar{\mu})^{-1}Q\bar{h}E(\bar{\nu}) = E(\bar{\mu})^{-1}h_1E(\bar{\nu})$, where $Q: H^{-\beta}(S^{n-1}) \to E(\bar{\mu})L_2^+(S^{n-1})$ is the projection operator. This means that for any function $v \in E(\bar{\nu})C_0^\infty(S_+^{n-1})$ the element $Q\bar{h}v - h_1 v \in E(\bar{\mu})L_2^+(S^{n-1})$ is the zero element. Hence for $w \in E(\mu)C_0^\infty(S_+^{n-1})$ we have

$$0 = (Q\bar{h}v - h_1 v, w) = (Q\bar{h}v - Qh_1 v, w) = \tag{3.5}$$
$$(\bar{h}v - h_1 v, Q^* w) = (\bar{h}v - h_1 v, w),$$

since Q^* is the projection operator $H^\beta(S^{n-1}) \to E(\mu)L_2^+(S^{n-1})$. (Here (f,g) denotes the extension of the scalar product in $L_2(S^{n-1})$.) In order to shorten the notation, for a function $\omega \mapsto v(\omega)$ on S^{n-1} we will write $v(\mathring{\omega},t)$ instead of $(v\circ\kappa)(\mathring{\omega},t)$, where $\mathring{\omega} = \omega'/|\omega'|$, $t = \omega_n/|\omega'|$. Equation (3.5) leads to the relation

$$\int_{S^{n-1}} (\bar{h} - h_1) v\bar{w}\, d\omega =$$

$$= \int_{S^{n-2}} d\mathring{\omega} \int_{-\infty}^{+\infty} (h(\mathring{\omega},t) - h_1(\mathring{\omega},t)) v(\mathring{\omega},t)\overline{w(\mathring{\omega},t)}(1+t^2)^{-n/2} dt = 0.$$

This implies that $\bar{h}(\mathring{\omega},t) = h_1(\mathring{\omega},t)$ for almost all $t \in \mathbb{R}$ and for fixed $\mathring{\omega} \in S^{n-2}$.

By the definitions of h, h_1 (and by theorem 1.6.5), the functions $t \mapsto h(\mathring{\omega},t)(1+t^2)^{(i\mu - i\nu)/2}$ and $t \mapsto h_1(\mathring{\omega},t)(1+t^2)^{(i\bar{\nu} - i\bar{\mu})/2}$ can be analytically extended to the halfplane $\operatorname{Im} t \leqslant 0$. Put

$$f(t) = \begin{cases} h_1(\mathring{\omega},t)(1+t^2)^{(i\bar{\nu} - i\bar{\mu})/2} & \text{for } \operatorname{Im} t \leqslant 0, \\ \overline{h(\mathring{\omega},\bar{t})(1+\bar{t}^2)^{(i\mu - i\nu)/2}} & \text{for } \operatorname{Im} t \geqslant 0. \end{cases} \tag{3.6}$$

Since $h, h_1 \in L_2(S^{n-1})$, it follows that f is locally integrable on the real line. Hence f is an entire function.

Recall that $h = v/u$, where $v = E(\nu)w$, $w \in L_2^+(S^{n-1})$, and u is the function from lemma 3.1 (as $\lambda = \mu$). By applying theorem 1.6.5 we obtain the estimate

$$\int_{\operatorname{Im} t=\tau} |h(\hat{\omega},t)|^2(1+|t|^2)^{-n/2}dt =$$

$$= \int_{\operatorname{Im} t=\tau} \frac{|v(\hat{\omega},t)(1+t^2)^{-(i\nu+n/2)/2}|^2}{|u(\hat{\omega},t)(1+t^2)^{-(i\mu+n/2)/2}|^2}(1+|t|^2)^{-n/2}dt \leqslant$$

$$\leqslant c_1 \int_{\operatorname{Im} t=\tau} \frac{|v(\hat{\omega},t)(1+t^2)^{-(i\nu+n/2)/2}|^2}{(1+|t|^2)^{\beta-n/2}}(1+|t|^2)^{-n/2}dt =$$

$$= c_1 \int_{\operatorname{Im} t=\tau} |v(\hat{\omega},t)(1+t^2)^{-(i\nu+n/2)/2}|^2(1+|t|^2)^{-\beta}dt \leqslant c_2$$

for all $\tau \leqslant 0$. Here $\beta = \operatorname{Im}\mu = \operatorname{Im}\nu$, and the constant c_2 is independent of τ. A similar estimate holds for h_1 too. Hence, for all real τ,

$$\int_{\operatorname{Im} t=\tau} |f(t)|^2(1+|t|^2)^{-n/2}dt \leqslant c = \text{const} < \infty.$$

Now Liouville's theorem implies that f is a polynomial. (In the form necessary for us, this theorem can be found in, e.g., [74].)

For U^{-1} we can derive

$$\mathrm{U}^{-1} = \mathrm{E}(\mu)^{-1}h'\mathrm{E}(\nu), \tag{3.7}$$

where $h' \in H^\beta(S^{n-1})$ (cf. the derivation of (3.3)). The function $t \mapsto h'(\hat{\omega},t)(1+t^2)^{(i\nu-i\mu)/2}$ can be analytically extended to the halfplane $\operatorname{Im} t \leqslant 0$. Formulas (3.3) and (3.7) lead to $hh' = 1$. Hence the function $t \mapsto h(\hat{\omega},t)(1+t^2)^{(i\mu-i\nu)/2}$ is zero free for $\operatorname{Im} t < 0$. A similar conclusion holds for h_1 too. Thus, the function f in (3.6) is zero free in the complex plane. (It has no real zeros since $h' = h^{-1}$ is locally integrable on the real line.) So, $f \equiv$ const. This means that $(h \circ \kappa)(\hat{\omega},t)(1+t^2)^{(i\mu-i\nu)/2} = q(\hat{\omega})$, i.e. $h(\omega) = |\omega'|^{i\nu-i\mu}q(\hat{\omega})$. By (3.3) we hence obtain $\mathrm{U} = \mathrm{E}(\nu)^{-1}|\omega'|^{i\nu-i\mu}q(\hat{\omega})\mathrm{E}(\mu)$.

Introduce the operator $\mathrm{U}_g = \mathrm{E}(\nu)^{-1}h(g\omega)\mathrm{E}(\mu)$, where $g \in SO(n-1)$. The operator $\mathrm{E}(\lambda)$ commutes with the operator $(g^*u)(\omega) = u(g\omega)$, and $\mathrm{U}_g = g^*\mathrm{U}(g^{-1})^*$. Thus, U_g is also unitary. Furthermore,

$$\mathrm{U}_g\mathrm{P}^+\mathfrak{U}(\mu) = g^*\mathrm{U}(g^{-1})^*\mathrm{P}^+\mathfrak{U}(\mu) = \tag{3.8}$$
$$= g^*\mathrm{U}\mathrm{P}^+(g^{-1})^*\mathfrak{U}(\mu) = g^*\mathrm{U}\mathrm{P}^+\mathfrak{U}_{g^{-1}}(\mu)(g^{-1})^*,$$

where $\mathfrak{U}(\mu) = \mathrm{E}(\mu)^{-1}\Phi(\omega)\mathrm{E}(\mu)$, $\mathfrak{U}_{g^{-1}}(\mu) = \mathrm{E}(\mu)^{-1}\Phi(g^{-1}\omega)\mathrm{E}(\mu)$. Since $\mathrm{P}^+\mathfrak{U}_{g^{-1}}(\mu) \in \mathfrak{S}_c^+(\mu)$, we have

$$g^*\mathrm{UP}^+\mathfrak{U}_{g^{-1}}(\mu)(g^{-1})^* = g^*\mathrm{P}^+\mathfrak{U}_{g^{-1}}(\nu)\mathrm{U}(g^{-1})^* = \mathrm{P}^+\mathfrak{U}(\nu)\mathrm{U}_g. \tag{3.9}$$

Formulas (3.8), (3.9) imply that $\mathrm{U}_g\mathrm{P}^+\mathfrak{U}(\mu) = \mathrm{P}^+\mathfrak{U}(\nu)\mathrm{U}_g$. The same is true if μ and ν are replaced by $\bar{\mu}$ and $\bar{\nu}$. Hence U_g is intertwining for the identity representations of the algebras $\mathfrak{S}_c^+(\mu)$ and $\mathfrak{S}_c^+(\nu)$. It is obvious that $\mathrm{U}^{-1}\mathrm{U}_g\mathrm{P}^+\mathfrak{U}(\mu) = \mathrm{U}^{-1}\mathrm{P}^+\mathfrak{U}(\nu)\mathrm{U}_g = \mathrm{P}^+\mathfrak{U}(\mu)\mathrm{U}^{-1}\mathrm{U}_g$, i.e. $\mathrm{U}^{-1}\mathrm{U}_g$ commutes with all elements of $\mathfrak{S}_c^+(\mu)$. Since $\mathfrak{S}_c^+(\mu)$ is irreducible (proposition 1.8), this implies that $\mathrm{U}^{-1}\mathrm{U}_g = c(g)\mathrm{I}$, where $c(g)$ is a complex number, $|c(g)| = 1$ (cf. Chapter 5, §1.4). So, $h(g\omega)/h(\omega) = c(g)$.

The map $\mathrm{SO}(n-1) \ni g \mapsto c(g)$ defines a one-dimensional representation of the group $\mathrm{SO}(n-1)$. For $n > 3$, $\mathrm{SO}(n-1)$ has only the trivial one-dimensional representation, i.e. $c(g) \equiv 1$, and for $n = 2,3$ we obtain $c(g) = e^{im\omega}$, where m is an integer and g is identified with rotation over the angle ω (if $n = 2$, then $\omega = 0, \pi$).

Suppose $n > 3$. Then $h(\omega) = |\omega'|^{i\nu - i\mu}$ and $\mathrm{U} = \mathrm{E}(\nu)^{-1}|\omega'|^{i\nu - i\mu}\mathrm{E}(\mu)$. It is obvious that for $\operatorname{Im}\mu = \operatorname{Im}\nu = \beta = 0$, U is an intertwining operator. We prove that for $\beta \neq 0$ this operator is not intertwining. Thus a contradiction arises, finishing the proof.

Define a unitary operator r on $L_2(S^{n-1})$ by $(ru)(\omega', \omega_n) = u(\omega', -\omega_n)$. Clearly, $r: L_2^{\pm}(S^{n-1}) \to L_2^{\mp}(S^{n-1})$, and $\mathrm{U} = r^{-1}\mathrm{U}r$, where U is regarded as an operator on $L_2(S^{n-1})$. Suppose that $\beta > 0$ and that $\mathrm{U}: L_2^+(S^{n-1}) \to L_2^+(S^{n-1})$ is unitary. Then $\mathrm{U}: L_2^-(S^{n-1}) \to L_2^-(S^{n-1})$ is also unitary, and so $\mathrm{U}: L_2(S^{n-1}) \to L_2(S^{n-1})$ is unitary. Applying the operator $\mathrm{U}^* = \mathrm{U}^{-1}$ to 1, we find

$$\mathrm{E}(\bar{\mu})^{-1}|\omega'|^{i\mu - i\nu}\mathrm{E}(\bar{\nu})1 = \mathrm{E}(\mu)^{-1}|\omega'|^{i\mu - i\nu}\mathrm{E}(\nu)1.$$

By formula (1.3.3) this can be rewritten as

$$\mu_0(\bar{\nu})\mathrm{E}(\bar{\mu})^{-1}|\omega'|^{i\mu - i\nu} = \mu_0(\nu)\mathrm{E}(\mu)^{-1}|\omega'|^{i\mu - i\nu}, \tag{3.10}$$

where $\mu_m(\lambda) = (-i)^m 2^{i\lambda}\Gamma((m + i\lambda + n/2)/2)\Gamma((m - i\lambda + n/2)/2)^{-1}$. Expand the function $\omega \mapsto |\omega'|^{i\mu - i\nu}$ in a series in spherical harmonics: $|\omega'|^{i\mu - i\nu} = \Sigma \gamma_{mk} Y_{mk}$. This function is not of class $C^\infty(S^{n-1})$, hence the series is infinite. Relations (3.10) and (1.3.3) imply that

$$\mu_0(\bar{\nu}) \sum_{m=0}^{\infty} \sum_{k=1}^{k_m} \gamma_{mk}\mu_m(\bar{\mu})Y_{mk} = \mu_0(\nu) \sum_{m=0}^{\infty} \sum_{k=1}^{k_m} \gamma_{mk}\mu_m(\mu)Y_{mk}.$$

Therefore $\mu_0(\bar\nu)\mu_m(\bar\mu) = \mu_0(\nu)\mu_m(\mu)$ for an infinite set of natural numbers m. However, as $m\to\infty$, the lefthand side of the last equations grows as $(m^2+|\mu|^2)^{\beta/2}$, while the righthand side decreases as $(m^2+|\mu|^2)^{-\beta/2}$ (cf. (1.5.23)). We have obtained a contradiction, implying that for $\beta\neq 0$ the operator cannot be unitary.

In case $n = 2$ or 3 a similar reasoning leads to the same conclusion. ■

Let $\mathcal{K}_0^+(l_\beta)$ be the algebra of continuous functions on the line l_β with values in $\mathcal{K}L_2^+(S^{n-1})$, and tending to zero at infinity, with norm

$$\|v;\mathcal{K}_0^+(l_\beta)\| = \sup_{\lambda\in l_\beta}\|v(\lambda);L_2^+(S^{n-1})\to L_2^+(S^{n-1})\|.$$

Lemma 3.3. *The following inclusion holds:* $\mathcal{K}_0^+(l_\beta)\subset\mathfrak{G}_c^+(l_\beta)$.

This assertion can be proved similarly to lemma 5.4.3; we only have to use propositions 2.1 and 2.2 instead of proposition 5.3.1 and 5.4.2.

Proposition 3.4. *The algebras* $\mathfrak{G}_c^+(l_\beta)/\mathcal{K}_0^+(l_\beta)$ *and* $\mathfrak{G}_c^+(l_0)$ *are isomorphic.*

Proof. By proposition 3.2.2, the operator-function $\lambda\mapsto \mathrm{T}(\lambda) = \mathrm{P}^+\mathfrak{U}(\lambda)-\mathrm{P}^+\mathfrak{U}(\sigma)$, where $\mathfrak{U}(\lambda) = \mathrm{E}(\lambda)^{-1}\Phi(\omega)\mathrm{E}(\lambda)$, $\sigma = \mathrm{Re}\,\lambda$, belongs to the ideal $\mathcal{K}_0^+(l_\beta)$. Thus,

$$\inf_{\mathrm{T}\in\mathcal{K}_0^+(l_\beta)}\sup_{\lambda\in l_\beta}\left\|\sum_j\prod_k \mathrm{P}^+\mathfrak{U}_{jk}(\lambda)+\mathrm{T}(\lambda)\right\|\leqslant \sup_{\sigma\in R}\left\|\sum_j\prod_k\mathrm{P}^+\mathfrak{U}_{jk}(\sigma)\right\|, \tag{3.11}$$

where $\mathfrak{U}_{jk}(\lambda)$ is an operator of the form $\mathrm{E}(\lambda)^{-1}\Phi_{jk}(\omega)\mathrm{E}(\lambda)$ (or $\mathrm{E}(\bar\lambda)^{-1}\Phi_{jk}(\omega)\mathrm{E}(\bar\lambda)$), and j,k run through a finite set of values. It remains to convince ourselves that equality holds in (3.11). For any $\lambda\in l_\beta$,

$$\inf_{\mathrm{T}\in\mathcal{K}L_2^+(S^{n-1})}\left\|\sum_j\prod_k\mathrm{P}^+\mathfrak{U}_{jk}(\lambda)+\mathrm{T}\right\| = \left\|\sum_j\prod_k\mathrm{P}^+\mathfrak{U}_{jk}(\sigma)\right\| \tag{3.12}$$

(a consequence of (1.11)). But every algebra $\mathfrak{G}_c^+(\sigma)$ is isomorphic to the algebra $\mathfrak{G}_c^+(l_0)$ (proposition 1.5). Hence the righthand side of (3.12) equals $\|\Sigma_j\Pi_k\mathrm{P}^+\mathfrak{U}_k;\mathfrak{G}_c^+(l_0)\|$, and the inequality opposite to (3.11) holds. ■

Let $\hat{\mathfrak{G}}_c^+$ be the space of theorem 1.7, and let $\hat{\mathfrak{G}}_c^+(l_\beta) = \hat{\mathfrak{G}}_c^+\cup l_\beta$. We introduce a topology on $\hat{\mathfrak{G}}_c^+(l_\beta)$. A fundamental system of neighborhoods in $\hat{\mathfrak{G}}_c^+(l_\beta)$ of a point $s\in\hat{\mathfrak{G}}_c^+$ is formed by sets $\mathfrak{A}\cup l_\beta$, where $\mathfrak{A}$ is a neighborhood of s in

$\hat{\mathfrak{S}}_c^+$, and a fundamental system of neighborhoods of a point $\lambda \in l_\beta$ is formed by open intervals on l_β containing λ.

Theorem 3.5. *Let* $\mathfrak{U}(\lambda) = E(\lambda)^{-1}\Phi(\psi)E(\lambda)$, $\beta \neq 0$, *and let* $l_\beta \ni \lambda \mapsto P^+\mathfrak{U}(\lambda)$ *be a function from the algebra* $\hat{\mathfrak{S}}_c^+(l_\beta)$. *The following assertions hold:*

1) *Together with the maps* $\pi(\lambda): P^+\mathfrak{U} \mapsto P^+\mathfrak{U}(\lambda)$, $\lambda \in l_\beta$, *the maps* $\pi(\psi)$, $\pi(t)$, $\tau(\psi)$ *defined in theorem* 1.7 *induce inequivalent irreducible representations of* $\mathfrak{S}_c^+(l_\beta)$.

2) *Every irreducible representation of* $\mathfrak{S}_c^+(l_\beta)$ *is equivalent to a representation listed in* 1).

3) *The correspondence* $\pi(\psi) \mapsto \psi \in S^{n-1}$, $\pi(t) \mapsto (0,t)$, $\tau(\psi) \mapsto \psi \in \partial S_+^{n-1}$, $\pi(\lambda) \mapsto \lambda \in l_\beta$ *is a bijection of the spectrum of* $\mathfrak{S}_c^+(l_\beta)$ *onto* $\hat{\mathfrak{S}}_c^+(l_\beta)$. *The topology on* $\hat{\mathfrak{S}}_c^+(l_\beta)$ *coincides with the Jacobson topology.*

Proof. The first assertion follows from theorem 1.12 and proposition 3.2.

We verify the second assertion. If π is an irreducible representation of the quotient algebra $\mathfrak{S}_c^+(l_\beta)/\mathcal{K}_0^+(l_\beta)$, then by proposition 3.4 and theorem 1.7 it is equivalent to one of the representations $\pi(\psi)$, $\pi(t)$, or $\tau(\psi)$ (for any $\lambda \in l_0 = \mathbb{R}$, the algebras $\mathfrak{S}_c^+(l_0)$ and $\mathfrak{S}_c^+(\lambda)$ are isomorphic). Suppose that π is an irreducible representation of $\mathfrak{S}_c^+(l_\beta)$ with nonzero restriction $\pi|_{\mathcal{K}_0^+(l_\beta)}$. Then $\pi|_{\mathcal{K}_0^+(l_\beta)}$ is an irreducible representation of $\mathcal{K}_0^+(l_\beta)$ (proposition 5.1.2). Thus, $\pi|_{\mathcal{K}_0^+(l_\beta)}$ is equivalent to a representation of the form $\pi(\lambda): v \mapsto v(\lambda)$, $v \in \mathcal{K}_0^+(l_\beta)$, $\lambda \in l_\beta$ (proposition 5.1.8). This implies that π is equivalent to the representation $\pi(\lambda)$ of $\mathfrak{S}_c^+(l_\beta)$.

The verification of the third assertion is left to the reader. We only show that every neighborhood in $\hat{\mathfrak{S}}_c^+(l_\beta)$ of a point $s \in \hat{\mathfrak{S}}_c^+$ contains the complete line l_β. By the description of the Jacobson topology (given above proposition 5.1.7), it suffices to prove that if $\|\pi(s)P^+\mathfrak{U}\| > 1$ for some $P^+\mathfrak{U} \in \mathfrak{S}_c^+(l_\beta)$, then $\|\pi(\lambda)P^+\mathfrak{U}\| > 1$ for all $\lambda \in l_\beta$. (Here $\pi(s)$ denotes one of the representations $\pi(\psi)$, $\pi(t)$, or $\tau(\psi)$.) By corollary 1.11, for all $\lambda \in l_\beta$,

$$\inf_{T \in \mathcal{K}\mathcal{L}_2(S^{n-1})} \|P^+\mathfrak{U}(\lambda) + T\| = \|P^+\mathfrak{U}(\sigma)\|, \quad \sigma = \operatorname{Re}\lambda.$$

Hence $\|\pi(\lambda)P^+\mathfrak{U}\| = \|P^+\mathfrak{U}(\lambda)\| \geqslant \|P^+\mathfrak{U}(\sigma)\| \geqslant \|\pi(s)P^+\mathfrak{U}\|$, i.e. $\|\pi(\lambda)P^+\mathfrak{U}\| > 1$. ∎

§4. The algebras $\mathfrak{G}^+(l_\beta)$

1. The algebras $\mathfrak{G}^+(l_\beta)$ and $\mathfrak{D}$. The algebra $\mathfrak{G}^+(l_\beta)$ is generated by the operator-functions $l_\beta \ni \lambda \mapsto \mathrm{P}^+\mathfrak{U}(\lambda)$, where $\mathfrak{U}(\lambda) = \mathrm{E}(\lambda)^{-1}\Phi(\phi,\omega)\mathrm{E}(\lambda)$, $\beta \neq \pm(k+n/2)$, $k = 0,1,\cdots$. The operations are pointwise, and the norm is

$$\|\mathrm{P}^+\mathfrak{U},\mathfrak{G}^+(l_\beta)\| = \sup_{\lambda\in l_\beta}\|\mathrm{P}^+\mathfrak{U}(\lambda);L_2^+(S^{n-1})\to L_2^+(S^{n-1})\|.$$

Thus, the algebra $\mathfrak{G}_c^+(l_\beta)$ considered in §3 is a subalgebra of $\mathfrak{G}^+(l_\beta)$.

P r o p o s i t i o n 4.1. *Let μ,ν be distinct points of l_β. Then the maps $\pi(\lambda)$:$\mathrm{P}^+\mathfrak{U}\mapsto\mathrm{P}^+\mathfrak{U}(\lambda)$ for $\lambda = \mu,\nu$ define inequivalent irreducible representations of $\mathfrak{G}^+(l_\beta)$.*

P r o o f. Irreducibility of the representations is ensured by proposition 2.1, and inequivalence is established verbatim as in proposition 5.4.2; moreover, for $\beta \neq 0$ inequivalence follows from proposition 3.2. ■

The verification of the following assertion is completely similar to the proof of lemma 5.4.3.

L e m m a 4.2. *The following inclusion holds: $\mathcal{K}_0^+(l_\beta) \subset \mathfrak{G}^+(l_\beta)$. (Here, $\mathcal{K}_0^+(l_\beta)$ has the same meaning as in §3.)*

Now we introduce the algebra $\mathfrak{D}$; in sequel it will be shown to be isomorphic to the quotient algebra $\mathfrak{G}^+(l_\beta)/\mathcal{K}_0^+(l_\beta)$. For a $\phi \in S_+^{n-1}$ we denote by $\mathfrak{D}(\phi)$ the algebra $C(S^{n-1})$ of continuous functions. For a $\phi \in \partial S_+^{n-1}$, $\mathfrak{D}(\phi)$ is understood to mean the algebra of continuous functions $\omega\mapsto\mathfrak{T}(\omega)$ on ∂S_+^{n-1}, where $\mathfrak{T}(\omega)$ is a Toeplitz operator on $\mathcal{H}^-(\mathbf{R})$ of the form $(\mathfrak{T}(\omega)u)(s) = \Pi_{t\to s}^-\Xi(\omega,t)u(t)$, and Ξ is a continuous function on $\partial S_+^{n-1}\times[-\infty,+\infty]$. Put $\|\mathfrak{T};\mathfrak{D}(\phi)\| = \sup_\omega\|\mathfrak{T}(\omega)\|$. Every function $\Psi \in C^\infty(S^{n-1}\times S^{n-1})$ generates a vector field, i.e. a map $\overline{S_+^{n-1}} \ni \phi\mapsto\mathfrak{D}_\Psi(\phi) \in \mathfrak{D}(\phi)$. By definition, for $\phi \in S_+^{n-1}$ we set $\mathfrak{D}_\Psi(\phi) = \Psi(\phi,\cdot)$, while for $\phi \in \partial S_+^{n-1}$, $\mathfrak{D}_\Psi(\phi)$ denotes the operator-function $\partial S_+^{n-1} \ni \mathring{\omega}\mapsto\Pi^-\Psi(\phi,\kappa(\mathring{\omega},\cdot))$. The algebra $\mathfrak{D}$ is generated by all such vector fields. Multiplication and involution in it are defined pointwise, and the norm is

$$\|\mathfrak{D}_\Psi(\cdot);\mathfrak{D}\| = \sup_{\phi\in\overline{S_+^{n-1}}}\|\mathfrak{D}_\Psi(\phi);\mathfrak{D}(\phi)\|.$$

2. The isomorphism between $\mathfrak{G}^+(l_\beta)/\mathcal{K}_0^+(l_\beta)$ and $\mathfrak{D}$. Let $\mathfrak{U}(\lambda) = E(\lambda)^{-1}\Phi(\phi,\omega)E(\lambda)$ and let $\mathfrak{D}_\Phi$ be the element of $\mathfrak{D}$ determined by Φ. Propositions 2.3 and 2.4 imply that the correspondence $\mathfrak{G}^+(l_\beta)/\mathcal{K}_0^+(l_\beta) \ni [P^+\mathfrak{U}] \mapsto \mathfrak{D}_\Phi \in \mathfrak{D}$ induces a map which preserves the algebraic operations. This section is devoted to proving that this map is isometric.

P r o p o s i t i o n 4.3. *Let $\mathfrak{U}_{jk}(\lambda) = E_{\omega\to\phi}(\lambda)^{-1}\Phi_{jk}(\phi,\omega)E_{\psi\to\omega}(\lambda)$, and $B(\lambda) = \Sigma_{j=1}^{M}\Pi_{k=1}^{N}P^+\mathfrak{U}_{jk}(\lambda)$, where $M,N<\infty$. Then, for any positive number ϵ and any positive $\delta < 1/4$, there is a constant c such that*

$$\|B(\lambda)u;L_2(S^{n-1})\| \leqslant \left\{\left\|\sum_j\prod_k \mathfrak{D}_{\Phi_{jk}};\mathfrak{D}\right\|+\epsilon\right\}\|u;L_2(S^{n-1})\|+ \tag{4.1}$$

$$+\|u;H^{-\delta}(\lambda,S^{n-1})\|$$

for all $u \in L_2^+(S^{n-1})$, $\lambda \in l_\beta$.

P r o o f. Let U_1, U_2 be open covers of the sphere S^{n-1} such that $\overline{U}_1 \subset S_+^{n-1}$, and such that the boundaries ∂U_q, $q = 1,2$, are located near the equator ∂S_+^{n-1}. Let also $\zeta \in C^\infty(S^{n-1})$, $\operatorname{supp}\zeta_q \subset U_q$, and $\zeta_1^2+\zeta_2^2 = 1$.

Rewrite the operator $\zeta_2 B(\lambda)$ as

$$\zeta_2 B(\lambda)u - T(\lambda)u =$$

$$= \zeta_2 E_{\omega\to\phi}(\lambda)^{-1}\left\{\left[J(\lambda)^{-1}\sum_j\prod_k\Phi_{jk}(\phi,\kappa(\hat\omega,\cdot))\times\right.\right.$$

$$\left.\left.\times J(\lambda)(E_{\psi\to\omega}(\lambda)u)(\kappa(\hat\omega,\cdot))\right]\circ\kappa^{-1}(\omega)\right\};$$

by remark 2.5 the operator T is subject to the inequality

$$\|T(\lambda)u;L_2(S^{n-1})\| \leqslant c(\sigma)\|u;H^{-\sigma}(\lambda,S^{n-1})\| \tag{4.2}$$

for $\sigma < 1/2$ and $\lambda \in l_\beta$. Put $\mathcal{F}(\phi,\hat\omega) = \Sigma_j\Pi_k\Phi_{jk}(\phi,(\kappa(\hat\omega,\cdot))$ and introduce the operator

$$M(\lambda) = E(\lambda)^{-1}\{[J(\lambda)^{-1}\zeta_2(\phi)^2\mathcal{F}(\phi,\hat\omega)^*\mathcal{F}(\phi,\hat\omega)J(\lambda)(E(\lambda)u)(\kappa(\hat\omega,\cdot)]\circ\kappa^{-1}(\omega)\},$$

where $\mathcal{F}(\phi,\hat\omega)^*$ is operator adjoint to $\mathcal{F}(\phi,\hat\omega)$ on $\mathcal{H}^-(\mathbb{R})$. Denoting by $(\zeta_2B(\lambda))^*$ the operator adjoint to $\zeta_2B(\lambda)$ on $L_2^+(S^{n-1})$, we find that (4.2) holds for the difference $T(\lambda) \equiv M(\lambda) - (\zeta_2 B(\lambda))^* \zeta_2 B(\lambda)$ (the letter T stands for several operators at the same time). Hence

$$\mathrm{Re}\,(M(\lambda)u,u) - \|\zeta_2 B(\lambda)u, L_2(S^{n-1})\|^2 \geqslant c\|u;H^{-\sigma/2}(\lambda, S^{n-1})\|. \quad (4.3)$$

Let $F = \|\Sigma_j \Pi_k \mathfrak{D}_{\Phi_{jk}};\mathfrak{D}\|$. Since the set $\mathrm{supp}\,\zeta_2 \cap \overline{S_+^{n-1}}$ lies in a small neighborhood of the equator ∂S_+^{n-1}, the operator-function $(\phi,\mathring{\omega}) \mapsto \Xi(\phi,\mathring{\omega})^2 = \zeta_2(\phi)^2(F^2 - \mathfrak{F}(\phi,\mathring{\omega})^*\mathfrak{F}(\phi,\mathring{\omega})) + \epsilon$ is positive for $(\phi,\mathring{\omega}) \in S_+^{n-1} \times S^{n-1}$. Introduce the operator

$$N(\lambda)u = E(\lambda)^{-1}\{[J(\lambda)^{-1}\Xi(\phi,\mathring{\omega})J(\lambda)(E(\lambda)u)(\kappa(\mathring{\omega},\cdot))]\circ\kappa^{-1}(\omega)\}.$$

The estimate (4.2) holds for the operator $T(\lambda) = \zeta_2^2 F^2 - M(\lambda) + \epsilon - N(\lambda)^* N(\lambda)$. (The last assertion, as well as the boundedness of the map $N(\lambda): L_2^+(S^{n-1}) \to L_2^+(S^{n-1})$, requires further substantiation; propositions 2.3, 2.4, and remark 2.5 are not immediately applicable in this situation. We will not dwell into this.) This implies the inequality

$$F^2\|\zeta_2 u;L_2(S^{n-1})\|^2 + \epsilon\|u;L_2(S^{n-1})\|^2 - \mathrm{Re}\,(M(\lambda)u,u) \geqslant$$
$$\geqslant c\|u;H^{-\sigma/2}(\lambda,S^{n-1})\|^2 + \|N(\lambda)u;L_2(S^{n-1})\|^2 \geqslant$$
$$\geqslant c\|u;H^{-\sigma/2}(\lambda,S^{n-1})\|^2.$$

Together with (4.3) this leads to the estimate

$$\|\zeta_2 B(\lambda)u;L_2(S^{n-1})\|^2 \leqslant \quad (4.4)$$
$$\leqslant F^2\|\zeta_2 u;L_2(S^{n-1})\|^2 + \epsilon\|u;L_2(S^{n-1})\| + c\|u;H^{-\sigma/2}(\lambda,S^{n-1})\|^2.$$

The last inequality is contained in theorem 3.6.2; also,

$$\|\zeta_1 B(\lambda)u;L_2(S^{n-1})\|^2 \leqslant \quad (4.5)$$
$$\leqslant F^2\|\zeta_1 u;L_2(S^{n-1})\|^2 + \epsilon\|u;L_2(S^{n-1})\|^2 + c\|u;H^{-\sigma/2}(\lambda,S^{n-1})\|^2.$$

Adding (4.4) and (4.5) we obtain

$$\|B(\lambda)u;L_2(S^{n-1})\|^2 \leqslant$$
$$\leqslant (F^2+\epsilon)\|u;L_2(S^{n-1})\| + c(\sigma,\epsilon)\|u;H^{-\sigma/2}(\lambda,S^{n-1})\|^2$$

for any $\sigma < 1/2$. ■

Proposition 4.4. *Let* B(λ) *be the operator from proposition 4.3. Then there is, for every positive number* ϵ, *an operator-function* $\mathrm{T} \in \mathcal{K}_0^+(l_\beta)$ *such that*

$$\|\mathrm{B}+\mathrm{T};\mathfrak{G}^+(l_\beta)\| \leqslant \left\|\sum_j \prod_k \mathfrak{D}_{\Phi_{jk}};\mathfrak{D}+\epsilon\right\|. \tag{4.6}$$

Proof. Denote by q_K a $C_0^\infty(\mathbf{R})$ function such that $q_K(t) = 1$ for $|t| < K$ and $q_K(t) = 0$ for $|t| > K+1$. Introduce the operator $\hat{q}_K(\lambda):L_2(S^{n-1}) \to L_2(S^{n-1})$ by

$$\hat{q}_K(\lambda) = \sum_{m=0}^{\infty} \sum_{k=1}^{k_m} q_K(m^2+|\lambda|^2)u_{mk}Y_{mk}$$

and put $\mathrm{T}(\lambda) = -\mathrm{B}(\lambda)\mathrm{P}^+\hat{q}_K(\lambda)$. By (4.1), for $u \in L_2^+(S^{n-1})$,

$$\|(\mathrm{B}(\lambda)+\mathrm{T}(\lambda))u;L_2(S^{n-1})\| = \tag{4.7}$$
$$= \|\mathrm{B}(\lambda)\mathrm{P}^+(\mathrm{I}-\hat{q}_K(\lambda))u;L_2(S^{n-1})\| \leqslant$$
$$\leqslant (F+\epsilon)\|(\mathrm{I}-\hat{q}_K(\lambda))u;L_2(S^{n-1})\| + c\|(\mathrm{I}-\hat{q}_K(\lambda))u;H^{-\delta}(\lambda,S^{n-1})\|,$$

where, as before, $F = \|\Sigma_j\Pi_k\mathfrak{D}_{\Phi_{jk}};\mathfrak{D}\|$. The definitions of the norm in $H^{-\delta}(\lambda,S^{n-1})$ (1.5.20) and of the function q_K imply the estimate $\|(\mathrm{I}-\hat{q}_K(\lambda))u;H^{-\delta}(\lambda,S^{n-1})\| \leqslant K^{-\delta}\|u;L_2(S^{n-1})\|$. Using (4.7), we obtain for K large, $\|\mathrm{B}(\lambda)+\mathrm{T}(\lambda))u;L_2(S^{n-1})\| \leqslant (F+\epsilon)\|u;L_2(S^{n-1})\|$ everywhere on l_β. ■

Proposition 4.5. *Put* $\mathfrak{U}_{jk}(\lambda) = \mathrm{E}_{\omega\to\phi}(\lambda)^{-1}\Phi_{jk}(\phi,\omega)\mathrm{E}_{\psi\to\omega}(\lambda)$ *and* $\mathrm{B}(\lambda) = \Sigma_{j=1}^M \Pi_{k=1}^N \mathrm{P}^+\mathfrak{U}_{jk}(\lambda)$, *where* $M,N < \infty$. *Then*

$$\inf_{\mathrm{T}\in\mathcal{K}_0^+(l_\beta)} \|\mathrm{B}+\mathrm{T};\mathfrak{G}^+(l_\beta)\| \geqslant \left\|\sum_j\prod_k \mathfrak{D}_{\Phi_{jk}};\mathfrak{D}\right\|. \tag{4.8}$$

Proof. Let $\phi \in S_+^{n-1}$ and $\omega_0 \in S^{n-1}$. Also, let $\zeta,\eta \in C_0^\infty(S^{n-1})$, $0 \leqslant \zeta,\eta \leqslant 1$, with $\zeta = 1$ in a neighborhood of ϕ_0 and $\zeta\eta = \zeta$. We have

$$\mathrm{B}(\lambda)\zeta = \eta\mathrm{E}(\lambda)^{-1}\sum_j\prod_k \Phi_{jk}(\phi,\omega)\mathrm{E}(\lambda)\zeta + \mathrm{T}(\lambda), \tag{4.9}$$

where $\mathrm{T} \in \mathcal{K}_0^+(l_\beta)$. Choose a function g, homogeneous of degree zero, such that $\nabla g(\phi_0)$ ($\neq 0$) lies in the plane spanned by the vectors ϕ_0 and ω_0. Suppose that sequences $\{\lambda_m\} \subset l_\beta$, $\{\tau_m\} \subset \mathbf{R}$ are subject to the requirements: $\lambda_m \to \infty$, $\tau_m \to \infty$, and that the direction of the vector $(\mathrm{Re}\lambda_m)\phi_0 + \tau_m\nabla g(\phi_0)$ tends to the direction of ω_0. Theorem 3.5.1 implies that for a function $u \in C_0^\infty(S_+^{n-1})$ supported in a small neighborhood of ϕ_0 the following inequality holds for m large:

$$\left| \mathfrak{U}(\lambda_m) e^{i\tau_m g} u - \sum_j \prod_k \Phi_{jk}(\phi_0, \omega_0) e^{i\tau_m g} u \right| < \epsilon,$$

where $\mathfrak{U}(\lambda) = \mathrm{E}(\lambda)^{-1} \zeta(\phi) \Sigma_j \Pi_k \Phi_{jk}(\phi, \omega) \mathrm{E}(\lambda)$ and ϵ is an arbitrary pre-given number. Together with (4.9) this leads to

$$\inf_{\mathrm{T} \in \mathfrak{K}_0^+(l_\beta)} \| \mathrm{B} + \mathrm{T}; \mathfrak{G}^+(l_\beta) \| \geqslant \max_{(\phi, \omega) \in S_+^{n-1} \times S^{n-1}} \left| \sum_j \prod_k \Phi_{jk}(\phi, \omega) \right|. \tag{4.10}$$

Let now $\phi_0 \in \partial S_+^{n-1}$. We may assume that

$$\sup_{\omega \in S^{n-1}} \left| \sum_j \prod_k \Phi_{jk}(\phi_0, \omega) \right| < \sup_{\dot{\omega} \in S^{n-2}} \left\| \sum_j \prod_k \Pi^- \Phi_{jk}(\phi_0, \kappa(\dot{\omega}, \cdot)) \right\|$$

(in the opposite case the estimate required already holds). Choose a point $\omega_0 \in S^{n-2}$ such that the following inequality holds:

$$\sup_{\omega \in S^{n-1}} \left| \sum_j \prod_k \Phi_{jk}(\phi_0, \omega) \right| < \| \mathfrak{F}(\phi_0, \omega_0); \mathcal{K}^-(\mathbf{R}) \to \mathcal{K}^-(\mathbf{R}) \|, \tag{4.11}$$

where, as before, $\mathfrak{F}(\phi_0, \omega_0) = \Sigma_j \Pi_k \Pi^- \Phi(\phi_0, \kappa(\omega_0, \cdot))$. Determine a function v just as in formula (2.3) (orthogonality of ϕ_0 and ω_0 is not assumed here). Denote by g a real-valued smooth function on S^{n-1}, homogeneous of degree zero, which is everywhere constant on the meridian passing through ω_0 except in small neighborhoods of the poles. Let also $\nabla g(\omega_0) \neq 0$. Choose $\sigma_0, \tau_0 \in \mathbf{R}$ such that σ_0 is large while the direction of the vector $\sigma_0 \omega_0 + \tau_0 \nabla g(\omega_0)$ coincides with the direction of ϕ_0. Put $w(\phi) = \mathrm{E}_{\omega \to \phi}(\sigma_0)^{-1} e^{-i\tau_0 g(\omega)} v(\omega)$. Let U be a domain on S^{n-1} containing a set $V = \{(\sigma_0 \omega + \tau_0 \nabla g(\omega)) / |\sigma_0 \omega + \tau_0 \nabla g(\omega)| : \omega \in (\operatorname{supp} v) \setminus \mathfrak{W}\}$, where $\mathfrak{W}$ is a small neighborhood of the poles $(0, \pm 1)$ (outside which g is constant on the meridian indicated above). We may assume that for a given $\delta > 0$ we have

$$\int_{S^{n-1} \setminus (U \cap S_+^{n-1})} |w(\phi)|^2 d\phi < \delta \tag{4.12}$$

(cf. the proof of proposition 2.6). Moreover, the definition of w implies

$$\left\| \sum_j \prod_k \mathrm{P}^+ \mathrm{E}(\sigma_0)^{-1} \Phi_{jk}(\phi_0, \cdot) \mathrm{E}(\sigma_0) \mathrm{P}^+ w; L_2(S^{n-1}) \right\| \geqslant \tag{4.13}$$

$$\geqslant (\|\mathcal{F}(\phi_0,\omega_0);\mathcal{K}^-(\mathbf{R})\to\mathcal{K}^-(\mathbf{R})\|-\delta)\|\mathrm{P}^+ w;L_2(S^{n-1})\|.$$

Let $\zeta,\eta \in C_0^\infty(U)$ with $\zeta = 1$ in a neighborhood of V and $\zeta\eta = \zeta$. Since $(1-\eta)\Sigma_j\Pi_k\mathrm{P}^+\mathrm{E}(\sigma)^{-1}\Phi_{jk}(\phi_0,\cdot)\mathrm{E}(\sigma)\zeta \in \mathcal{K}_0^+(l_0)$, by taking into account (4.12) and (4.13) we obtain

$$\left\|\left|\eta\sum_j\prod_k\mathrm{P}^+\mathrm{E}(\sigma_0)^{-1}\Phi_{jk}(\phi_0,\cdot)\mathrm{E}(\sigma_0)\right|\right\| \geqslant \|\mathcal{F}(\phi_0,\omega_0)\|-\delta \tag{4.14}$$

(on the left the norm is the operator norm in $L_2^+(S^{n-1})$, on the right that in $\mathcal{K}^-(\mathbf{R})$).

Suppose that domains U_1, U_2 cover the sphere S^{n-1}, $\overline{U}_1 \subset S_+^{n-1}$, and that the boundaries ∂U_q lie in a small neighborhood of the equator ∂S_+^{n-1}. Denote by χ_1, χ_2 two $C^\infty(S^{n-1})$ functions such that $\operatorname{supp}\chi_q \subset U_q$, $\chi_1^2+\chi_2^2 = 1$. Let also $\zeta_q \in C_0^\infty(U_q)$, $\zeta_q\chi_q = 1$, $q = 1,2$. Put $\mathrm{A}(\lambda) = \eta\Sigma\Pi\mathrm{P}^+\mathrm{E}(\lambda)^{-1}\Phi_{jk}(\phi_0,\cdot)\mathrm{E}(\lambda)$. We have

$$\|\mathrm{A}(\sigma)u\|^2 = \sum_{j=1}^{2}\|\chi_j\mathrm{A}(\sigma)u\|^2 = \tag{4.15}$$

$$= \sum_{j=1}^{2}\|\zeta_j(\mathrm{A}(\sigma)+\mathrm{T}_j(\sigma))\chi_j u\|^2 \leqslant$$

$$\leqslant \max_j\{\|\zeta_j(\mathrm{A}(\sigma)+\mathrm{T}_j(\sigma));L_2^+(S^{n-1})\to L_2^+(S^{n-1})\|^2\}\|u\|^2,$$

where $\mathrm{T}_j \in \mathcal{K}_0^+(l_0)$. Proposition 4.4 implies that for large $|\sigma|$ and any $\delta>0$,

$$\|\zeta_2(\mathrm{A}(\sigma)+\mathrm{T}_2(\sigma))\| < \sup_\omega\left|\sum_j\prod_k\Phi_{jk}(\phi_0,\omega)\right|+\delta,$$

and (4.11) leads to the estimate

$$\|\zeta_2(\mathrm{A}(\sigma)+\mathrm{T}_2(\sigma))\| < \|\mathcal{F}(\phi_0,\omega_0)\|. \tag{4.16}$$

Therefore

$$\|\zeta_1\mathrm{A}(\sigma)\| \geqslant \|\mathcal{F}(\sigma_0,\omega_0)\| \tag{4.17}$$

(for $|\sigma|$ large). Indeed, in the opposite case we would have, by (4.15), (4.16), $\|\mathrm{A}(\sigma)\| < \|\mathcal{F}(\phi_0,\omega_0)\|$. This contradicts (4.14), in which $|\sigma_0|$ may be regarded arbitrarily large and δ arbitrarily small. The formula $\mathrm{A}(\lambda) = \mathrm{A}(\sigma)+\mathrm{T}(\lambda)$, where $\sigma = \operatorname{Re}\lambda$, $\mathrm{T} \in \mathcal{K}_0^+(l_\beta)$, and inequality (4.17) imply

$$\inf_{T\in\mathcal{K}_0^+(l_\beta)} \sup_{\lambda\in l_\beta} \left\| \left| \zeta_1\eta \sum_j \prod_k P^+ E(\lambda)^{-1}\Phi_{jk}(\phi_0,\cdot)E(\lambda)+T(\lambda)\right| \right\| \geqslant \tag{4.18}$$
$$\geqslant \|\mathcal{F}(\phi_0,\omega_0)\|.$$

Since ζ,η are supported near ϕ_0, this implies that

$$\inf_{T\in\mathcal{K}_0^+(l_\beta)} \sup_{\lambda\in l_\beta} \left\| \left| \sum_j \prod_k P^+ E(\lambda)^{-1}\Phi_{jk}(\phi_0,\cdot)E(\lambda)+T(\lambda)\right| \right\| \geqslant \tag{4.19}$$
$$\geqslant \|\mathcal{F}(\phi_0,\omega_0)\|.$$

(Assume this to be not true. Then, for some $T_0 \in \mathcal{K}_0^+(l_\beta)$ we have

$$\sup_{\lambda\in l_\beta} \left\| \sum_j \prod_k P^+ E(\lambda)^{-1}\Phi_{jk}(\phi_0,\cdot)E(\lambda)+T(\lambda)\right\| < \|\mathcal{F}(\phi_0,\omega_0)\|.$$

and thus

$$\sup_{\lambda\in l_\beta} \left\| \zeta_1\eta \left[\sum_j \prod_k P^+ E(\lambda)^{-1}\Phi_{jk}(\phi,\cdot)E(\lambda)\right] + T_0(\lambda)\right\| < \|\mathcal{F}(\phi_0,\omega_0)\|.$$

Clearly, in view of the smallness of the support of $\zeta_1\eta$ the latter inequality also holds if ϕ is replaced by ϕ_0. The result contradicts (4.18).) Applying proposition 2.3, formula (4.19) implies the estimate

$$\inf_{T\in\mathcal{K}_0^+(l_\beta)} \|B+T;\mathfrak{G}(l_\beta)\| \geqslant \sup_{(\phi,\omega)\in\partial S_+^{n-1}\times\partial S_+^{n-1}} \|\mathcal{F}(\phi,\omega);\mathcal{K}^-(\mathbf{R})\to\mathcal{K}^-(\mathbf{R})\|.$$

This, together with (4.10), gives (4.8). ■

P r o p o s i t i o n 4.6. *The algebras* $\mathfrak{G}^+(l_\beta)/\mathcal{K}_0^+(l_\beta)$ *and* $\mathfrak{D}$ *are isomorphic.*

The p r o o f can be obtained by combining propositions 2.3, 2.4, and 4.4, 4.5. ■

3. The spectrum of the algebra $\mathfrak{G}^+(l_\beta)$. Denote by $\hat{\mathfrak{G}}^+(l_\beta)$ the disjoint union of the sets $S_+^{n-1}\times S^{n-1}$, $\partial S_+^{n-1}\times\partial S_+^{n-1}$, $\partial S_+^{n-1}\times[-1,1]$, and the line l_β. Identify $\partial S_+^{n-1}\times\{\pm 1\}\subset\partial S_+^{n-1}\times[-1,1]$ with $\partial S_+^{n-1}\times\{(0,\pm 1)\}\subset\subset\overline{S_+^{n-1}}\times S^{n-1}$, i.e. the interval $[-1,1]$ is represented by the diameter with end points at the poles $(0,\pm 1)$ of the second factor in the product $\overline{S_+^{n-1}}\times S^{n-1}$.

We introduce a topology on $\hat{\mathfrak{G}}^+(l_\beta)$. The typical neighborhoods (i.e. neighborhoods forming a fundamental system) of a point $(\phi_0,\omega_0) \in \overline{S_+^{n-1}} \times S^{n-1}$, under the assumption that $\phi_0 \in S_+^{n-1}$, $\phi_0\omega_0 = 0$, are unions $\mathcal{U}(\phi_0,\omega_0) \cup l_\beta$; here $\mathcal{U}(\phi_0,\omega_0)$ is a neighborhood of (ϕ_0,ω_0) on $\overline{S_+^{n-1}} \times S^{n-1}$.

Let $(\phi_0,\omega_0) \in \overline{S_+^{n-1}} \times S^{n-1}$, with also $\phi_0 \in \partial S_+^{n-1}$, $\phi_0\omega_0 = 0$, and suppose ω_0 is not one of the poles $(0,\pm 1)$. Typical neighborhoods are in this case unions $\mathcal{U}(\phi_0,\omega_0) \cup \mathcal{V}(\phi_0,\omega_0) \cup l_\beta$, where $\mathcal{V}(\phi_0,\omega_0)$ is a neighborhood of $(\phi_0,\dot{\omega}_0)$ on $\partial S_+^{n-1} \times \partial S_+^{n-1}$ (recall that if $\omega = (\omega',\omega_n)$, $\omega' \neq 0$, then $\dot{\omega} = \omega'/|\omega'|$). For a point $(\phi_0,\omega_0) \in \partial S_+^{n-1} \times \partial S_+^{n-1}$ satisfying $\phi_0\omega_0 = 0$, a fundamental system of neighborhoods is formed by sets $\mathcal{V}(\phi_0,\dot{\omega}_0) \cup l_\beta$.

If the requirement $\phi_0\omega_0 = 0$ is replaced by $\phi_0\omega_0 > 0$ (respectively, $\phi_0\omega_0 < 0$), then in the description of typical neighborhoods the line l_β must be replaced by a set $\{\lambda \in l_\beta : \mathrm{Re}\lambda > N\}$ (respectively, $\{\lambda \in l_\beta : \mathrm{Re}\lambda < N\}$), where N is an arbitrary real number.

A fundamental system of neighborhoods of a point $(\phi_0,\omega_0) \in \partial S_+^{n-1} \times (-1,1)$ is formed by unions $\mathcal{W}(\phi_0) \times \mathcal{T}(t_0) \cup \mathcal{W}(\phi_0) \times \partial S_+^{n-1} \cup l_\beta$, where $\mathcal{W}(\phi_0)$ is a neighborhood of ϕ_0 on ∂S_+^{n-1} and $\mathcal{T}(t_0)$ is a neighborhood of t_0 on $[-1,1]$.

Typical neighborhoods of a point (ϕ_0,ω_0), where $\phi_0 \in \partial S_+^{n-1}$ and ω_0 is one of the poles $(0,\pm 1)$, are sets $\mathcal{W}(\phi_0) \times \mathcal{T}(\omega_0) \cup \mathcal{W}(\phi_0) \times \partial S_+^{n-1} \cup \mathcal{U}(\phi_0,\omega_0) \cup l_\beta$. A neighborhood in $\hat{\mathfrak{G}}^+(l_\beta)$ of a point $\lambda \in l_\beta$ is an open interval on l_β containing λ.

T h e o r e m 4.7. *Let* $\mathfrak{U}(\lambda) = \mathrm{E}(\lambda)^{-1}\Phi(\phi,\omega)\mathrm{E}(\lambda)$, $\Phi \in C^\infty(S^{n-1} \times S^{n-1})$. *Then the operator-function* $l_\beta \ni \lambda \mapsto \mathrm{P}^+\mathfrak{U}(\lambda)$ *belongs to the algebra* $\mathfrak{G}^+(l_\beta)$, *and the following assertions hold:*

1) *The maps*

$$\pi(\phi,\omega):\mathrm{P}^+\mathfrak{U} \to \Phi(\phi,\omega), \quad (\phi,\omega) \in S_+^{n-1} \times S^{n-1};$$

$$\pi(\phi,t):\mathrm{P}^+\mathfrak{U} \mapsto \frac{1-t}{2}\Phi(\phi,(0,-1)) + \frac{1+t}{2}\Phi(\phi,(0,1)),$$

$$(\phi,t) \in \partial S_+^{n-1} \times (-1,1);$$

$$\tau(\phi,\omega):\mathrm{P}^+\mathfrak{U} \mapsto \prod{}^{-} \Phi(\phi,\kappa(\omega,\cdot)), \quad (\phi,\omega) \in \partial S_+^{n-1} \times \partial S_+^{n-1};$$

$$\pi(\lambda):\mathrm{P}^+\mathfrak{U} \mapsto \mathrm{P}^+\mathfrak{U}(\lambda), \quad \lambda \in l_\beta,$$

induce pairwise inequivalent irreducible representations of $\mathfrak{G}^+(l_\beta)$*; the representations* $\pi(\phi,\omega)$, $\pi(\phi,t)$ *are one-dimensional, the* $\tau(\phi,\omega)$ *are representations in* $\mathcal{K}^-(\mathbf{R})$, *and the* $\pi(\lambda)$ *are representations in* $L_2^+(S^{n-1})$.

2) *Every irreducible representation of* $\mathfrak{G}^+(l_\beta)$ *is equivalent to a representation listed in* 1).

3) *The correspondence* $\pi(\phi,\omega)\mapsto(\phi,\omega)$, $\pi(\phi,t)\mapsto(\phi,t)$ $\tau(\phi,\omega)\mapsto(\phi,\omega)$, $\pi(\lambda)\mapsto\lambda$ *defines a bijection of the spectrum of* $\mathfrak{G}^+(l_\beta)$ *onto* $\hat{\mathfrak{G}}^+(l_\beta)$. *The topology on* $\hat{\mathfrak{G}}^+(l_\beta)$ *coincides with the Jacobson topology.*

Proof. The first assertion follows from propositions 4.1, 4.6, and theorem 2.7. The second can be verified similarly to the corresponding part of theorem 3.5. The topology on the spectrum (third assertion) is clarified by a reasoning similar to that in the proof of theorem 5.4.6. ■

§5. The spectrum of an algebra of pseudodifferential operators on a manifold with boundary

In this paragraph we describe the spectra of algebras generated by pseudodifferential operators on a smooth manifold with boundary. The transition from spectra of the 'local' algebras studied in the preceding paragraphs of this Chapter to the spectrum of an algebra of pseudodifferential operators is performed similarly to the case of a closed manifold. This transition is not discussed here; we confine ourselves to statements only. We will study three types of algebras: the algebra of pseudodifferential operators in a halfspace (this case is distinguished for its simplicity of description); the algebra generated by pseudodifferential operators with continuous symbols; and the algebra of pseudodifferential operators in whose symbols discontinuities 'of the first kind' are allowed (the algebra from theorem 5.4.8).

1. Algebras on a halfspace. Let $L_2(\mathbf{R},|x|^{2\beta})$ $(=H_\beta^0(\mathbf{R}^n))$ be the space of functions on $\mathbf{R}^n$ with finite norm

$$\|u\| = \left[\int_{\mathbf{R}^n} |x|^{2\beta}\,|u(x)|^2dx\right]^{1/2}, \quad \beta\in\mathbf{R}^n,$$

let $L_2^+(\mathbf{R}, |x|^{2\beta})$ be the subspace of functions with support in $\overline{\mathbf{R}^n_+}$, and let $\mathrm{P}^+ : L_2(\mathbf{R}^n, |x|^{2\beta}) \to L_2^+(\mathbf{R}^n, |x|^{2\beta})$ be the projection operator. Let also $\mathfrak{U}(\lambda) = \mathrm{E}_{\omega\to\phi}(\lambda)^{-1}\Phi(\phi,\omega)\mathrm{E}_{\psi\to\omega}(\lambda)$, $\Phi \in C^\infty(S^{n-1}\times S^{n-1})$, and

$$(\mathrm{A}u)(x) = \tag{5.1}$$

$$= \frac{1}{\sqrt{2\pi}} \int_{\mathrm{Im}\lambda=\beta} r^{i(\lambda+in/2)} \mathrm{E}_{\omega\to\phi}(\lambda)^{-1}\Phi(\phi,\omega)\mathrm{E}_{\psi\to\omega}(\lambda)\tilde{u}(\lambda+in/2,\psi)d\lambda,$$

where $r = |x|$, $\phi = x/|x|$, $\beta = \pm(k+n/2)$, $k = 0, 1, \cdots$. Recall that the map $\mathrm{A}: L_2(\mathbf{R}^n, |x|^{2\beta}) \to L_2(\mathbf{R}^n, |x|^{2\beta})$ is continuous, and that on a dense set the operator (5.1) coincides with an operator of the form $\mathrm{F}^{-1}_{\xi\to x}\Phi(\phi,\xi/|\xi|)\mathrm{F}_{y\to\xi}$, $\phi = x/|x|$ (cf. theorem 2.3.5). Denote by $\mathcal{A}$ the algebra generated on $L_2^+(\mathbf{R}^n, |x|^{2\beta})$ by the operators $\mathrm{P}^+\mathrm{A}$, and by $\mathcal{A}_c$ the subalgebra of $\mathcal{A}$ generated by the operators whose symbols Φ depend on $\omega \in S^{n-1}$ only. Parseval's equality (1.2.7) for the Mellin transform means that $\mathcal{A}$ and $\mathcal{A}_c$ are isomorphic, respectively, to the algebras $\mathfrak{G}^+(l_\beta)$ and $\mathfrak{G}_c^+(l_\beta)$. Thus, a description of the spectrum of $\mathcal{A}$ is given by theorem 4.7, and of $\mathcal{A}_c$ - by theorem 3.5 (if $\beta \neq 0$). If $\beta = 0$, then $\mathcal{A}_c$ is isomorphic to any of the algebras $\mathfrak{G}_c^+(\lambda)$, $\lambda \in \mathbf{R}$. The spectrum of $\mathfrak{G}_c^+(\lambda)$ is elucidated in theorem 1.7.

2. The algebra of pseudodifferential operators with continuous symbols. Let $\mathfrak{M}$ be a Riemannian manifold with C^∞ boundary $\partial\mathfrak{M}$; let $S^*(\mathfrak{M})$ be the bundle of cotangent unit vectors. Denote by $\mathcal{A}$ the algebra generated on $L_2(\mathfrak{M})$ by the pseudodifferential operators of order zero whose symbols are smooth functions on $S^*(\mathfrak{M})$. Introduce the space $\hat{\mathfrak{G}}$ (which will turn out to be homeomorphic to the spectrum of the quotient algebra $\mathcal{A}/\mathcal{K}L_2(\mathfrak{M}))$ as the disjoint union of the sets $S^*(\mathfrak{M})$, $S^*(\partial\mathfrak{M})$, and $\mathfrak{N}(\partial\mathfrak{M})$, where $\mathfrak{N}(\partial\mathfrak{M})$ is the bundle of vectors with length at most one, cotangent to $\mathfrak{M}$, and normal to $\partial\mathfrak{M}$. The fiber $\mathfrak{N}(\partial\mathfrak{M})_x$ is the diameter of the sphere $S^*(\mathfrak{M})_x$; vectors $\pm n_x \in \mathfrak{N}(\partial\mathfrak{M})_x$, $|n_x| = 1$, are identified with the corresponding poles of $S^*(\mathfrak{M})_x$ (n_x is the inner normal vector to $\partial\mathfrak{M}$). We define a topology on $\hat{\mathfrak{G}}$. Below, neighborhoods of a point of $\hat{\mathfrak{G}}$ are understood to be sets forming a fundamental system of neighborhoods.

The neighborhoods in $\hat{\mathfrak{G}}$ of a point $(x_0,\theta_0) \in S^*(\mathfrak{M})$, $x_0 \in \mathfrak{M}$, coincide with its ordinary neighborhoods in $S^*(\mathfrak{M})$.

Let $(x_0,\theta_0) \in S^*(\mathfrak{M})$, $x_0 \in \partial\mathfrak{M}$, while θ_0 does not coincide with any of the poles of $S^*(\mathfrak{M})_{x_0}$, i.e. end points of the diameter $\mathfrak{N}(\partial\mathfrak{M})_{x_0}$. Denote by U a neighborhood of x_0 in $\mathfrak{M}$. We will represent the bundles $S^*(\mathfrak{M})|_U$ and $S^*(\partial\mathfrak{M})|_{U\cap\partial\mathfrak{M}}$ as the products $U\times S^*(\mathfrak{M})_{x_0}$ and $(U\cap\partial\mathfrak{M})\times S^*(\mathfrak{M})_{x_0}$. Let also $W(\theta_0)$ be a neighborhood of θ_0 on $S^*(\mathfrak{M})_{x_0}$, and let $\mathfrak{W}(\mathring{\theta}_0)$ be a neighborhood of $\mathring{\theta}_0$ on $S^*(\partial\mathfrak{M})_{x_0}$ (recall that for $\omega=(\omega',\omega_n)$, $\omega'\neq 0$, the point $\mathring{\omega}$ is defined as $\omega'/|\omega'|$). A neighborhood of (x_0,θ_0) in $\hat{\mathfrak{S}}$ now is a union of sets $U\times W(\theta_0)$ and $(U\cap\partial\mathfrak{M})\times\mathfrak{W}(\mathring{\theta}_0)$.

Now suppose that $(x_0,\theta_0)\in S^*(\mathfrak{M})$, $x_0\in\partial\mathfrak{M}$, while θ_0 is a pole. A neighborhood of (x_0,θ_0) in $\hat{\mathfrak{S}}$ is a union of sets $U\times W(\theta_n)$ and $(U\cap\partial\mathfrak{M})\times(\mathfrak{T}(\theta_0)\cup S^*(\partial\mathfrak{M})_{x_0})$, where $\mathfrak{T}(\theta_0)$ is a neighborhood of θ_0 on $\mathfrak{N}(\mathfrak{M})_{x_0}$.

A neighborhood in $\hat{\mathfrak{S}}$ of a point $(x_0,\theta_0)\in S^*(\partial\mathfrak{M})$ is an ordinary neighborhood of (x_0,θ_0) in $S^*(\partial\mathfrak{M})$.

Finally, the neighborhoods of a point $(x_0,t_0)\in\mathfrak{N}(\partial\mathfrak{M})$, $|t_0|<1$, are sets $(U\cap\partial\mathfrak{M})\times(\mathfrak{T}(t_0)\cup S^*(\partial\mathfrak{M}))$, where $\mathfrak{T}(t_0)$ is a neighborhood of t_0 on $\mathfrak{N}(\partial\mathfrak{M})_{x_0}$.

T h e o r e m 5.1. *Let* A *be a pseudodifferential operator from the algebra* $\mathfrak{A}$, *and let* Φ *be its symbol. The following assertions are true:*

1) *The maps*

$$\tau(x,\theta):\mathrm{A}\mapsto\prod{}^{-}\Phi(x,\kappa(\theta,\cdot)):\mathcal{H}^-(\mathbf{R})\to\mathcal{H}^-(\mathbf{R}),\quad x\in\partial\mathfrak{M},\quad \theta\in S^*(\partial\mathfrak{M})_x,$$

$$\pi(x,\theta):\mathrm{A}\mapsto\Phi(x,\theta),\quad x\in\mathfrak{M},\quad \theta\in S^*(\mathfrak{M})_x,$$

$$\pi(x,t):\mathrm{A}\mapsto\frac{1-t}{2}\Phi(x,-n_x)+\frac{1+t}{2}\Phi(x,n_x),\quad x\in\partial\mathfrak{M},\ |t|<1,$$

induce inequivalent irreducible representations of the quotient algebra $\mathfrak{A}/\mathcal{KL}_2(\mathfrak{M})$.

2) *Every irreducible representation of* $\mathfrak{A}/\mathcal{KL}_2(\mathfrak{M})$ *is equivalent to a representation listed in* 1).

3) *The correspondence* $\tau(x,\theta)\mapsto(x,\theta)\in S^*(\partial\mathfrak{M})$, $\pi(x,\theta)\mapsto(x,\theta)\in S^*(\mathfrak{M})$, $\pi(x,t)\mapsto(x,t)\in\mathfrak{N}(\partial\mathfrak{M})$ *defines a bijection of the spectrum of* $\mathfrak{A}/\mathcal{KL}_2(\mathfrak{M})$ *onto* $\hat{\mathfrak{S}}$. *The topology on* $\hat{\mathfrak{S}}$ *is the Jacobson topology.*

3. The algebra of pseudodifferential operators with discontinuous symbols. Denote by $S_+(\mathfrak{M})$ the bundle of tangent unit vectors above $\mathfrak{M}$; the sign + signifies the fact that the fiber $S_+(\mathfrak{M})_x$ above a point $x \in \partial\mathfrak{M}$ contains only vectors directed to the side of $\mathfrak{M}$; the fiber $\overline{S_+}(\mathfrak{M})_x$ contains, in addition, also the vectors directed along $\partial\mathfrak{M}$. Let $\mathfrak{M}$ be the set of functions Φ with discontinuities of the first kind (it was introduced at the beginning of Chapter 5, §2; see also Chapter 4, §4.1). A function $\Phi \in \mathfrak{M}$ can be regarded as being given on the Whitney sum $\overline{S_+}(\mathfrak{M})\oplus S^*(\mathfrak{M})$.

In this section, $\mathcal{Q}$ denotes the algebra generated on $L_2(\mathfrak{M})$ by the pseudodifferential operators of order zero whose symbols belong to $\mathfrak{M}$.

Introduce the space $\hat{\mathfrak{G}}$ as the disjoint union of the sets $\overline{S_+}(\mathfrak{M})\oplus S^*(\mathfrak{M})$, $S(\partial\mathfrak{M})\oplus S^*(\partial\mathfrak{M})$, $S(\partial\mathfrak{M})\oplus\mathfrak{N}(\partial\mathfrak{M})$, and $\mathfrak{L}$, where $\mathfrak{L}$ is the disjoint union of lines l_x indexed by the points of $\mathfrak{M}$. Moreover, $S(\partial\mathfrak{M})_x\oplus\{\pm n_x\}$ is identified with the subset of $(\overline{S_+}(\mathfrak{M})\oplus S(\mathfrak{M}))_x$, and the fiber $\mathfrak{N}(\partial\mathfrak{M})_x$ is regarded as the diameter of $S^*(\mathfrak{M})_x$. We define a topology on $\hat{\mathfrak{G}}$ by indicating for every point the neighborhoods forming a fundamental system.

A) Neighborhoods of a point $\lambda \in l_x$. Every line l_x is represented as $\mathbf{R}$. A neighborhood in $\hat{\mathfrak{G}}$ of $\lambda \in l_x$ is an open interval on l_x containing λ.

B) Neighborhoods of a point $(x_0,\phi_0,\theta_0) \in S_+(\mathfrak{M})\oplus S^*(\mathfrak{M})$, $x_0 \in \partial\mathfrak{M}$. These are defined as in the case of a closed manifold $\mathfrak{M}$. We recall this definition. The part of $S_+(\mathfrak{M})\oplus S^*(\mathfrak{M})$ above a neighborhood U of a point x_0 in $\mathfrak{M}$ is regarded as a set of triples (x,ϕ,θ), where $x \in U$, and ϕ,θ are tangent and cotangent unit vectors. Let $V(\phi_0)$ and $W(\theta_0)$ be neighborhoods of ϕ_0, respectively θ_0, on the unit spheres, let $K(\phi_0)$ be a subset of U whose image under some coordinate map $\kappa: U \to \mathbf{R}^n$ is the intersection of an open cone with vertex at $\kappa(x_0)$ and containing all directions from $\kappa'(x_0)(V(\phi_0))$, and an open n-dimensional ball with center at $\kappa(x_0)$. In case $\phi_0\theta_0 = 0$, a neighborhood of (x_0,ϕ_0,θ_0) is a union of sets $\{x_0\}\times V(\phi_0)\times W(\theta_0)$, $K(\phi_0)\times S^{n-1}\times W(\theta_0)$, and of the lines l_x $(x \in K(\phi_0))$ and l_{x_0}. If $\phi_0\theta_0 \lesseqgtr 0$, a neighborhood is also such a union, but with l_{x_0} replaced by a set $\{\lambda \in l_{x_0} : \lambda \lesseqgtr N\}$, where N is an arbitrary real number.

C) Neighborhoods of a point $(x_0,\phi_0,\theta_0) \in \overline{S_+}(\mathfrak{M})\oplus S^*(\mathfrak{M})$, $x_0 \in \partial\mathfrak{M}$. If $\phi_0 \notin \partial S_+(\mathfrak{M})_{x_0}$, neighborhoods are defined as in case B); of course, κ is understood to mean a map $\kappa: U \to \overline{\mathbf{R}^n_+}$, and $V(\phi_0)$ is a neighborhood of ϕ_0 in the open hemisphere $S_+(\mathfrak{M})_{x_0}$.

Suppose that $\phi_0 \in \partial S_+(\mathfrak{M})_{x_0}$ and that θ_0 is not a pole of $S^*(\mathfrak{M})_{x_0}$, i.e. an end point of the diameter orthogonal to $\partial\mathfrak{M}$. Let, also, $V(\phi_0)$ be a neighborhood of ϕ_0 in the closed hemisphere $\overline{S_+(\mathfrak{M})_{x_0}}$, and let $K(\phi_0)$ be defined as in B). Finally, denote by $\mathcal{V}(\phi_0)$ and $\mathcal{W}(\mathring{\theta}_0)$ neighborhoods of ϕ_0, $\mathring{\theta}_0$ on $S(\partial\mathfrak{M})_{x_0}$ and $S^*(\partial\mathfrak{M})_{x_0}$, respectively. If $\phi_0\theta_0 = 0$, a neighborhood of (x_0,ϕ_0,θ_0) is a union of the following sets: the sets of lines l_x, $x \in K(\phi_0)$, and l_{x_0}: $\{(\{x_0\} \times V(\phi_0)) \cup (\overline{S_+(\mathfrak{M})}|_{K(\phi_0)})\} \times W(\theta_0)$; and $\{(\{x_0\} \times \mathcal{V}(\phi_0)) \cup (S(\partial\mathfrak{M})|_{K(\phi_0)\cap\partial\mathfrak{M}})\} \times \mathcal{W}(\mathring{\theta}_0)$. If $\phi_0\theta_0 \neq 0$, then l_{x_0} must be modified as in B).

Let, as before, $\phi_0 \in \partial S_+(\mathfrak{M})_{x_0}$, but now θ_0 is a pole of $S^*(\mathfrak{M})_{x_0}$. A neighborhood of (x_0,ϕ_0,θ_0) is in this case a union of the sets l_{x_0}: $\{l_x\}_{x\in K(\phi_0)}$; $\{(\{x_0\} \times V(\phi_0)) \cup (\overline{S_+(\mathfrak{M})} \times K(\phi_0))\} \times W(\theta_0)$; and $\{(\{x_0\} \times \mathcal{V}(\phi_0)) \cup (S(\partial\mathfrak{M})|_{K(\phi_0)\cap\partial\mathfrak{M}})\} \times (\mathcal{T}(\theta_0) \cup S^*(\partial\mathfrak{M})_{x_0})$ (here, $\mathcal{T}(\theta_0)$ is a neighborhood of θ_0 on $\mathfrak{N}(\partial\mathfrak{M})_{x_0}$).

D) Neighborhoods of a point $(x_0,\phi_0,\theta_0) \in S(\partial\mathfrak{M}) \oplus S^*(\partial\mathfrak{M})$. A neighborhood in this case is a union of sets $\{(\{x_0\} \times \mathcal{V}(\phi_0)) \cup (S(\partial\mathfrak{M})|_{K(\phi_0)\cap\partial\mathfrak{M}})\} \times \mathcal{W}(\theta_0)$, the sets of lines $\{l_x\}_{x\in K(\phi_0)\cap\partial\mathfrak{M}}$, and l_{x_0} in case $\phi_0\theta_0 = 0$. If $\phi_0\theta_0 \neq 0$, then l_{x_0} is modified as in B).

E) Neighborhoods of points $(x_0,\phi_0,t_0) \in S(\partial\mathfrak{M}) \oplus \mathfrak{N}(\partial\mathfrak{M})$, $|t_0| < 1$. Let $\mathcal{T}(t_0)$ be a neighborhood of t_0 on $\mathfrak{N}(\partial\mathfrak{M})_{x_0}$. A neighborhood of (x_0,ϕ_0,t_0) is a union of sets $\{(\{x_0\} \times \mathcal{V}(\phi_0)) \cup (S(\partial\mathfrak{M})|_{K(\phi_0)\cap\partial\mathfrak{M}})\} \times (\mathcal{T}(t_0) \cup S^*(\partial\mathfrak{M})_{x_0})$ and of the lines l_x, $x \in K(\phi_0) \cap \partial\mathfrak{M}$, and l_{x_0}.

The topology defined on $\hat{\mathfrak{G}}$ by these neighborhoods is not separable; $\hat{\mathfrak{G}}$ is a T_0-space. The topology induced on the subspace $\overline{S_+(\mathfrak{M})} \oplus S^*(\mathfrak{M})$ is the weakest of all topologies in which all functions from $\mathfrak{M}$ are continuous.

Let A be a pseudodifferential operator of order zero, and let $\Phi \in \mathfrak{M}$ be its symbol. Introduce the maps

$$\pi(x,\lambda): \mathrm{A} \mapsto \mathfrak{U}(x,\lambda) = \tag{5.2}$$

$$= \mathrm{E}_{\theta\to\phi}(\lambda)^{-1}\Phi(x,\phi,\theta)\mathrm{E}_{\psi\to\theta}(\lambda): L_2(S(\mathfrak{M})_x) \to L_2(S(\mathfrak{M})_x),$$

where $x \in \mathfrak{M} \setminus \partial\mathfrak{M}$, $\phi,\psi \in S(\mathfrak{M})_x$, $\theta \in S^*(\mathfrak{M})_x$, $\lambda \in l_x = \mathbf{R}$;

$$\pi(x,\lambda): \mathrm{A} \mapsto \mathrm{P}^+\,\mathfrak{U}(x,\lambda): L_2^+(S(\mathfrak{M})_x) \mapsto L_2^+(S(\mathfrak{M})_x), \tag{5.3}$$

where $x \in \partial\mathfrak{M}$, $\lambda \in l_x = \mathbf{R}$, $L_2^+(S(\mathfrak{M})_x)$ is the subspace of $L_2(S(\mathfrak{M})_x)$ consisting of the functions with support in the 'upper' halfsphere $\overline{S}_+(\mathfrak{M})_x$, and $\mathrm{P}^+: L_2(S(\mathfrak{M})_x) \to L_2^+(S(\mathfrak{M})_x)$ is the projection operator;

$$\tau(x,\phi,\theta): \mathrm{A} \mapsto \prod{}^- \Phi(x,\phi,\kappa(\theta,\cdot)): \mathcal{H}^-(\mathbf{R}) \to \mathcal{H}^-(\mathbf{R}), \tag{5.4}$$

where $(x,\phi,\theta) \in S(\partial\mathfrak{M}) \oplus S^*(\partial\mathfrak{M})$;

$$\pi(x,\phi,\theta): \mathrm{A} \mapsto \Phi(x,\phi,\theta), \quad (x,\phi,\theta) \in \overline{S_+}(\mathfrak{M}) \oplus S^*(\mathfrak{M}), \tag{5.5}$$

$$\pi(x,\phi,t): \mathrm{A} \mapsto \frac{1-t}{2}\Phi(x,\phi,-n_x) + \frac{1+t}{2}\Phi(x,\phi,n_x), \tag{5.6}$$

where $(x,\phi,t) \in S(\partial\mathfrak{M}) \oplus \mathfrak{N}(\partial\mathfrak{M})$, $|t| < 1$.

T h e o r e m 5.2.

1) *The maps* (5.2) - (5.6) *induce pairwise inequivalent irreducible representations of the quotient algebra* $\mathfrak{A}/\mathcal{K}L_2(\mathfrak{M})$.

2) *Every irreducible representation of* $\mathfrak{A}/\mathcal{K}L_2(\mathfrak{M})$ *is equivalent to a representation* (5.2) - (5.6).

3) *The correspondence*

$$\pi(x,\lambda) \mapsto \lambda \in l_x, \quad x \in \mathfrak{M};$$

$$\tau(x,\phi,\theta) \mapsto (x,\phi,\theta) \in S(\partial\mathfrak{M}) \oplus S^*(\partial\mathfrak{M});$$

$$\pi(x,\phi,\theta) \mapsto (x,\phi,\theta) \in \overline{S_+}(\mathfrak{M}) \oplus S^*(\mathfrak{M});$$

$$\pi(x,\phi,t) \mapsto (x,\phi,t) \in S(\partial\mathfrak{M}) \oplus \mathfrak{N}(\partial\mathfrak{M})$$

defines a bijection of the spectrum of $\mathfrak{A}/\mathcal{K}L_2(\mathfrak{M})$ *onto* $\hat{\mathfrak{S}}$. *The topology on* $\hat{\mathfrak{S}}$ *coincides with the Jacobson topology.*

This theorem implies that $\mathfrak{A}$ is a (non-separable) type I algebra.

Bibliographical sketch

Chapter 1. This chapter is based on Plamenevskii [44]. All results from the theory of generalized functions used in it are contained in [11] or in [13]. Formula (1.3.3) may be derived from (1.4.7) and results in [7]. Expansion in spherical harmonics from the point of view necessary for us was considered in [1], [39], [40]. The proof of theorem 1.5.1 originates from [39]. Parameter-dependent norms were introduced, in an appropriate setting, in [2]. A.O. Derviz helped in the preparation of §6. A sketch of results concerning the operator E(λ) on L_p spaces can be found in [63].

Chapter 2. It seems that V.A. Kondrat'ev [33] was the first who systematically used the spaces $H^s_\beta(\mathbf{R}^n)$. The results concerning the Fourier transform given in §2 are, to the best of my knowledge, new, and are published for the first time. The theorem asserting boundedness of a one-dimensional singular integral operator on a space $L_2(\mathbf{R}, |x|^{2\beta})$, if $|2\beta| < 1$, was proved by K.I. Babenko [4]. The multidimensional analog of this theorem is due to E.M. Stein, who established that a singular integral operator is bounded on $L_2(\mathbf{R}, |x|^{2\beta})$ if $|2\beta| < n$, [66]. Results concerning boundedness of singular integral operators with an exponent 2β outside the interval $[-n, n]$ were announced in [43] and proved in [44]. The latter also contains a representation of the form (2.3.6). Another manner of proving an estimate of the form (2.3.8) was given in [70]. In this connection we point to the recent article [38], which is concerned with the one-dimensional case. The spaces $H^s_\beta(\mathbf{R}^m, \mathbf{R}^{m-n})$ and $\mathfrak{S}^s_\beta(\mathbf{R}^n)$ were used in [35], [36] (for nonnegative integers s). In §4 - §6, results from Plamenevskii [53] are expounded.

Chapter 3. This Chapter contains (in reworked form) results from Plamenevskii [45], [46], [50], [51]. The symbolic calculus for meromorphic pseudodifferential operators is an analog of the symbolic calculus of Kohn-Nirenberg [32] for classical pseudodifferential operators. Theorem 3.5.1 is close to the corresponding theorem for classical pseudodifferential operators (cf., e.g., [67], Vol. 2).

Chapter 4 is based on Plamenevskii [45], [47], [49]. In **Chapter 5** results of joint work of Plamenevskii and V.N. Senichkin [54] - [57] are studied, and in **Chapter 6** - such results from [58] - [60].

For one-dimensional singular integral operators, other operator symbols and the various reflections based on them see [18] (smooth contour, discontinuous coefficients) and [20] (compound contour). In the case of a smooth contour the symbols in [18] are unitarily equivalent to the symbols in Chapter 4, §4.5 (formula (4.4.10)). If singularities are allowed on the contour, the symbols from [20] and those defined in Chapter 4, §4.5 (in accordance with [56]) essentially differ. The symbols in [20] do not realize, in general, irreducible representations of the algebra of singular integral operators, and the algebra of symbols is not *-isomorphic to the quotient of the algebra of singular integral operators by the ideal of compact operators. The literature devoted to one-dimensional singular integral operators with discontinuous coefficients is very extensive. We yet mention the work of R.V. Duduchava [26], [27], and the fundamental articles [64], [65] of I.B. Simonenko. One-dimensional operators play the role of examples in the present book, and our list of references is not intended to serve as a reliable orientation in this direction.

References

[1] M.S. AGRANOVICH; *Elliptic singular integro-differential operators,* Uspekhi Mat. Nauk SSSR 20, No. 5 (1965), 3-120. (Translation: Russian Math. Surveys 20, No. 5 (1965), 1-121.)

[2] M.S. AGRANOVICH, M.I. VISHIK; *Elliptic problems with a parameter and parabolic problems of general type,* Uspekhi Mat. Nauk SSSR 19, No. 3 (1964), 53-161. (Translation: Russian Math. Surveys 19, No. 3 (1964), 53-157.)

[3] F.V. ATKINSON; *Normal solvability of linear equations in normed spaces,* Mat. Sb. 28, No. 1 (1951), 3-14 (in Russian).

[4] K.I. BABENKO; *On dual functions,* Dokl. AN SSSR 62, No. 1 (1948), 157-160 (in Russian).

[5] H. BATEMAN, A. ERDÉLY, ET AL.; *Higher transcendental functions,* Vol. 1, McGraw-Hill, 1953.

[6] H. BATEMAN, A. ERDÉLY, ET AL.; *Higher transcendental functions,* Vol. 2, McGraw-Hill, 1953.

[7] S. BOCHNER; *Theta relations with spherical harmonics,* Proc. Nat. Acad. Sci. USA 37 (1951), 804-808.

[8] N.L. VASILEVSKII; *On the algebra generated by two-dimensional integral operators with Bergman kernel and piecewise continuous coefficients,* Dokl. AN SSSR 271, No. 5 (1983), 1041-1044. (Translation: Soviet Math.-Dokl. 28, No. 1 (1983), 191-194.)

[9] N.P. VEKUA; *Systems of singular integral equations and certain boundary value problems,* Noordhoff, 1967 (translated from the Russian).

[10] N.YA. VILENKIN; *Special functions and the theory of group representations,* Amer. Math. Soc., 1968 (translated from the Russian).

[11] V.S. VLADIMIROV; *Generalized functions in mathematical physics,* Nauka, 1976 (in Russian).

[12] F.D. GAKHOV; *Boundary value problems,* Pergamon, 1966 (translated from the Russian).

[13] I.M. GEL'FAND, G.E. SHILOV; *Generalized functions,* Vol. 1, Acad. Press, 1964 (translated from the Russian).

[14] I.M. GEL'FAND, N.YA. VILENKIN; *Generalized functions,* Vol. 4, Acad. Press, 1965 (translated from the Russian).

[15] I.M. GEL'FAND; *Eigenfunction expansions of equations with periodic coefficients,* Dokl. AN SSSR 73, No. 6 (1950), 1117-1120 (in Russian).

[16] I.Ts. GOKHBERG, M.G. KREIN; *Introduction to the theory of linear nonselfadjoint operators,* Amer. Math. Soc. 1969 (translated from the Russian).

[17] I.Ts. GOKHBERG, N.YA. KRUPNIK; *Einführung in die Theorie der eindimensionalen singulären Integraloperatoren,* Birkhäuser, 1979 (translated from the Russian).

[18] I.Ts. GOKHBERG, N.YA. KRUPNIK; *Algebra generated by one-dimensional singular integral operators with piecewise continuous coefficients,* Funkts. Anal. i Prilozhen. 4, No. 3 (1970), 26-36. (Translation: Funct. Anal. Appl. 4, No. 3 (1970), 193-201.)

[19] I.Ts. GOKHBERG, N.YA. KRUPNIK; *On the algebra generated by Toeplitz matrices,* Funkts. Anal. i Prilozhen. 3, No. 2 (1969), 46-56. (Translation: Funct. Anal. Appl. 3, No. 2 (1969), 119-127.)

[20] I.Ts. GOKHBERG, N.YA. KRUPNIK; *On singular integral operators on a compound contour,* Soobshch. AN GSSR 64, No. 1 (1971), 21-24 (in Russian).

[21] I.Ts. GOKHBERG, E.I. SIGAL; *An operator generalization of the logarithmic residue theorem and the theorem of Rouché,* Mat.-Sb. 84, No. 4 (1971), 607-629. (Translation: Math. USSR-Sb. 13 (1971), 603-625.)

[22] V.V. GRUSHIN; *Pseudodifferential operators in* $\mathbf{R}^n$ *with bounded symbols,* Funkts. Anal. i Prilozhen. 4, No. 3 (1970), 37-50. (Translation: Funct. Anal. Appl. 4, No. 3 (1970), 202-212.)

[23] A.O. DERVIZ; *Boundary-value problems for meromorphic pseudodifferential operators,* Izv. Vyzov. Mat. 3 (1985), 84-86 (Translation: Soviet Math.-Izv. 29, No. 3 (1985), 108-111).

[24] J. DIXMIER; *C^*-algebras,* North-Holland, 1977 (translated from the French).

[25] R.G. DOUGLAS; *Banach algebra techniques in the theory of Toeplitz operators,* Amer. Math. Soc., 1973.

[26] R.V. DUDUCHAVA; *On integral operators of convolution type with discontinuous symbols,* Math. Nachr. 79 (1977), 75-98 (in Russian).

[27] R.V. DUDUCHAVA; *On bisingular integral operators with discontinuous coefficients,* Mat. Sb. 30, No. 4 (1976), 584-609. (Translation: Math. USSR-Sb. 30, No. 4 (1976), 515-537.)

[28] V.P. IL'IN; *On inequalities between the norms of partial derivatives of functions of several variables,* Tr. Mat. Inst. Steklov. 84 (1965), 144-173 (Translation: Proc. Steklov Inst. Math. 84 (1965), 161-193 (edited by V.N. Faddeeva).)

[29] A.P. CALDERON, A. ZYGMUND; *Singular integral operators and differential equations,* Amer. J. Math. 79, No. 4 (1957), 901-921.

[30] L.V. KANTOROVICH, G.P. AKILOV; *Functional analysis in normed spaces,* Pergamon, 1964 (translated from the Russian).

[31] A.I. KOMECH; *Elliptic boundary value problems for pseudodifferential operators on manifolds with conical points,* Mat. Sb. 86, No. 2 (1971), 268-298. (Translation: Math. USSR-Sb. 15, No. 2 (1971), 261-297.)

[32] J.J. KOHN, L. NIRENBERG; *An algebra of pseudo-differential operators,* Comm. Pure Appl. Math. 18, No. 1/2 (1965), 269-305.

[33] V.A. KONDRAT'EV; *Boundary problems for elliptic equations in domains with conical or angular points,* Tr. Moskov. Mat. Obshch. 16 (1967), 209-292. (Translation: Trans. Moscow Math. Soc. 16 (1967), 227-314.)

[34] H.O. CORDES, D.A. WILLIAMS; *An algebra of pseudo-differential operators with non-smooth symbols,* Pacific J. Math. 78, No. 2 (1978), 279-291.

[35] V.G. MAZ'YA, B.A. PLAMENEVSKII; *L_p-estimates of solutions of elliptic boundary value problems in domains with edges,* Tr. Moskov. Mat. Obshch. 37 (1978), 49-93. (Translation: Trans. Moscow Math. Soc. 37 (1978), 49-98.)

[36] V.G. MAZ'YA, B.A. PLAMENEVSKII; *Elliptic boundary value problems on manifolds with singularities,* Probl. Mat. Anal. 6 (1977), 85-142 (in Russian).

[37] V.G. MAZ'YA, T.O. SHAPOSHNIKOVA; *Theory of multipliers in spaces of differentiable functions,* Pitman, 1985 (translated from the Russian).

[38] B. MUCKENHOUPT, R. WHEEDEN, WO-SANG YOUNG; *L_2-multipliers with power weights,* Adv. Math. 49, No. 2 (1983), 170-216.

[39] N.M. MIKHAILOVA-GUBENKO; *Singular integral equations in Lipschitz spaces II.* Vestn. Leningr. Gos. Univ. Ser. Mat. Mekh. Astr. 21, No. 7 (1966), 45-57 (in Russian).

[40] S.G. MIKHLIN; *Multidimensional singular integrals and integral equations,* North-Holland, 1970 (translated from the Russian).

[41] N.I. MUSKHELISHVILI; *Singular integral equations,* Noordhoff, 1953 (translated from the Russian).

[42] S.A. NAZAROV; *Elliptic boundary value problems with periodic coefficients in a cylinder,* Izv. AN SSSR Ser. Mat. 45, No. 1 (1981), 101-112. (Translation: Math. USSR-Izv. 18, No. 1 (1982), 89-98.)

[43] B.A. PLAMENEVSKII; *Singular integral equations in a cone,* Dokl. AN SSSR 179, No. 5 (1968), 1057-1059. (Translation: Soviet Math. Dokl. 9, No. 2 (1968), 509-511.)

[44] B.A. PLAMENEVSKII; *On the boundedness of singular integrals in weighted spaces,* Mat. Sb. 76, No. 4 (1968), 573-592 (in Russian).

[45] B.A. PLAMENEVSKII; *On an algebra of pseudodifferential operators in spaces with weighted norms,* Mat. Sb. 106, No. 2 (1978), 296-320. (Translation: Math. USSR-Sb. 34 (1978), 841-865.)

[46] B.A. PLAMENEVSKII; *On meromorphic pseudodifferential operator-functions,* Izv. Vyzov. Mat. 4 (1978), 79-90 (in Russian).

[47] B.A. PLAMENEVSKII; *On algebras generated by pseudodifferential operators with isolated singularities of symbols,* Dokl. AN SSSR 248, No. 2 (1979), 297-302. (Translation: Soviet Math.-Dokl. 20, No. 5 (1979), 1013-1017.)

[48] B.A. PLAMENEVSKII; *On boundary value problems for meromorphic pseudodifferential operators,* Izv. Vyzov. Mat. 4 (1980), 69-78. (Translation: Soviet Math.-Izv. 24, No. 4 (1980), 77-87.)

[49] B.A. PLAMENEVSKII; *On algebras generated by pseudodifferential operators with isolated singularities of symbols,* Probl. Mat. Fiz 10 (1982), 209-241. (Translation: Selecta Math. Sov. 5, No. 1 (1986), 77-100.)

[50] B.A. PLAMENEVSKII; *On pseudodifferential operators on* $\mathbb{R}^n$ *with discontinuities of the second kind in the symbols,* Izv. Vyzov. Mat. 12 (1982), 30-38 (in Russian).

[51] B.A. PLAMENEVSKII; *On traces of pseudodifferential operators,* Izv. Vyzov. Mat. 12 (1982) 30-38. (Translation: Soviet Math.-Izv. 26, No. 12 (1982), 27-37.)

[52] B.A. PLAMENEVSKII; *On the index of pseudodifferential operators with isolated singularities of symbols in* $\mathbb{R}^n$, Dokl. AN SSSR 263, No. 5 (1982), 1062-1065.

(Translation: Soviet Math.-Dokl. 25, No. 2 (1982), 494-497.)

[53] B.A. PLAMENEVSKII; *Estimates for a convolution operator in spaces with weighted norms,* Dokl. AN SSSR 286, No. 1 (1986), 36-39. (Translation: Soviet Math. Dokl. 33, No. 1 (1986), 25-29.)

[54] B.A. PLAMENEVSKII, V.N. SENICHKIN; *On the spectrum of C^*-algebras generated by pseudodifferential operators with isolated singularities of symbols,* Dokl. AN SSSR 261, No. 6 (1981), 1304-1306. (Translation: Soviet Math.-Dokl. 24, No. 3 (1981), 686-689.)

[55] B.A. PLAMENEVSKII, V.N. SENICHKIN; *On the spectrum of C^*-algebras generated by pseudodifferential operators with discontinuous symbols,* Izv. AN SSSR Ser. Mat. 47, No. 6 (1983), 1263-1284. (Math. USSR-Izv. 23, No. 3 (1984), 525-544.)

[56] B.A. PLAMENEVSKII, V.N. SENICHKIN; *On C^*-algebras of singular integral operators with discontinuous coefficients on a complex contour I,* Izv. Vyzov. Mat. 1 (1984), 25-33; *II,* Izv. Vyzov. Mat 4 (1984), 37-46. (Translations: Soviet Math.-Izv. 28, No. 1, (1984), 28-37; Soviet Math.-Izv. 28, No. 4 (1984), 47-58).

[57] B.A. PLAMENEVSKII, V.N. SENICHKIN; *On the spectrum of a C^*-algebra of pseudodifferential operators with singularities in the symbols,* Math. Nachr. 121 (1985), 231-268 (in Russian).

[58] B.A. PLAMENEVSKII, V.N. SENICHKIN; *The spectrum of an algebra of pseudodifferential operators with discontinuous symbols on a manifold with boundary,* Dokl. AN SSSR 277, No. 6 (1984), 1327-1330. (Translation: Soviet Math.-Dokl. 30, 1 (1984), 286-289.)

[59] B.A. PLAMENEVSKII, V.N. SENICHKIN; *On the spectrum of a C^*-algebra of pseudodifferential operators with discontinuous symbols on a manifold with boundary,* Probl. Mat. Fiz. 11 (1985), 178-209 (in Russian).

[60] B.A. PLAMENEVSKII, V.N. SENICHKIN; *On the spectrum of an algebra of pseudodifferential operators in a half-space,* Probl. Mat. Anal. 10 (1986), 160-179 (in Russian).

[61] B.A. PLAMENEVSKII, V.N. SENICHKIN; *Spectra of C^*-algebras of pseudodifferential operators with multidimensional singularities in symbols,* Funkts. Anal. i Prilozhen. 20, No. 4 (1986), 85-86. (Translation: Funct. Anal.

Appl. 20, No. 4 (1986), 328-329.)

[62] S.G. SAMKO; *Hypersingular integrals and their applications,* Izd. Rostok. Univ., 1984 (in Russian).

[63] S.G. SAMKO; *Singular integrals over the sphere and construction of the characteristics with respect to the symbols,* Izv. Vyzov. Mat. 4 (1983), 28-42. (Translation: Soviet Math.-Izv. 27, No. 4 (1983), 35-52.)

[64] I.B. SIMONENKO; *A new general method for studying linear operator equations of the type of singular integral equations, I,* Izv. AN SSSR Ser. Mat. 29, No. 3 (1965), 567-586 (in Russian).

[65] I.B. SIMONENKO; *On global and local factorability of a measurable matrix function and the Noetherian property induced by its singular operator,* Izv. Vyzov. Mat. 4 (1983), 81-87. (Translation: Soviet Math.-Izv. 27, No. 4 (1983), 99-106.)

[66] E.M. STEIN; *Note on singular integrals,* Proc. Amer. Math. Soc. 8, No. 2 (1957), 250-254.

[67] F. TRÈVES; *Introduction to pseudodifferential and Fourier integral operators,* Vol. 1-2, Plenum, 1980.

[68] M. TAYLOR; *Pseudo-differential operators,* Springer, 1974.

[69] K.O. FRIEDRICHS; *Pseudo-differential operators,* Courant Inst. Math. Sci. Lecture Notes, 1970.

[70] YU.E. KHAIKIN; *On operators of convolution type in weighted spaces,* Vestn. Leningr. Gos. Univ. Ser. Mat. Mekh. Astr. 13 (1969) (in Russian).

[71] B.V. KHVEDELIDZE; *The method of Cauchy-type integrals in the discontinuous boundary-value problems of the theory of holomorphic functions of one complex variable,* Itogi Nauk. i Tekhn. Sovr. Probl. Mat. 7 (1975), 5-162. (Translation: J. Soviet Math. 7 (1977), 309-415.)

[72] L. HÖRMANDER; *Pseudo-differential operators,* Comm. Pure Appl. Math. 18 (1965), 501-517.

[73] M.A. SHUBIN; *Pseudodifferential operators and spectral theory,* Springer, 1987 (translated from the Russian).

[74] G.I. ESKIN; *Boundary value problems for elliptic pseudodifferential equations,* Amer. Math. Soc., 1981 (translated from the Russian).

Index